D1058519

PLAYWRITING: THE STRUCTURE OF ACTION

PRENTICE-HALL SERIES IN
THEATRE AND DRAMA

Oscar G. Brockett, *Consulting Editor*

Robert Benedetti
THE ACTOR AT WORK

Charlotte Kay Motter
THEATRE IN HIGH SCHOOL: PLANNING, TEACHING, DIRECTING

Sam Smiley
PLAYWRITING: THE STRUCTURE OF ACTION

Francis Hodge
PLAY DIRECTING: ANALYSIS, COMMUNICATION, AND STYLE

Oscar G. Brockett and Robert R. Findlay
CENTURY OF INNOVATION: A HISTORY OF AMERICAN THEATRE & DRAMA 1870-1970

PRENTICE-HALL INTERNATIONAL, INC., London
PRENTICE-HALL OF AUSTRALIA, PTY. LTD., Sydney
PRENTICE-HALL OF CANADA, LTD., Toronto
PRENTICE-HALL OF INDIA PRIVATE LIMITED, New Delhi
PRENTICE-HALL OF JAPAN, INC., Tokyo

PLAYWRITING: THE STRUCTURE OF ACTION

SAM SMILEY

PRENTICE-HALL, INC., *Englewood Cliffs, New Jersey*

C 13-684548-7
P 13-684530-4

Library of Congress Catalog Card Number: 78-125077

Printed in the United States of America

Current printing (last number):
10 9 8

For

HUBERT C. HEFFNER

PREFACE

The artists of today can no longer afford total solitude because forces of dehumanization challenge the very possibility of art. Naturally, each writer must work alone, but he needs to share knowledge and inspiration with others. Whatever energy this book may contain, that idea is its source.

This book explores principles of writing plays. It presents ideas about selecting and arranging dramatic materials for an artistic purpose. While explaining both theory and practice, it focuses on the structuring of a drama. It reveals dramatic structure as something more dynamic than mere form, because drama has to do with the turbulence of human relationships and actions.

Although this book describes structural principles, it does not *prescribe* their use. Neither does it argue for the well-made play as a literary genre, but it definitely contends that every play should be made well. It reflects a playwright's viewpoint, not that of a critic. And it is intended for all playwrights—beginners, interns, and professionals.

Strangely, in the seven fine arts some information traditionally handed from one artist to another has never been written down. Within the following pages are many of the techniques that for generations older writers have personally disclosed to younger ones. This work also contains some principles described in earlier books. But each generation deserves to hear in its own idiom the ancient bits of knowledge and to understand how they relate to current society and theatre. Further, it explains some new techniques of discovery, organization, and execution in drama. All the principles and practices herein are, however, only potential aids. Each playwright must discover naturally, for himself, as he writes, how and when to use them.

The book offers theoretical, practical, and productive information. Specifically, it describes the writer's vision and its possible components—intellectual and emotional, aesthetic and ethical. It traces the writing process from germinal idea through scenario and draft to final version. It suggests a manuscript format and a procedure for getting a production. By emphasizing certain dramatic artists, it depicts contemporary theatre and the relationships within it. Most importantly, the book explains ideas about the internal nature of plays. But it proclaims no laws. It simply offers help to any playwright who wants in his own way to create unique drama.

Although many plays furnished examples, the following provided the most: *Oedipus the King* by Sophocles, *Lysistrata* by Aristophanes, *Hamlet* by Shakespeare, *The Miser* by Molière *The Ghost Sonata* by August Strindberg, *Arms and the Man* by George Bernard Shaw, *A Streetcar Named Desire* by Tennessee Williams, *The Good Woman of Setzuan* by Bertolt Brecht, *Waiting for Godot* by Samuel Beckett, *The Chinese Wall* by Max Frisch, and *The Homecoming* by Harold Pinter.

The playwrights who encouraged this book were Dene Hammond, Eugene Hochman, James Lineberger, Richard Maibaum, David McFadzean, Frank Moffett Mosier, Clifford Odets, Barrie Stavis, Megan Terry, and William Weber. Other professional writers who furnished ideas and inspiration were: Stephen Birmingham, Jean Poindexter Colby, Lucille Fletcher, David Fortney, Jesse Hill Ford, Philip Hamburger, Hal Higdon, William Kozlenko, Anton Myrer, Jeanne Suhrheinrich, Douglass Wallop, Maia Wojciechowska, John Weston, and William Zinsser.

In a personal sense, the book is for them and for Hubert Heffner, Harold Crain, and Lawrence Tucker—three great teachers; plus Eric Bentley and Elder Olson, whose books are superb teachers. A group of theatre artists encouraged the book too. These friends wanted some functional dramatic theory, something useful to actors, directors, and designers: Gary Bayer, Marcia Bennett, Duane Campbell, Larry Clark, Fredrik deBoer, David Emge, Henderson Forsythe, Ron Glass, Joe Hazzard, Robert Hazzard, Douglas Hubbell, John David Lutz, David Shelton, George Touliatos, Dudley Thomas, and a host of excellent students.

The editorial advice of Oscar Brockett improved the book immeasurably. Theodore Hoffman provided inspiration for the whole book and ideas for the sections on contemporary drama. Patricia Madsen, for several years chairman of the AETA Playwrights Program, furnished information about the needs of young playwrights. Several parts of the book were put together for the Indiana University Writers' Conference at the invitation of Robert Mitchner. Byron Schaffer, Jr., and Donald E. Polzin, co-editors of *Players Magazine*, gave concrete encouragement by printing two segments of this book before publication. Howard Cady and Kirk Polking have helped too with gracious and intelligent advice. Without the editorial and design work of Paula Cohn and the personal assistance of Arthur Rittenberg the entire project would have been impossible. And the book would not have come into being at all without the manuscript labors of Joan Ballard and Sherry Noyes. Ann Smiley continually provided intellectual, practical, and emotional support.

For several decades, a growing number of fine college teachers have been helping young people learn about playwriting. Such men as George Pierce Baker, E. P. Conkle, Frederick Koch, E. C. Mabie, Kenneth Macgowan, and Kenneth T. Rowe were the pioneers. Today, a surprising number of people are extending the work of those men by laboring for hours over student manuscripts and holding discussions of genuine value to young writers. These teachers work effectively despite small budgets, constricted curricula, and the lethargy of institutions. For the teacher of playwriting, praise is rare and criticism frequent. It is time that someone recognized the contributions that playwriting teachers are making to theatre—such people as Joseph Baldwin, Kenneth Cameron, Joel Climenhaga, Sherwood Collins, Roger Cornish, Norman Fedder, Marian Gallaway, Eugene McKinney, Richard Moody, William Monson, William Reardon, James Rosenburg, George Savage, James Schevill, Webster Smalley, Howard Stein, and many others.

The publishing companies noted below were kind enough to allow the use of quotations from the following authors and books: Aristotle, *Poetics,* trans. Ingram Bywater, in *The Basic Works of Aristotle,* ed. Richard McKeon (New York: Random House, Inc., 1941); Antonin Artaud, *The Theater and Its Double,* trans. Mary Caroline Richards (New York: Grove Press, Inc., 1958. Copyright © 1958 by Grove Press, Inc. By permission of Grove Press, Inc. and Calder & Boyars, Ltd., in Great Britain, trans. by Victor Corti; Jean-Louis Barrault, *Reflections on the Theatre,* trans. Barbara Wall (London: Rockliff Publishing Corporation Limited, 1951); Bertolt Brecht, "A Short Organum for the Theatre," in *Brecht on Theatre,* ed. and trans. John Willett (New York: Hill and Wang, 1966); Albert Camus, *The Rebel,* trans. Anthony Bower (New York: Alfred A. Knopf, 1956); Benedetto Croce, *Guide to Aesthetics,* trans. Patrick Romanell (Indianapolis: The Bobbs-Merrill Company, Inc., 1965); Friedrich Duerrenmatt, "Problems of the Theatre," in *Playwrights on Playwriting,* ed. Toby Cole (New York: Hill and Wang, 1961); William Gibson, *The Seesaw Log* (New York: Alfred A. Knopf, Inc., 1959); Tyrone Guthrie, *In Various Directions* (New York: The Macmillan Company, 1965), by permission of The Macmillan Company and Michael Joseph, Ltd.; Ernest Hemingway, "Monologue to the Maestro: A High Seas Letter," from *By-Line: Ernest Hemingway,* ed. William White (New York: Charles Scribner's Sons, 1967), by permission of Charles Scribner's Sons and William Collins Sons & Co., Ltd.; Friedrich Nietzsche, *The Birth of Trag-*

edy, trans. by Clifton P. Fadiman, in *The Philosophy of Nietzsche,* The Modern Library (New York: Random House, Inc., 1927); John Osborne, "Declaration," in *Playwrights on Playwriting,* ed. Toby Cole (New York: Hill and Wang, 1961); by permission of John Osborne; Jean-Paul Sartre, *What Is Literature?,* trans. Bernard Frechtman, published by Harper & Row, Publishers, Inc., 1965, copyright by Philosophical Library, Inc., 1949; August Strindberg, Foreword to *Miss Julie,* trans. by Elizabeth Sprigge, in *Six Plays of Strindberg,* A Doubleday Anchor Book (New York: Doubleday & Company, Inc., 1955). Reprinted by permission of Collins-Knowlton-Wing, Inc. Copyright © 1955 by Elizabeth Sprigge, and by permission of A. P. Watt & Son.

S. S.

University of Missouri
Columbia, Missouri

TABLE OF CONTENTS

PLAYWRITING: THE STRUCTURE OF ACTION

PART I: *A PLAYWRIGHT'S SOLITARY WORK*

In upholding beauty, we prepare the way for the day of regeneration when civilization will give first place—far ahead of the formal principles and degraded values of history—to this living virtue on which is founded the common dignity of man and the world he lives in, and which we must now define in the face of a world that insults it.

ALBERT CAMUS
The Rebel

1: *THE WRITER'S VISION*

...each one, by inventing his own issue, invents himself. Man must be invented each day.

JEAN-PAUL SARTRE
What Is Literature?

Before a writer plunges into the productive activity of creating a play, he needs to consider what he is going to write about and how he will proceed. What does he think about life—its order, its meaning, and its potential? He must have something to say. Any writer naturally discovers part of what he wants to say while he writes, but a part too must exist in his mind before he begins. To discover reasons for man's existence and to perfect a vision of life are fascinating activities for anyone, especially a writer. The search requires awareness, sensitivity, and thought. Each work reflects a writer's life attitudes, but he must in some way identify those attitudes in himself before they will ever energize a work of value.

SEEING INTO LIFE

The artist affirms, and the artist denies. He affirms life by selecting materials and by shaping them to make an object that reflects his vision.

But denial rises as strongly in him as affirmation. They stand together in his nature. He exults in the possibilities of life but damns its chaos and suffering. By the very act of creation, he destroys the world's irrationality, incomprehensibility, and inhumanity. The artist is above all human. Who more than he concentrates at once on creation, love, faith, thought, and work? He assents to life's potential but rebels against its terror. Both actions depend upon his vision. A writer's vision provides him with something to say.

Vision consists of a complex of emotional and intellectual conditions within a writer. A writer as artist searches for unity in the disorder of existence, a unity involving mind and heart. In life such unity is impossible, and so he rejects what he sees and reconstructs through his personal vision a substitute universe in the controllable actuality of his art. When he finishes his art object, he destroys for that material and space and time some of the world's confusion. He thus belongs to a group which attacks disorder. The artists will not be the ones to end the world; they want to create it.

Some possible components of an artist's vision are his awareness, his perspective, his good and bad dreams, his intoxication with life, and his battles—with society, himself, and the universe.

Working through the personal to get to the general, he progresses only by maintaining intense consciousness of the world around him. Such *awareness* simultaneously goads him and acts as his weapon. It impels him to question the human condition and the social insanities of his time. Eventually, he perceives mankind's recurrent questions: To what purpose does man exist? Why does man suffer? What is the essence of life? What meaning can there be in death? Where do all man's struggles lead? Such problems sting the artist, but he also reacts to the life issues of his immediate time: How can nations resolve their conflicts and end all wars? What is the solution to the schism between the capitalist world and the socialist world? If man continues to reproduce his own kind so rapidly, can he survive the population flood? The issues of a writer's country, too, will concern him. In the United States they have to do with mechanization, interdependence, affluence, race, success, violence, and established authority.

Without some *perspective,* however, an artist cannot help but produce art that is private, arbitrary, disordered, and distorted. Proceeding through a lifetime of isolation, disillusion, and discontinuity—the inevitable norm of experience—he must, in order to work, establish for himself a functional perspective. He faces the paradox of living within a threatening and degenerating society and yet remaining vibrant and humane.

Perspective, a combination of diverse attitudes, dictates for the playwright the sort of action he chooses to formulate in his drama. It filters the materials he selects and affects his use of time and place, character and thought. Every artist's perspective develops in the interaction between his personal life and his historical period.

An artist, especially a writer, often creates a projection of his *good and bad dreams.* All men dream, and to some extent all try to make their dreams come true. The artist perfects a medium for the expression of his dreams. These may be short or long, whole or partial, anormal or universal, but they always possess signification. Stylistically, they may be illusory or non-illusory, objective or subjective. But dreams always reflect the dreamer. As each artist employs his imagination, intelligence, and skill to build an object, he often calls upon his sweet dreams and his nasty ones.

Intoxication with life also stimulates art. Regardless of what social milieu an artist lives in, he remains an individual, a one among the many. He is alone always, as is every man. But for the artist, loneness often gives a heightened sense of life. Loneliness may make him sad, but loneness means inner freedom. Alone, a man faces the terror and rejoices in the ecstasy of life. One driving force in any artist is intoxication with being, with living. Loneness and liveness furnish the individual with the energy to create.

Every dramatist deals, to some degree, with conflict. Most playwrights employ conflict in their dramas, and they no doubt learn about it from life. Like most artists, a writer often experiences *battles* with collective forces in society, with the desires and frustrations in himself, and with the natural facts of life and death. The intense artist is seldom willing to accept without question what others tell him to feel or believe. Rather, he is likely to insist on examining things for himself and reaching his own conclusions. His engagement in social, personal, and even political conflicts catapults him toward freedom of choice. He may perceive new patterns of behavior, or he may reaffirm traditional values. In either case, the battles of life help an artist capture universal experience in his work.

In order to create art works of any worth, each artist must have something to say, some values, some attitudes, some store of experience—a vision. The persistence of great art reflects the endurance of man in life and of man at work. Vision is the power in an artist of seeing into life and of making something of what he sees.

IDEAS ABOUT ART

To discern the principles of any art, one must first develop some ideas for reflecting about aesthetic knowledge. And it is advantageous to heed some professional thinkers. For example, Aristotle and Croce presented differing but useful approaches to knowledge and aesthetics. Since the artist should always be eclectic, he can freely draw ideas from such theorists and blend them with his own.

Aristotle divided all knowledge into three types. He considered the objectives of the various kinds of knowledge, their nature, the human faculties involved in knowing them, and the purpose of each kind. He examined knowledge as cause rather than experience. Inductively, he discerned three bodies of knowledge that are crucial to three corresponding categories of human activity, or "sciences." The three types of knowledge (and activity) are distinct in subjects, problems, methods, and functions.

The three areas of human knowledge are the theoretical, the practical, and the productive. The three related activities can best be characterized by the verbs to know, to do, and to make. The first type, theoretical science, involves knowledge for its own sake. It includes all philosophic studies—ideas, theories, and concepts. Knowledge itself is the object. Propositions, probabilities, and their supporting facts comprise the materials for the theoretical type. The method is dialectic, or logical argument. And the goal is knowing, or truth. The theoretical sciences include metaphysics, theology, aesthetics, theoretical mathematics, physical theory, and the like. A thorough study of any field leads ultimately to theoretical study in that field, the study of principles. For example, this chapter sets forth an approach to the theoretical study of drama, and is thus involved in the subject of aesthetics, the theory of art.

Practical science, the second type, has to do with human activity. In the practical sciences, knowledge is subordinate to action. Ethics, politics, rhetoric, and applied mathematics stand as examples of this group of sciences. The object is human behavior, for instance the practice of speechmaking. The materials are any means appropriate to performing the action. The method is the manner of the activity as performed by a given individual. And the goal of any practical science is the successful completion of the chosen action—the good in behavior.

Third, the productive sciences are the arts, both useful and fine—carpentry and glassblowing, dance and music. With these, the goal is production of an object, such as a brick wall by the useful art of bricklaying or a poem by the fine art of poetry. The end is beauty in the fine arts or utility in the useful arts. In this category, product dominates knowledge or action. For the arts, the object of study is the essence, or idea, of the thing to be created. The materials are those items of physical matter proper to the given product. And the method is the style with which an artist handles his materials.

Each of the fine arts, then, belongs in the realm of productive science and appropriately involves productive knowledge. This is not to say, however, that there is no theory or activity involved. All three types of knowledge—theoretical, practical, and productive—are interrelated. Drama, as an instance of fine art, should be understood theoretically within the realm of aesthetics, practically as a kind of human activity, and productively as it involves the construction of a concrete object called a play. Parts I and III of this book deal with theoretical knowledge for its own sake and with practical knowledge about the activities of the playwright. Most properly, Part II is the largest portion of the book, and it exclusively treats the internal nature of drama as an art product. Every playwright needs to know about theory, practice, and product in his art.

The functions of art focus upon man himself. All fine arts have to do with human life in its most intense states. Although most people have some experiences with art, the discerning are forever amazed that objects exist which enhance life, objects made by human beings and enjoyed by other human beings. Art astonishes. The specific functions of art are as infinite as the number of artists multiplied by the number of their works and multiplied again by the number of the people who come into contact with those works.

Art has some identifiable general functions. First, art objects produce specific pleasure in human beings. The generation of that quality alone makes the labor that goes into art worthwhile; there are never too many pleasures in any man's life. Art can also furnish knowledge—about man. It always signifies something about human life, even if only a sight, a feeling, or an awareness of "nothing." Man constantly drives himself to know significations. Art also satisfies the imagination of both the person who makes it and the person who observes it. And the human imagination—that amalgam of intellect, emotion, and instinct—is forever whirling and seeking. Finally, art functions as a special kind of *order* in the chaos of life. It offers controlled and permanent beauty. An artist *creates* by giving diverse materials a controlled form in a skilled manner for a pleasurable end. With each art object, he creates an image possessing unity, harmony, and balance. The artist as creator can give life increased value.

All human beings live most minutes of most days in an habitual or semiconscious manner. The noteworthy moments in any individual's life are the few experiences of intense consciousness. Psychologists now explain that people are fully awake only a few minutes each day. People live for those stimuli which bring about total awareness of life. Such stimuli come from many sources and cause varying reactions. To look at the brilliance of a million stars at night, to feel the surge of sexual love, to watch the face of your child during a happy time—such common experiences may be memorable, live moments. And art too can provide them. At its best, or whenever it achieves its potential for any individual, it causes an intense awareness of life.

The formal term for the full human response to art is aesthetic reaction. When art causes an aesthetic reaction, then it has caused live moments. This heightened consciousness of life consists of emotional, intellectual, and physical responses. Art produces some emotion and stimulates some thoughts in viewers, and as each person feels or thinks, physical response is inevitable. Such aesthetic responses are often more covert than overt. But if the artist watches and listens to his audiences, he soon learns to identify their physical responses, the movements and noises which reveal their emotional and intellectual states.

The artist also has an aesthetic reaction to his own work. Although it too is composed of emotional, intellectual, and physical responses, his reaction is different from that of an audience. It can come anytime. But it usually occurs when he completes his art work. An audience does not even have to see a playwright's drama for him to get a sense of aesthetic fulfillment.

An artist also gets pleasure from work itself. He finds worth in life by using his intuition to contemplate, to imagine, to invent, and to make an unified and beautiful object. Art brings ideals into reality, but it is more than a set of physical facts in space and time. Ideality is a part of the essence of art. Although an artist must think while constructing a work and although each work is an organized whole, art is usually non-logical—that is non-scientific, non-dialectic. It is not, in the common sense, true or false, moral or immoral. Although perceived by the human sensors, it is more than sensuous. Art is not the result of a scientific or practical act, but of an intuitive one. Such an act does not exclude thought; it requires a different sort of thought. Call it imagination, fancy, dream, ideality. Intuition implies that art never distinguishes between reality and unreality. Art is both.

An artist's intuition produces an image, at once concrete and abstract. Thus, the image at the heart of every art object is an ideal—an image as its own essence. Vision and intuition fuse within each artist. They permit him to discover the essence which is the existence of art. Hence, art requires more imagination than logic, more taste than judgment, and more genius than intellect. The artist's vision depends on yearning and feeling. These generate in him the energy to blend the worlds of fantasy and fact to make a universe more real in association with the object he produces. Life, unity, compactness, and fullness in art proceed from the intuitive vision of the artist.

DRAMA AS A FINE ART

The fine arts are seven highly developed ways for man to make the abstractions of his fancy assume the concrete reality of physical objects. The traditional seven fine arts are architecture, dance, drama, music, painting, poetry, and sculpture.

Architecture is the creative planning and construction with physical substance of a human milieu. Dance is the imaginative and organized execution of human activity in three-dimensional space. Music consists of sounds in the form of melody, harmony, and rhythm. Painting, a spatial art, involves visual images organized with line, mass, color, and texture. Poetry has to do with the construction of verbal images with words. Sculpture is the formulation of material into a three-dimensional image.

A drama is a constructed object which exists in a given time span and which can be repeated. Its materials are words and physical activities. Its form is human action, human change. The manner of its presentation requires live performance through acting.

All the fine arts encompass multiple purposes. Some art objects are functional, some decorative, and others didactic. But most are beautiful in some respect. And nearly all serve to make an image, stimulate emotion, or intensify experience.

Many people have discussed drama as merely a combination of other fine arts. Not true! It may share features with others or involve some of the same human skills, but at best it employs all these differently. Drama relates to poetry because at its core is a play, a verbal object made by a writer who utilizes many principles and materials also used by a poet or novelist. But a dramatist always writes present action for live actors; most other authors usually write past action for the printed page. Drama resembles dance with its human performers acting in space. But the forms of the two are different, as are the specific executions of movements. Architecture often affects the drama because one possible human milieu an architect may create is a theatre. Some of the spectacle of drama also corresponds to architectural practice. But the two arts remain separated in basic materials, forms, styles, and effects. Painting and sculpture relate more obviously to the work of the director and scene designer who stage a drama. Set designs and scenic pieces are frequently painted, but a flat picture never makes a satisfactory setting for drama. In fact, most graphic artists who think of two dimensional pictures fail as theatrical designers. Sculpture and drama both use space, but in widely divergent ways. Although some plays utilize music as a part of their spectacle, the music thus employed is simply material to the total drama and never an end in itself, as it properly is in its own artistic realm.

Even though drama most often begins as a written play, it never exists until that play is performed. A play, by itself, is not art but only one, albeit the most important, ingredient for the creation of drama. Drama does not come into being except when performed live on stage; in script form, a play is merely a *potential* drama.

THE NATURAL VS. THE ARTIFICIAL

The fine arts do not exist in nature. They are not even a necessary attribute of man. They come into being only as individual products, and their principles are determined by these products themselves, not by any physical, biological, or chemical laws. The principles of art are not natural ones, such as gravity or photosynthesis, but are artificial. Objects in nature—rocks, trees, animals—are obviously not man-made. Such objects are *natural.* A tree sprouts, grows, and dies in a natural process. It is an organized whole, containing certain elements and parts, and it possesses natural beauty, insofar as it is a fulfillment of its normal existence.

Art objects—paintings, dances, or dramas—are artificial. They are man-made. They too are organized wholes, many individual parts and elements welded together in aesthetic unity. But these objects can never occur naturally. A block of marble could stand in a garden for a thousand centuries, but it will never turn itself into a statue anything like Michelangelo's "David." To call art *artificial* is not to label it phony or bogus. The terms *natural* and *artificial* simply permit the division of all things into two classes, those which come into being through a natural process or those which come into being because of man. The playwright, like other artists, should know that the parts he puts together—such as the story, the characters, the words—do not go together naturally. Consciously and subconsciously, he must do the composing. All the parts blend only because of him and never of themselves. For the artist to realize this total control is both a joy and a curse. The responsibility for his final art product belongs to him alone.

The principles of art are not principles of activity but of product. Although a playwright gets aesthetic pleasure from practicing his art, the activity of writing is not an end in itself. His goal is the play. He must concern himself, therefore, with principles of the object. How does a play, as an object, come to be?

Aristotle provided four causes for the coming into being of an artificial product: material cause, formal cause, efficient cause, and final cause. The material cause is the means of the formulation. It consists of the concrete materials, or matter, used. In a play, the material is words. Formal cause refers to the object of the formulation. Simply explained, it is the form. Form is a common term among artists and critics, but not everyone understands its two precise meanings. Form is the organization of any art work; form is also the controlling idea in the work itself. This idea—or essence, or conception—is not necessarily the same as the idea in the mind of the artist as he makes the work. In a mimetic play, the form, or formal cause, is usually a human action. (In a didactic play, the formal cause is a thesis.) An action is a pattern of human life involving changing relationships between and within individuals. Efficient cause is the manner of the formulation. This cause refers to the artist and how he brings a play, as an object, into being. For every writer, it is the style of putting the words together. For each dramatist, it is the overall style he applies to a whole play, from his use of words in the dialogue to the implications he gets into the script about how it should be produced. The final cause is the end to which the whole is organized. It is the purpose, goal, or function of the art object. No matter how abstract or concrete the purpose of a given work may be, the artist needs an awareness of the basic nature of the work, to whom it is directed, and what response it should elicit.

To clarify the four causes, two simply sketched illustrations should suffice, the coming into being of a chair and of *A Streetcar Named Desire*. In the useful art of chairmaking, the material cause is wood, metal, plastic, padding, and the like. The formal cause is the idea of what the finished chair should look like and feel like. The efficient cause is the style of the chair in design, decoration, and workmanship. The final cause includes the twin functions of the chair being useful for sitting and pleasant to look at. In the fine art of playmaking, the material cause for *A Streetcar Named Desire* consists of the words of the play, both dialogue and stage directions. The formal cause *in* the play amounts to the combined activity of Blanche, Stanley, Stella, Mitch, and all the other characters. Their serious action is the effort to find and hold onto a happy and secure place in life. The efficient cause is the manner Tennessee Williams employed in writing the play as evinced in the arrangement of the words. He used American, realistic, and somewhat poetic prose. The style for the whole production, in fact, is a poetic realism. The final cause of the play is the creation of an object of beauty, in the special sense of contemporary tragedy, that will stimulate an aesthetic response in a twentieth-century audience. All four causes are crucial to the playwright's full understanding of a comprehensive method of play construction.

Another term which needs some explanation for the playwright is *imitation*. Aristotle used it in his *Poetics* in about 335 B.C. Imitation, as a functional term for an artist, does not mean producing a photographic, or representational, copy of what exists in nature. Aristotle employed imitation to mean man-made; he meant the term to epitomize the coming into being of any art object; and he intended that the idea of imitation would include the artist's control of the four causes. Imitation means that an artist imagines a conception (formal cause), uses (efficient cause) matter (material cause) to fulfill that conception, for a certain purpose (final cause). When an artist creates a work, which conforms to his imagination and which has an aesthetic purpose, then such a work of art is in its nature *mimetic*. Mimetic art does not mimic nature so much as it reflects an artist's vision.

An artist's idea is the basis of the form of a work. Only after he conceives an idea does he marshal materials to carry out his conception. Every art object depends on a form-matter principle to the elements of drama. The qualitative, functional elements of drama are: plot, character, thought, diction, sounds, and spectacle. They are comprehensive and exclusive. All items in a play can be related to one of the six. Although the six most important chapters of this book, in Part II, respectively treat each of the six elements, their form-matter connections should be understood here. The following arrangement reveals their relationships:

Form ↓   Plot ↑
Character
Thought
Diction
Sounds
Spectacle   Matter

The two arrows indicate how the parts work together in the formulation of a play. Reading down the list, each element is form to those below it, and reading upwards, each element is material to those above. Plot is the organization of the whole; it is not synonymous with story. Story is *one* of many ways of organizing a plot. Plot stands as the form to all the elements listed below it. Character is the most important material to plot. When all the sayings and doings of the characters are taken as an organic whole, that whole is the plot. Then, character is form to all things below it, especially thought. Thought is everything that goes on within a character—emotions, qualities, and ideas. All these taken together are the material to character. Thought is the form of the diction. Some thought operates as the organization of every series of words, and those words are the material of the thought. The words, i.e., diction, are made up of individual sounds. Thus, diction is the form of sounds, and sounds are the matter of diction. The music of the human voice, e.g., the use of emphasis and emotive coloring, gives words their meaning. Finally, spectacle, the physical actions that accompany the sounds and words, is the most simple material of all. For the playwright, plot has the greatest importance, character second, thought third, and so on. As actors, directors, and designers prepare a production, the list is turned around in order of importance. Theatre artists normally consider spectacle first, and then each of the other elements in ascending order. If the dramatist understands the form-matter relationship, he can adroitly utilize all the possible elements of drama as he formulates his play.

OF SUBJECTS AND SOURCES

The life of every artist amounts to a continual search. His quest is both expedient and ethical. He looks for material and seeks meaning. By living a life of scrutiny, the playwright constantly explores attitudes, fields of knowledge, experiences, and ideas. His sources are everywhere, and his subjects may be anything. The vision of the playwright, then, not only embodies an awareness of life issues, the study of ancient and current art, and knowledge about aesthetic principles, but also it consists of the kind of sources he utilizes and the subjects he chooses to treat through his art.

A playwright's greatest problem regarding subjects is for him to learn to write about what *he* wishes. In a sense, he always chooses his subjects. Otherwise, how could he proceed to work? But too often he selects a subject that seems popular, that someone else foists upon him, or that he thinks will please audiences. If a playwright makes selections for such reasons, he cripples his own creativity, and he will write lesser plays.

The playwright should break most well-known rules! He should question the teachers or textbooks that say: "Write about what you know." "Write a popular play." "Write something for the market." He will always write what he knows, and *knowing* in this connection depends as much on his imagination as on his firsthand experience. In fact, it is better to write something from imagination, conscious thought, or reading than to write about personal experiences that have occurred within one's own home, block, or city. The writer will automatically use experiences from those locales, but they will not ordinarily provide much thrust for creation. It is not easy, however, for anyone to discern what he truly wants to write about or what he truly feels. Ernest Hemingway, along with many other writers, said that his greatest difficulty was finding out what he really felt and distinguishing that from what he was supposed to feel. Each playwright must solve the same problem when selecting subjects for a play.

Theorists frequently identify a play's subject in a variety of ways: the informational field, the type of people, the school of philosophy, and so on. The following working definition functions well for a playwright.

The subject of a play amounts to the total activity of the characters as they respond to their immediate surroundings. It is affected by time and place, and it involves their social, professional, and personal relationships. The subject should be simple and clear. The concrete subject often implies a broader subject. The obvious subject of Shaw's *Arms and the Man* is how a Bulgarian-Serbian war of 1885 affects a specific family, but the ultimate subject is the romantic attitude about war. In *No Exit,* Sartre's simple subject is hell, but the broader implications have to do with man's personal responsibility for his actions. Often a person decides to read a given novel because of his threshold interest in an informational area. Eric Remarque's *All Quiet on the Western Front* is about World War I; *U. S. A.* by John Dos Passos treats American life during the 1930's; and *Moby Dick* by Herman Melville deals with whaling. But plays are too brief to present anywhere near the same amount of historical information. In plays, the information appears mostly in the action.

A playwright's ideas about a drama usually relate to a subject, and the playwright should clearly know his true subject before he begins. His ideas are imaginative notions about subject, incidents, characters, and thoughts. Where does a playwright find his ideas? The kinds of sources he consciously inspects also reflect his vision.

A playwright's total life, of course, is his source for inspiration and material. But some phases and activities are likely to be more consciously useful than others. *Reading is a major source.* Every writer must be a professional reader. No matter what educational degrees he earns, continual reading is his continuing process of education. Reading is a conscious search for knowledge, experience, and technique. Connected patterns of reading, rather than miscellaneous sampling, lead to more value for anyone.

Reading gives the playwright knowledge, experience, and technical proficiency. He can glean facts and concepts from books in such fields as philosophy, history, biography, sociology, and psychology. Nietzsche, Toynbee, Riesman, and Freud, for instance, have helped countless writers.

American playwrights have not yet made much use of the history of this country. *The Crucible* by Arthur Miller and *Harpers Ferry* by Barrie Stavis are examples of the sort of plays that can be made from American history. Periodical and daily journalism is also important reading matter, although too many writers have overemphasized its usefulness. Magazines and newspapers help a writer know his own world and give him specific germinal ideas. Also, the playwright who wishes to write speciality plays for specific organizations, occasions, or markets must read particular associated matter.

More important than factual knowledge, reading can provide experience. If an author knows how to write fiction, he can take the reader to the place of his book, and give him the experiences of the book. Such experience is secondhand only in the most superficial sense. Many reading experiences are more vivid, memorable, and affective than most everyday personal experiences. Art is life intensified, and fiction is an art that describes life in detail. To read John Steinbeck's *Grapes of Wrath* is to have lived with the Okies during the Depression. Albert Camus' *The Stranger* can provide the near madness of an Absurdist hell in Algiers. Reading Hemingway's *For Whom the Bell Tolls* permits one to participate in the Spanish Civil War. Additionally, to read lyric poetry will not only increase a writer's emotive experience, but also it will teach him about the tensions of words. John Berryman's poems are excellent examples. Also, every dramatist must read plays. If he finds writing plays necessary to his life, then naturally and compulsively he will read them, both to experience the life they set forth and to learn how others have constructed their dramas. Of course, to experience the life in works of fiction, lyric poetry, and drama, the reader must give himself to each work. And he must read slowly and reflectively.

To glean information about writing technique, a playwright can beneficially read the basic books on the theory of drama, such as Aristotle's *Poetics*, Elder Olson's *Tragedy and the Theory of Drama,* and Eric Bentley's *The Life of the Drama.* Many special theories will be useful too, such as those of Antonin Artaud or Michel de Ghelderode. Some books of criticism can increase his consciousness—for example, H. D. F. Kitto's *Greek Tragedy; The Heart of Hamlet* by Bernard Grebanier; and *Brecht: The Man and His Work* by Martin Esslin. A few playwriting books would also help: George Pierce Baker's *Dramatic Technique,* Marian Gallaway's *Constructing a Play,* and John Howard Lawson's *Theory and Technique of Playwriting.* Every playwright could benefit from occasionally reviewing a basic English grammar and from skimming such technical books as *Creative Writing* by George G. Williams. Books about technique abound, but the writer must find the few valuable ones and not waste time on the others.

In connection with a playwright's formal and informal reading, another source of vitality and ideas are the heroes he takes into his life. Everyone has them. Some remain important for years, some only for a short time. But drawing upon the ideas, vision, and spirit of a great man has always been a way for an individual to raise himself. Heroes of this sort may come from any age and any field. Every writer can help himself by being aware of the few people, other than parents and teachers, who have most affected his life. One's heroes might be artists, thinkers, or men of activity. The more one knows about each, the more they will help. It is beneficial, too, to identify the heroes of men of genius themselves. They also have them and have referred to them in much of what they have said or written.

In any discussion of sources for the playwright, the subject of adaptation ought to be mentioned. This is also connected with reading. Sometime every playwright has the impulse to adapt one of his favorite works of fiction or history into a drama. For the young playwright, it is a rewarding exercise and may help him to produce a first-rate play. For the more experienced, it may lead him to a carefully wrought work. Adaptation is not necessarily uncreative. The dramatist should, of course, be faithful to the spirit or factual truth of the original, but he can still write with freedom. A novel or a biography is simply life material, in a well organized form. The playwright must exercise fine selectivity to discern what to include in his dramatization. Since he cannot employ everything from a longer work, he usually chooses the major crises.

Regarding permission for adaptation, there are several ways to work. The non-established writer should first make his adaptation and then write to the publisher about how to secure permission to adapt. The publisher will give an answer or refer the playwright to the original author, his agent, or his estate. If the young playwright fails to secure permission, he must simply realize he has had a useful experience and then pack the play into his trunk of unproduceable works. The established dramatist has greater persuasive appeal. Since he probably has more writing projects underway, he is likely to inquire first. Most playwrights are surprised, however, at the eagerness of other authors to have their works dramatized. Any work older than fifty-six years can be adapted without permission. Adaptation can be a rewarding experience so long as the dramatist does not feel fettered by the original and so long as writing adaptations does not become his only creative activity.

Reading many kinds of works for varying purposes is, therefore, one of the two major sources for the playwright. A writer is a reader. He can draw from books the knowledge of many other men. From reading plays, he can learn what is best in drama, and he can thereby establish a scale of value for judging his own work.

One cautionary note is that a writer can also destroy himself by reading. Inordinate concern with the classics sometimes generates the fear that they are unbeatable. Awe can destroy the impulse to create. Or a writer can contract the disease of reading the superficial pieces of commercial story tellers and become so taken with them that all his work comes out as mimicry of hollow writing. He should vary his reading from classic to common, from fact to fiction, and from the lyric to the dramatic. A writer must be a self-confident reader, giving himself to what he reads, but never losing his creative identity in the process.

The second major source for any artist, the playwright included, is his own direct experience. This is more immediate than his reading. Although what he reads is pertinent to his art, what he lives is more important. Each writer's personal experiences are so infinite, complex, and unique that generalizations are not very helpful. This book gives more space to using reading as a source, but that should not belittle the significance of the incidents and relationships of a writer's own life. A few things can be said in this context that a playwright might well tuck permanently into his mind.

The artist should not try to control his experience for purposes of his art. To throw oneself into some situation simply for the experience is to live falsely. A writer should become involved in what is natural to him. Not everyone needs to live on the left bank in Paris, bum around the country, or smoke marijuana. A writer cannot perfectly control all his experiences anyway. He should choose his life day by day and try to remain honest with himself. Most importantly, he should look within himself and at life around him as consciously as possible. When he experiences an emotion, he ought at the same time to observe it, to touch it, and to store it for future use.

The most vivid circumstances, the ones affecting the artist deeply, will be the ones he will depend upon the most, even though he may never use them as subjects. Of course, such occurrences may come from any phase of life—loves or hates, desires or rejections. He will draw material from contacts with other individuals, institutions, art works, and even from what he hears. Whenever his feelings are roused, whenever he has emotions, that is when something is happening that will affect his work. He should look to portions of life that move him. An artist is emotional man.

The other phase of direct personal experience as source, one often overlooked by non-artists, is the artist's imagination. His imaginative life probably provides more source material than all else together. His imagination compresses the totality of the rest of his life; it is the amalgam of his direct and indirect experience. It intermixes all he has read, heard, thought, and experienced. His fantasies and his daydreams become his works of art. The playwright's imagination, under control and consciously employed, *is* his vision.

Certain attitudes toward subjects and sources will assist the playwright. He should write what *he* wants and never cater to what someone else suggests or is willing to put on stage. He should despise maturity and love wisdom—if maturity means the process of losing the sharp edges of personality, and if wisdom means not just collecting facts but using the mind as a probe. He should live with frenzy, if that means overcoming natural human lethargy and the impulse to treasure the security of one place and a narrow circle of relationships. He should cultivate friendships; people are his major subjects. He should participate in or learn about as many different activities—physical, social, intellectual—as possible; he will use every one in his work. He should read. He should make a philosophy of life. He should daydream and imagine. His search in life for significance, order, and meaning is also his search through subjects, materials, and ideas. He must live his art.

CREATIVITY

Institutional representatives on all sides assert that creativity is an indefinable quality, rare in human beings and something everyone wants. They call it "a capacity predictive of high individual production potential." They spend thousands of man-hours trying to find those—freshmen, technological trainees, or administrators—who have it. Perhaps modern psychologists will someday discover the exact nature of creativity. But until scientists isolate it and give everyone a shot of it, attempts ought to be made to define it. Enough psychoanalysts, critics, and educationists have tried. Artists ought to stop laughing at those definitions and reveal their views about what it is to be creative. Each practicing artist knows it in himself, but maybe he cannot explain it verbally.

Artistic creativity consists of certain components in the artist, attitudes of the artist, and some external conditions for the artist. The creative facet of a personality is basic to an aesthetic view of life and ultimately to the establishment by each artist of a conscious-subconscious vision. The creative ingredients in a man stimulates his inspiration and also impels him into long work processes. It kindles his joy and gives him pain. It lets an artist love humanity and despise the world. And despite these glittering abstractions—that are not at all abstract to an artist—creativity always includes some specific elements.

The three natural components in the artist are intellect, talent, and compulsion. He must have at least minimal intellectual powers. Neither a two-year-old child nor a moron can be an artist. Equally, intellectual genius does not seem necessary for great creativity. The artist can use as much intelligence as he possesses, and his intelligence must be developed by education. The artist's education should provide him with above average awareness. He needs to develop aesthetic taste. He has to cultivate powers of selectivity. He uses, and hence ought to exercise, powers of reason. He must learn to see likenesses. Although artists seldom create from intellectual themes, they need the ability to make unity with things. All these items amount to intellectual skill and knowledge.

The artist must also possess talent. This necessary component in the artist frightens most people. They shudder at defining it. Non-artists usually think about talent only superficially and conclude that it is unknowable. Artists are wary about mentioning it. They are superstitious, perhaps, and do not wish to tamper with, or know too much about, their own portion of it. Talent is not the same as creativity; it is one component of creativity. For an artist, talent is a compound of imagination and motor skill. It depends especially upon sensitivity, and sensitivity describes a volatile set of emotions combined with a heightened sensory awareness. Talent, then, for a playwright or any other artist, involves the intellectual, emotional, and motor capabilities appropriate to his art.

The other natural component of creativity is compulsion. It can be defined as the artist's inner drive, his power of volition. He *must* practice his art in order to fulfill himself or find any harmony in existence. When a person discovers this drive in himself, then he knows the compulsion to art.

The two qualities most artists find necessary are discipline and love. The artist, insofar as he produces regularly, behaves with volition. He needs will-power, the ability to make things happen. And no one can help. When he works, his control must be a fist. Some like to call it dedication, but it is more the mental and emotional capacity for concentration. The artist concentrates on his dream, or his frenzy, and uses materials to give it reality. Thus, he forces things to happen. His self-discipline applies not only to his best creative projects but also, in a slightly different sense, to the necessary processes of learning his craft, practicing it, and establishing regular habits of work. A writer of plays can write only when he wants to, but if he seldom wants to, then he is no playwright. It is always startling to discover the crowds of pretenders in the arts. They automatically disqualify themselves, regardless of their talent, when they do not *make* art. Wanting is not the same as doing. The artist is always volitional. Secondly, most artists find that love of people is necessary in them. Some would rather call it interest or concern. Whatever the label, an artist must care about others. This attitude will help in his life, and it will show in his work. The man is always seen in the piece of art he makes, and if there is no love in him, his work will be the less.

Creativity also depends upon a degree of freedom for the artist. The condition of freedom applies internally as well as externally. The circumstances in which the artist works may be superficially pleasant and yet imprison his creativity. Or he may be severely limited in his exterior life and yet have great inner freedom. A young American may have a happy home and ample money but be stultified by middle-class pressures for him to be successful and well liked. He may need to live as an expatriate to find freedom, or he may be able to make certain compromises, live conventionally, and still retain his own autonomy. A Jean Genêt can find creative freedom in prison, or a James Baldwin may discover aesthetic stimuli in the socio-economic bars enclosing his race. Freedom for each artist is always relative to the inner state and the viewpoint of his own self. He will undoubtedly be affected by restrictions that are political, economic, social, and personal. The greatest threats to his creative freedom are likely to be those connected with responsibility, security, and time. Freedom of time is freedom of self.

An artist is subordinate to his vision. It consists of his intuition, his bridge of labor, and the works of art themselves. Only by means of a vision—partly conscious, partly intuitive, and fully creative—can the artist pursue his virtue and fulfill his potential. A writer must have something to say. But in order to have something to say, he must see, feel, and think.

As modern artists view human existence, their collective vision of Man vs. Nothingness has its effect on individuals in their midst. Modern man appears to be cursed with inner poverty. Artists exhibit this in paintings, plays, and novels. But, ironically, as an artist formulates such living statements, he disproves the inner poverty of man. An artist creates from inner plenty, even if that plenty is concerned with the grotesque. An artist's will to create reflects his life force. To the artist, the startling thing about man is not that he is a rational animal nor that he may, or may not, have a soul. The marvelous in man is his creativity.

2: *THE PROCESS OF PLAYWRITING*

Whenever I sit down to write, it is always with dread in my heart.

JOHN OSBORNE
"Declaration"

Each writer must devise his own system of developing a work from germinal idea to completed manuscript. Young playwrights often fail to understand the process, or they discover a reasonable system only after many discouraging starts. Experienced writers always seek to improve their compositional process. This chapter describes a pattern of creative activity which should help any dramatist perfect his craft. Its purpose is creative efficiency. Of the fourteen possible steps in the process, the first twelve are the most necessary for bringing a drama into being. Most of the process works equally well for other kinds of writers. Naturally, each individual should make use of this universalized process as it suits his fancy. For example, instead of writing out everything in the entire process, some writers will prefer to retain some factors mentally. And every author will take each new piece of writing through a slightly different process. Nevertheless, a playwright can ignore the steps described here only at the peril of his plays. This process, then, is ideal, but ideals are often necessities.

THE BEGINNING

Experienced writers know that one crucial factor in the compositional process precedes anything most non-writers would conceive as an initiating element. A play usually begins not with an idea but with a feeling. This feeling is a rising *creative compulsion.* The playwright lives for a time without the compulsion, and suddenly he realizes that the act of composition is day by day becoming a necessity. His view of life sharpens. His senses and intuition become more lively. He realizes that he will soon find an initiating idea for a play. This creative compulsion is like a field of rich soil ready for seeding. The playwright must be internally ready before an idea will ever strike him as right and thus set him working on a play. The first step, then, in the process of writing a play is the inner readiness of the playwright, a rising creative compulsion.

The second step in the process is the *germinal idea.* Some imaginative idea to which the playwright applies his craft always provides the basis for the formulation of a play. It is germinal because nothing in it is developed, and yet it contains the potential for a total drama. Although the perceptive playwright notices many possible ideas for plays, a true germinal idea strikes him rarely. The ones usually resulting in completed works meet at least three qualifications. A good germinal idea strongly commands the conscious interest of the writer. It must be one that he can live with daily for months or years. In such an idea, he should sense the potential for the dramatic. He cannot, of course, recognize dramatic possibility unless he owns a knowledge of the parts of drama. Another qualification for a germinal idea is that the playwright should consider it capable of making life excitement. That is, it should contain the potential for a work that intensifies his life, the lives of the characters eventually included, and the lives of the audience members it ultimately contacts. These qualifications and the recognition of the idea itself depend upon the extent of the artist's vision and upon the intensity of his creative compulsion.

Each true germinal idea will be unique. If it is not different from all others, the playwright will ignore it anyway. Germinal ideas vary in quality, kind, and frequency.

When a germinal idea may occur is obviously unpredictable, and the source of any particular idea is equally so. About the time of occurrence, an idea may make itself immediately apparent, or it may be recognizable only in retrospect. It probably will not occur in the moments while a writer goes to sleep. A playwright ought to ignore the popular myth that the best ideas suddenly pop forth as sleep begins. Few writers jump out of bed to jot ideas into a handy notebook.

What are the best sources of ideas? Each writer will learn to look to those which have proven fruitful in the past. A playwright can consciously search for and successfully find germinal ideas, though not usually within a predictable time limit. The major source categories are: present experience, conversation, reading, memory, and imagination.

A writer carries about many ideas, some mental and some written. Most authors do not write down ideas that are merely notions. Sometimes a playwright even forces himself to carry a full-blown germinal idea mentally because he wants it to remain changeable for a time so his imagination can enrich it. The best ideas, however, should always be recorded. They slip out of memory with surprising ease. Despite the faith some authors place in the activity of the subconscious, the writer cannot really work on an idea unless it is on paper. Besides, the accumulation of written ideas, although they seem disconnected, can amount to the construction of a rough scenario.

With most good germinal ideas, the writer discovers the idea, carries it mentally for some time, and then writes it in short form. By putting it down concretely, he sets it. The play is fully underway. From that time on, he adds notes to it and helps it grow. A germinal idea may originally be only a sentence or two, but it must increase quantitatively before the writer can start shaping it. It is important to keep an idea fluid as long as possible and especially to avoid considering any audience for it. Most significantly, a germinal idea never achieves reality until written.

Germinal ideas most often appear as one of six types. Each stands in a relative order of frequency for most playwrights. The categories described are not so important as the indicated range of recognition.

The first type of germinal idea is a person or a character. Because human action forms the core of drama and because human behavior fascinates, writers naturally use people as the most obvious starters for plays. Every person probably could furnish material for some kind of drama. But a playwright must learn consciously to choose the persons who most interest him and who best fit the kind of drama he wants to make. Although unusual individuals often attract attention, more familiar persons or *imagined* characters more usually provide germinal ideas. To write character studies as exercises not only permits the writer to practice his craft but also assists him in sorting through potential materials. The people whom the playwright best remembers probably lead him most rapidly to realizing a character as a germinal idea.

The second most frequent type of germinal idea is place. Certain kinds of human actions tend to happen in predictable kinds of places. And different places attract differing kinds of people. What occurs in a prison camp is not ordinarily what comes to pass in an enchanted forest. And the people found in a hobo jungle are not usually the same as those on board a private yacht. Additionally, place relates to milieu. To a playwright, a place is a delimited space with apparent physical features; milieu refers to a broader world surrounding any specific place. While the writer may well consider place and milieu simultaneously, place often makes a germinal idea, but milieu seldom does. Also, the concept of place is more useful than that of setting. Setting refers to a stage with scenery. A playwright should avoid worrying about staging considerations while he is developing an idea. Some locales attractive to beginners and professionals alike are trite. No matter how unique the details or how startling the contained action, these places may impair a play's originality and impact: bars of every type, living rooms, kitchens, courtrooms, small restaurants, Greenwich Village apartments, bachelor pads, throne rooms, law offices, and porches. Human beings surely meet and do things that cannot be done in those places. When a playwright wants to get an idea about place, he can ask: "Where do things happen of the sort I want to write about? What kind of place makes things happen?" Because place gives impetus to activity and because it relates to mood, writers can use place to generate germinal ideas.

Incidents also make productive germinal ideas. An incident is a rapid change in circumstances for one or more people. It usually has to do with events and deeds, either active or passive; somebody does something to others, or something happens to him. Although incidents can be gentle or violent, the most productive ones are catastrophic, i.e., the change involves highly contrasting conditions. The following examples come from well-known modern plays: an old traveling salesman commits suicide; a man turns into a rhinoceros; a woman stabs a man while he sits in a bathtub. Random examples from one daily newspaper illustrate the easy availability of materials: a young Black sets fire to a corner grocery store whose owner had demeaned his old parents; an AWOL paratrooper surrenders to authorities and shoots himself in the stomach; the five-week-old first child of a couple in their forties dies; a young man unsuccessfully tries to keep his girl friend from piercing her nostril in order to wear Indian nose jewelry. Events suitable for plays abound in everyone's life. Incidents always contribute to story in a play; in fact, story can be defined simply as a sequence of incidents.

Another sort of germinal idea frequently used is conceptual thought. If a writer conceives essences or if he enjoys discovering universals by reading philosophy, thoughts are likely to stimulate him. To use a reflective thought as a spark for a play does not necessarily mean the resulting drama must be didactic. An initiating thesis need not control the entire structure, but it can be suggestive of the other elements, such as plot and character. The writer should take care, however, to avoid thoughts that have been overworked by others or ones that he does not fully understand. Some playwright, for example, might see a germinal idea in Kierkegaard's thoughts about the simultaneous advantage and misfortune of despair, or in Nietzsche's ideas about the necessity for all good men to laugh. Sometimes, writing conceptual thoughts as aphorisms can yield both initiating ideas and material thoughts for later germinal ideas.

Situations are a fifth kind of germinal idea. These can be defined as sets of human relationships between people and with things. The relationships that most usually stimulate writers consist of interlocking emotional attitudes. Contrasting individuals make possible colorful situations. Samuel Beckett's *Waiting for Godot* rests on a fascinating situation. And certainly the interlocking attitudes in *No Exit* by Jean-Paul Sartre and the contrasting individuals—a lesbian, a coward, and a woman with tendencies toward nymphomania who has killed her own baby—make a dramatic situation. The eternal love triangle and a youth's rebellion against parents are two typical, overused situations. Each person lives through set after set of changing relationships each day as he moves from home to work, to play, to sex, and so on. Many playwrights have made frequent use of situations as germinal ideas.

The sixth and final type of germinal idea that occurs to writers with frequency is that of informational area. A writer might decide to write a play dealing in some way with bowling, drinking, Pontius Pilate, state mental institutions, a summer riot in an American city, birth control, Camelot, Ramus, the PTA, a love-in, scuba diving, bull fighting, or Jesus. The decision to use informational material is itself the germinal idea, and then a process of research follows. This can be an excellent way to start a play, and far too few American playwrights take advantage of the store of historical materials available.

A germinal idea, of whatever type or size, sets the writer working. He depends upon both imaginative inspiration and intellectual selectivity in hunting one. It will not ordinarily be a total conception for a play. Nevertheless, the initiating idea is likely to color the entire play which stems from it, impelling it to be centered on character, story, or thought. Most important, the germinal idea gives actuality to the playwright's creative compulsion. He must have one before he can proceed with the process of writing a play.

The third step in the overall process is *the collection*. This is not so singular as a germinal idea. Encompassing a working period, it is the development of the germinal idea. Although this can be the time when the playwright checks such itemized lists of factors as he has amassed, the collection refers to rough notes of all sorts. Each note should add to the germinal idea, even if some do not make obvious connections with it. As the playwright piles up working notes, he will begin to give attention to other ideas he has and to combine them with his basic conception. A working title usually comes to light early. Other characters appear. The writer puts down notes about focus, contrast, and originality. He considers the conventions necessary for the theatre experience he wants to make, and he samples their limitations. At this time, a writer often feels like talking to other people about his work because his excitement is growing.

Most writers discover, however, that to talk about partially formulated work is to ruin it. There is nothing mystic about this. Logically, if imagination and excitement are consumed in conversation, they will come harder in writing sessions.

A typical collection includes such items as these: a cast of potential characters with qualifying traits, a description of place and time, a brief sketch of relationships, a mass of miscellaneous notes, and a statement about the organizational form of the play.

The collection always requires imagination and usually involves research. The playwright must discipline himself to sit at his desk and dream the materials for the play or get to work finding them. Anything can be included in the collection: situations, incidents, conflicts, characters, thoughts, bits of dialogue. Many of the notes he makes are addressed to himself. Some writers keep a journal for this period instead of collecting notes. Depending on the play's type and subject, research may also be necessary. Historical plays require research, but so also do many plays of contemporary life. The writer may need to study locales, types of people, and subject areas. Whether the notes come from imagination or from formal research, everything should be written down. So much material should accumulate that it could not be retained mentally. Ideas remain abstract until written. To compile a collection that functions well, a playwright should get as much as possible onto paper—rapidly.

THE SCENARIO

The scenarios of practicing dramatists assume many shapes and various lengths. The fourth step in the process of writing a play is the composition of a *rough scenario.* This is more than a collection of potentially useful notes. Whereas the collection consists of a mass of quickly written bits, the rough scenario is the first stage in the organization of the materials. With the collection, a playwright concentrates on materials, and in the rough scenario he struggles with form. The rough scenario should not and cannot be so well shaped as the final scenario, but it ought to contain most of the materials, at least in miniature. It should include sections treating all the qualitative parts of a drama. The materials—ideas, incidents, characters—normally retain their sketchy appearance, but with this step a dramatist arranges them in a viable format. The following contains the minimal elements of a rough scenario:

ROUGH SCENARIO

1. Working title.
2. Action: A statement describing what activity the characters as a group are engaged in, most usefully stated as an infinitive verb and modifiers; also an explanation of who changes and how they change.

3. Form: An identification of the comprehensive organization of the play—tragedy, comedy, melodrama—and how the play's action relates to appropriate emotional qualities.

4. Circumstances: Time and place of the action, plus other given circumstances of importance.

5. Subject: A definition of informational area.

6. Characters: A list of impressionistic descriptions; central character identified; relationships noted.

7. Conflict: An explanation of what people or forces are fighting, an identification of obstacles to the major characters, or a description of the disruptive factors in the situation; the basis of tension explained.

8. Story: At least a sequential list of the incidents, but better a detailed outline of the entire story; also notes about how the play begins and ends.

9. Thought: A discussion of meaning, a description of point of view, a list of key thoughts for the whole play, and perhaps for each major character.

10. Dialogue: A statement about the style of the dialogue and the manner of its composition.

11. Schedule: A time plan for the writing and completion of the play.

Such a rough scenario is nearly impossible to compose and retain mentally. If not written, it probably will never exist. It permits the writer to physicalize the materials of his play, and it clears his mind for further creative work.

The scenario, the full and formal treatment of the play, is the fifth stage of the compositional process. Its formulation is an insurmountable task without the previous steps, and it is crucial to the writing of the dialogue. Again, no writer should expect to construct an adequate one without putting it on paper. The scenario need not restate everything mentioned in the rough scenario. It should, however, contain at least the following essentials:

SCENARIO

1. Title.

2. Circumstances: A prose statement of time and place, as these are to appear in the script.

3. Characters: Descriptions of every major and minor character in as much detail as appropriate, using outline form to cover the six character traits for each.

4. Narrative: A prose narration of the play scene-by-scene, concentrating on plot and story, brief yet admitting all necessities.

5. Working outline: A detailed outline of the play, beat-by-beat.

The title, if possible, should no longer be a working reference but a final one. The statement of time and place should be refined from the rough scenario, probably shorter, and certainly more well written. If the play is to be full-length and to have more than one setting, this statement should include a description of each place. The character studies in the scenario should be extensions of those in the rough scenario; they should be fuller, and the major ones should include many traits. Possibly other minor characters may be added during the drafting, but such extras will probably harm the compactness of the whole.

The prose narration comes from the story and conflict segments of the rough scenario. It should consist of one paragraph for every French scene in the play. A French scene is a portion of dialogue demarked by the entrance or exit of a major character. To compose this scene-by-scene narrative, a playwright first makes a rough list of scenes, and then he writes a brief explanation of each. The final narrative results from extended preliminary work, and it depends upon the care with which the playwright wrote the story treatment in the rough scenario.

The beat-by-beat outline should be extremely long and full of particulars. A beat of dialogue is similar to a paragraph of prose. This outline of beats is the most important portion of the scenario. It requires the major structural work in the composition of the play. Its significance even surpasses that of the germinal idea and the writing of the dialogue. The playwright must apply to it all the adroitness and creativity he can. The more detail he packs into the scenario—on paper—the more chance the play will have of being good.

Although the scenario contains only five parts, it will be quite long. Not just because of its length but because of the arduous work of formulation, the scenario demands discipline. A writer who avoids or makes a quick job of the scenario simply demonstrates his laziness or stupidity. Without a scenario, a play will probably be only a dialogue. And conversation is not drama. With a scenario, the dialogue streams clearly and freshly from the playwright into the script. It saves months of revisions. A first draft written without a scenario stands merely as an opaque scenario in dialogue form and usually requires total rewriting. Dramatic composition demands economy and requires that every bit in the play be compact and multiplex. Such writing can happen only through the sort of planning a scenario requires.

The pre-drafting steps—germinal idea, collection, rough scenario, and scenario—usually require more time than the drafting of the dialogue. This is always true if the scenario is adequate.

Each of the pre-drafting items often needs more than one version. The period of putting them together is the playwright's thinking time. During this period, he evaluates the possibilities for each material item put into the play and for the structural principles applicable to the play. He must expect false starts, changes, and developments. The pre-drafting stage in the compositional process demands patience, endurance, and discipline as the companions to creativity. To write a play, a dramatist must discover a germinal idea, make a collection of materials, put together a rough scenario, and formulate a final scenario. Only then is he ready to write dialogue. During all these steps rather than thinking of writing a play, a playwright should think of constructing a drama.

DRAFTING AND REVISIONS

Drafting as a segment of the writing process contains three particular steps: first draft, revision, and "final" draft. The writer should recognize the possibility of all three, but he should compose the first and each succeeding draft as though it were his last. If while writing he depends too much on later revision to correct errors, his product will be shoddy. Or, if he is too conscious of the coming corrective work, he can easily become discouraged. Besides, if he concentrates, each draft will produce some fine particulars that will always remain in the script.

The sixth step, then, in the writing process is the *first draft.* It is easily the most maligned of all steps in the process. How often have the words "a play is not written but rewritten," or some such assertion, been jammed into the mind of every author? The adage carries a seed of truth, but it is time to say that the first draft is by far the most important draft. It is often the *only* draft. Even writers themselves forget this fact or fail to realize it. If the first draft is poor, it means that the scenario was badly conceived or that the author failed to concentrate while drafting. No amount of rewriting or doctoring will cure a bad first draft, only a whole new draft can work the miracle.

A draft of a play is the total wording of it from a scenario into dialogue, stage directions, and other necessary explanations. The rewriting and polishing of existing dialogue cannot produce another draft. A second draft of a play would mean a *complete* rewording of the entire manuscript, with perhaps a few bits of dialogue retained. All writers revise, but suprisingly few make more than one draft of any kind of literary work.

An author writes a draft by knowing what the characters will do and say and then by putting into actual diction those doings and sayings. Many playwrights place the beat-by-beat scenario beside the paper on which the draft will be written. They read a beat in the scenario, then write it in dialogue. A playwright should not let the dialogue wander too far from what the scenario stipulates, but he must be free and imaginative in investing the planned action with verbal reality. Most writers who plan thoroughly and then compose with assurance are surprised to discover how well the dialogue comes out. They command their inspiration because it consists of preparation on paper and inside them much more than it depends on their accidental mood. The professional writer is not necessarily one who makes a lot of money, but one who can control his inspiration.

The time required for drafting a one-act play averages about one to three weeks; a long play usually needs about three to nine weeks. If the writer can turn out two pages of dialogue a day, his writing is going well. Obviously, many exceptions to these averages occur, but they should provide the young writer an idea about writing rate.

Many people have written about when and where to write. But every writer must discover for himself how he *can* work. Once he discovers his possible, not necessarily his best, working circumstances then he should cultivate them. Another significant factor in the discipline of writing is regularity. Most writers *can* write anyplace any time, if they *will*.

How the writer puts words on paper is also a matter of preference. It usually amounts to a habit. Whether a writer uses pencil, pen, typewriter, or a combination is not so important as is his finding the tool that hinders him least. Those with typing facility usually get words down faster, but the longhand writers ordinarily find revision easier. Hemingway's principle of writing longhand and then getting a new look at the words when typed is a workable method. The author who writes in longhand and subsequently types his own manuscript benefits from having several fresh views of his work.

When a playscript goes into typed form, the length of the play becomes apparent. A manuscript format assumes importance partly because by using it the writer can know a script's approximate playing time. Thus, it is well to employ originally a good format. Additionally with regard to length, most writers find overwriting useful. It is easier to cut words than to add them. Compactness in a play comes partially from judicious cutting. A few other considerations can also be helpful. The scenario should be executed thoroughly. Also, the playwright ought to compose not just story beats but ones dealing with character, mood, and thought as well. The novice often fails at this. When working from a scenario, a playwright often overlooks the importance of motivated transitions between beats. He should handle these with care. It is also better to stop writing each day in the middle of a beat rather than at the end of one. If a writer stops when the words are coming easily, he will find that they will probably flow easily the next day too.

Certain attitudes toward drafting make possible a first version of quality. A writer should thoroughly understand his material before starting. He should not try to write a great play, but simply *his* play. It should be simple and personal. The dialogue should run freely, and never be whipped along. When the writer feels as though he is grinding it out, he had best stop for a while and do something else. This should never become an excuse, however, for avoiding the labor of writing. During every minute of his writing time, a writer should believe in his own ability to write. Finally, with the scenario lying near, the writer should not worry with rules and check lists, but he should write with natural fluency. He should stir his energy and write!

When the first draft is completed, most playwrights feel some pleasure. For the first time, the play has a concrete reality. There it is, a physical object! This is likely to be one of the two times a playwright experiences the unique aesthetic reaction of his art.

Immediately upon the completion of the draft, the seventh step of the writing process begins. This is the period of *revisions*. It incorporates ripening, testing, readings, and rewriting. The playwright, if he wants, can at this point more safely talk about his play and show it to others. Functionally, however, he should set the draft aside for a period of time so that when he looks at it next he will have lost some of his glow of accomplishment and be able to view it more technically. The appropriate ripening period will vary from writer to writer and from project to project. A draft should be left alone long enough for the writer to see it afresh when he renews work on it, but he should get to it before he loses emotional intimacy with it. Often a week or two is enough; a month is usually too long.

Several possible methods of testing a draft can be helpful. A playwright could employ them all. The purpose of testing is to spot points for revision. An attentive, but not a plodding, reading comes first. During this reading, the writer should correct obvious errors and make simple changes. Any uncertainties—such as word choice, punctuation, spelling, and development—should be resolved early. The writer who does not command such details will never have the confidence necessary to go on writing, and if he cannot handle such small items he will never be able to control more important ones.

Next, he can beneficially read the speeches of one character at a time. This involves choosing one character, and later each of the others, and sequentially reading only the speeches of that character from the beginning of the play to the end. In such a manner, he can revise for consistency, development, change, and climax in each individual.

During the third time through the script, the playwright should mark the beginning and end of every beat, segment, and scene. He can then leaf through the play and check to see if each of these pieces is whole, focused, and vital. During this reading, the majority of revisions occur. This beat-by-beat analysis of the play is the most important activity in the entire revision process.

If the writer knows someone who understands playwriting and whom he trusts, he may benefit from that person's comments. People of this kind are rare. Nearly everyone who might be willing to read a draft will make useless if not dangerous comments. Even a qualified person may not really understand the play. Usually such persons, even close friends, have neither the time nor the interest to read the draft with the necessary attention. Most writers learn by unfortunate experiences not to let anyone read a draft. Still, some like to have an intimate friend read their work. Although this person usually is non-expert, they like the ego boost they get from a happy reaction. A writer should learn to depend on his own judgment alone for a revision critique.

One sort of contact between others and the draft can be helpful to the playwright. For many it amounts to a necessity, and all proven playwrights ought to have such contacts available. This is a reading by good actors under the tutelage of an intelligent director. Of course, only those playwrights associated with a theatre group or those who live in a theatre center and have many close friends among theatre artists are likely to get a competent reading. Such a reading should come only after the revision process. If the actors are willing to give the draft merely a cold, unprepared reading, the playwright will benefit little. The rehearsal period for a reading need not be long—one or two read-throughs will do—but the actors must be familiar with the piece. The company should read for a selected, not a general audience. The playwright should follow the reading in a manuscript and make his own notes during and after the reading. The comments of all the others involved have little worth, except as a community of opinion. The playwright should ignore most compliments and criticisms. Those which arise in his own mind will probably be the most valid. Sometimes when several people with sound judgment are dissatisfied with a scene, the playwright can benefit from looking for reasons. He should pay attention to their dissatisfaction, not to their reasons why nor their suggestions for change. The reasons a scene may be unsatisfactory frequently have nothing to do with that scene, but with some earlier one. Also during a reading, the writer should observe to what degree the play stimulates emotion in the actors. His own perceptions, the community of response, and the actor emotions are the chief reasons for a reading. An additional function of a reading may also be that it might interest someone in producing the play later.

Another method by which theatre companies help playwrights develop plays is improvisation. A growing number of experimental companies, such as The Open Theatre, assist playwrights by means of improvisational sessions. The working methods of such companies differ so much that it is difficult to generalize about this sort of imaginative cooperation. Sometimes a playwright simply gets germinal ideas from listening to the actors. With other groups, the writer acts as a kind of organizer and transcriber of material. And some ensembles work with a partially developed script and help expand it by having the actors explore the possibilities by joining their imaginations with those of the playwright and the director.

At present, theatre groups in nearly every American city could offer a playwright a useful reading, and a few can help with improvisational sessions. Happily, an increasing number of groups are willing to help. Most, however, are more likely to be interested in reading a finished play than in aiding a playwright with a draft.

Whether a playwright views his play by reading it, by getting advice from an expert, by hearing actors read it, or by combining these methods, he must ultimately discern for himself the needed changes. How much time he spends with revisions is unpredictable. But in order to accomplish a finished play with alacrity, he should work a regular, minimum amount of time daily. Playwrights average about four to six weeks for revising a one-act play and about three to six months for a long play. These are, of course, only generalized averages. Some playwrights rework plays for several years, while others never change a word between the first draft and rehearsals.

If a playwright discovers the first draft to be unsatisfactory, he may well destroy it. When this happens, the scenario is usually at fault, and it too must be reworked. Another typical way of re-drafting is to use a beat of dialogue in the first draft as the scenario for a newly worded beat. Most playwrights rewrite beats and scenes; some recast acts; but few compose whole new drafts of long plays. Nevertheless, a second, third, or fourth draft will sometimes be necessary.

A playwright should learn to judge when the revision process has sufficiently altered the play. He will recognize when the script is nearing production state. Naturally, this power of recognition increases with experience. But with most writers it comes easily. This reaction toward the play indicates the start of the eighth step in the compositional process, *the "final" draft*. "Final" belongs in quotation marks because a play physically consists of so many alterable, minute particles—words—that no play ever really reaches an absolutely final draft until the playwright decides to become a mailman, to join a police force, or to die.

The "final" draft is more than a clean manuscript. It involves the special work of *polishing*. Polishing a play primarily requires meticulous *reading*. A writer should consider each word and revolve it in his mind until he knows it to be the exact word for that one location in the script. He must give the same attention to the individual sounds in each word, and to the play's phrases, clauses, punctuation, individual speeches, stage directions, and beats of dialogue. Few changes in plot, character, or thought occur at this stage. Polishing usually involves only the diction. To polish a one-act play normally requires a week and a long play about two or three. A writer can spend too much time with this. He should work through the manuscript once or twice and then stop. He ought not to change words merely for the sake of changes. Too much word juggling can waste the playwright's time and dull the play's luster.

After the play is polished, the final typing should take place, including an original and at least one carbon copy. When the playwright eventually gets this "final" manuscript completed, he should read it quickly but attentively and correct typing errors. Even the best typists make some. These mistakes can be corrected in ink on the manuscript. This work ought not occupy more than a day. Many producers believe that a play manuscript which has not been proofread represents the work of a playwright who does not really care about his play. Proofreading is essential.

With the completion of the "final" draft, the major portion of the playwright's solo creative work is over. He now has a completed playscript ready for the staging part of the total process. He should turn his activity to getting a production for the play.

As a ninth step in the process of playwriting, the business of *submission* is as important as any other, even though it is not creative. Circulating the finished play is part of the playwright's professional and solo work. Although this step does not affect a play's internal nature, no play ever achieves completion until produced. To get a production, every playwright works as a salesman in the activity of submission.

A dramatist should send out his play the day he proofreads the final manuscript. And he should continue sending it out, for years if necessary. Most new writers are too sensitive about submission. After a few rejections, they quit. A good plan involves compiling a list of a dozen appropriate submissions with names and addresses, and sending the play to each until it gets an acceptance. On the day a play returns, it should go out immediately to a new place. If all twelve "customers" turn it down, the playwright ought to reread the piece, revise it, and begin the submission process again. Or maybe he should lock it in a trunk forever. If a playwright combines talent and craftsmanship with a business attitude, a reasonable method of submission, and an understanding of markets, he will get his play produced.

REWRITING AND PRODUCTION

A play is merely a script until it steps onstage as a performed drama. Even then it is a drama only during the period of time that actors give it life. The next four steps, therefore, in the process of creating a play have to do with production. They are pre-rehearsal work, rehearsals, performances, and post-production revision.

One day during the period of submissions a letter or a phone call will notify the playwright that some producer or group likes his play. A theatre company wants to produce it. The writer will feel like framing the letter and calling everyone he knows. Immediately, the tenth step in the creation of a play begins, *the pre-rehearsal work.* The playwright should try to arrange personal association with the first production. The time and money he may spend will bring him unlimited benefits not only with the completion of that particular play but also by giving him knowledge about the theatre for all his future plays. Most theatre organizations will furnish some assistance so that the playwright may attend rehearsals and performances. They can assist with travel allowance, room, board, and at least a modest royalty. Also on that first beautiful day the playwright knows his play will be produced, he ought to begin a production log. Many writers even maintain a journal beginning with the germinal idea. A playwright needs a functional record of the production procedures, the names and addresses of people involved, and a permanent container for notes about revision.

After receiving an offer for production, the playwright should accept or decline promptly. Usually, he then enters into some sort of contract, if only an informal one. It is important to note that he should correspond and handle all details with dependability and dispatch. Further, the entire production will benefit if everyone strives to be pleasant and pliable.

Usually, the first contact will be with the director, but in some cases, especially in the professional theatre, the initial notice comes from the producer. Whenever the playwright learns the director's identity, he should expect and encourage a detailed correspondence or a series of conferences. The playwright's work with the director will normally be extensive. Before rehearsals begin, any competent director will carefully analyze the play. At this point, the director will have many questions and probably at least a few suggestions for dialogue revision. If the distance is not prohibitive, pre-rehearsal conferences will help the progress of the play. When the playwright is "permanently" associated with a theatre company or if the production is a professional one, the pre-rehearsal conferences are likely to be numerous. The pre-rehearsal period is one of reactions and decisions about specific items and possible changes in the play. The director will be thinking theatrically, and the playwright can probably sharpen the play's theatrical qualities by heeding the director, *if* the latter is a good one.

When the playwright joins the theatre company for *the rehearsal period*, he takes his eleventh step toward creating the play. Ideally, he will be able to attend tryouts and all rehearsals. But this is not always possible for the young playwright and the amateur company. In any case, his first job in rehearsals is to evaluate the production situation and get acquainted with his co-workers. He should attune himself to the company and try to discover how he can help most and thus best improve his play. He should avoid interfering with the work of the company. In fact, he should be more sensitive to the company's needs than to his own. His fellow artists will broaden and deepen his play and give it actuality. Hence, he should, above all, be human.

If in the director's opinion the play is too long, the playwright should cut it to an acceptable length. If he has employed the appropriate manuscript format and if he has limited the number of pages to about thirty to forty-five for a one-act play and ninety to one hundred for a full-length play, cutting will probably not be necessary. In any case, the playwright should be agreeable to cutting. But even though a director may have some good suggestions, the playwright alone should do the cutting. The playwright must face the fact that most productions require cuts as well as changes.

During each rehearsal he should listen for passages that fail and for new dimensions of meaning. Seeing the characters come alive is more important than just watching the actors. He can readily learn to distinguish between the successes and failures of the actors or of the play. He will receive suggestions for changes, and he will spot many errors himself. With each no matter how minor, he must carefully consider whether to alter the play or not. Each change during rehearsals ought to be mutually acceptable to playwright and director.

Not all rehearsals are alike. The rehearsal period consists of at least seven segments. First comes reading rehearsals; the actors read the play aloud several times and discuss interpretations with the director. The playwright often enters the discussion. During readings the playwright should note the passages that are unclear to the actors, and he should consider altering the phrases that specific actors cannot articulate. Second, during blocking rehearsals, the director designates stage areas and floor patterns the actors will use throughout the play. The playwright can derive new stage directions for his manuscript from these rehearsals and from the next set too. Third, businessing rehearsals are when the director and actors work out the other physical movements. Stage business consists of the detailed physicalization of the play. Such business might be as insignificant as an actor scratching his cheek or as important as an actor opening a treasure chest. If the company is competent and imaginative, its members will amaze the playwright with the life-like actions they devise. Fourth, although there are many names for it, the next set of rehearsals can best be called the patience period. During this time the actors struggle to memorize lines, perfect moves, and generate emotions. Properties, scenery, and costumes are taking shape. The director must be everywhere at once. He prods, corrects, demands, and sympathizes with everyone. The production seems to be an impossible mess. Like the director, the playwright must endure this developmental period with caution and calm. By the end of it, about two-thirds of the rehearsals will be over, and the major changes in the script should be completed and incorporated into the production. From this point on, only minor alterations in diction can be made with ease. Fifth, the polishing rehearsals come next. Whereas earlier each rehearsal dealt with only a few scenes, polishing rehearsals usually require run-throughs of acts or the entire play. The drama now truly begins to live. Many moments in these rehearsals may even surpass performances in the matters of actor inspiration and intensity. The playwright will probably be doing little work on the play at this time, but he can be central to the publicity program. Sixth, the addition of costumes, makeup, lights, sound effects, properties, and scenic pieces occurs during the technical rehearsals. Again, the production seems to go into nearly total confusion. If theatre is ever hectic, it is now. The playwright should stay out of the way, but be available to help as he can. The dress rehearsals are the seventh and final segment. If he gets along well with the director, the playwright can helpfully make notes about errors or improvements during each dress rehearsal. He should always give these notes to the director and *never* to the people to whom they apply.

The playwright can learn more about dramatic art from going through a rehearsal period of one of his plays than he can by watching fifty performances or reading a hundred books. But to learn as much as possible, he must get involved and concentrate while seeing and listening. Some rehearsals with some companies will be horrible and discouraging, but most will send the playwright's imagination and inspiration soaring. Every production experience will be unique and contain varying problems. The playwright, regardless of his level of expertise, can profit from joining the rehearsals of his play as carried out by different sorts of groups—amateur, semi-professional, and professional.

A playwright's twelfth step in the total process is attending *performances* of his drama. He should be present at the cast call before each performance, and he should avoid a lobby display of himself before the play begins. As the play starts, the temptation to pace the lobby or stand backstage will be great, but the playwright's place is in the audience. Witnessing all possible performances from beginning to end will enable him to judge each item of material and every bit of form in his play. And especially, he will have the opportunity to observe audience responses. Unless he has the chance to study his play and audience reaction to it *in performance*, he can never know whether or not it works. Indeed, he will never know it as a completed work. Only in performance is a play alive and complete. Only when acted does it become a work of art.

The playwright, like the director, has no place backstage during any performance. A competent stage manager never permits anyone there but actors and crews. At intermission, the playwright ought once again to stay away from the actors. This is the appropriate occasion for him to walk among audience members in the lobby. He will probably get more of an impression from their general mood than their individual comments. To judge the success or failure of the play as produced, he must analyze the audience response during performances.

Also during the performance period, a playwright must learn to discern the difference between the quality of a production and the quality of his play. He needs to be able to separate the performances from the characters. This knowledge grows within most playwrights only by seeing more than one production of each of their plays.

Further, a playwright should realize that amateur companies ordinarily make few, if any, changes in a production after it opens, while professional companies frequently introduce modifications of all sorts until the director and producer are satisfied.

The final concern of a playwright during the performance period is to make copious notes about possible revisions after each evening's performance. He may get these into the production immediately if it is a professional company, or if he is involved with an amateur group he may hold the notes and rewrite later. In any case, the performances will probably suggest many advantageous changes, and if the dramatist does not record these immediately, he will forget most of them.

The next step, the thirteenth, in the process is the execution of the changes suggested by the production. The playwright should compose a *post-production version* of the play. Just keeping an understandable manuscript is sometimes difficult in the course of a rehearsal and performance period. The playwright's job with the play is unfinished until he composes a new version. The quantity of alterations is irrelevant, but be they few or many, he should turn out a clean manuscript. The sources for this new version most probably will be: (1) changes made during rehearsals and included in the first production, (2) notes made during dress rehearsals and performances, and (3) post-production conferences with the director, the designers, and the chief actors. Although open sessions of discussion following one or more performances may turn out to be interesting, they are seldom useful to the playwright.

Most plays benefit from the discoveries and revisions a dramatist makes during the production period. The experienced playwright does not think of a play as finished until he comes up with a post-production version. It is, incidentally, accepted procedure to send at least one copy of this version to the director—the man who probably helped by suggesting numerous improvements. He is also a person who can suggest other directors or theatre companies who might be interested in producing the play. After the run of the play or whenever the playwright must leave the company, part of his job is to be certain that he is leaving only warm contacts among all those associated with the production. He should abide by an old theatre saying: Never make an enemy in the business; your contacts are your next job. The final item of post-production work is for the playwright to begin his submission activities once more with the new version of the play.

The fourteenth and final step in a playwright's work process ought to occur one way or another, but may not. It is the *publication* of the play. Concerning the playwright's work if he has a publication opportunity, several things are important. First, a playwright should never solicit or permit publication of his play until he has seen it produced and subsequently revised it. Publication usually results either from an approach by a publisher after a professional production or from the sales effort of an agent. Occasionally, with the play-leasing companies or with specialized theatre magazines, submission by the playwright will work. When an editor accepts a play for publication, he may also suggest some changes. Ordinarily, these will be details concerning manuscript format, punctuation, quantity of stage directions, and expanded pre-play explanations. If he agrees, the playwright should execute such changes promptly. After the requisite time passes, he will receive the printer's galley proofs, and he should read them meticulously. The playwright should correct every printer's error, but he should originate no changes at this point.

The published script, when it appears, is like a new baby. One has the impulse to carry it close to the heart and to show it to everyone. The proud playwright should remember, however, that a printed version is still subject to change and undoubtedly will be changed with each future production. The most fortunate circumstance about publication is the potential of wide distribution. Perhaps it will elicit many other productions.

Playwriting requires an extended time span. Such a long period of regular work demands unusual patience, craftsmanship, and discipline. Although every kind of writing is difficult, playwriting is for most writers the most arduous. Additionally, a playwright depends far more on others for the completion of the individual work. Because of the extreme problems involved, all playwrights soon accumulate an embarassing number of plays which sit frozen in various stages of the process. Practicing playwrights continually consider the questions: What happened? Did I lack the discipline to finish the play, or was it from the beginning a poor idea? Should I take up an unfinished piece and complete it? Or should I look only to new ideas? Most old projects are best left alone. A new germinal idea, however, should stimulate the playwright's resolve to work it through to completion.

In summary, the fourteen most important stages in the process of creating a play are: (1) the creative compulsion, (2) the germinal idea, (3) the collection, (4) the rough scenario, (5) the scenario, (6) the first draft, (7) revisions, (8) the "final" draft, (9) submissions, (10) pre-rehearsal work, (11) the rehearsal period, (12) performances, (13) the post-production version, and (14) publication. If a writer ignores any phase, his play is likely to suffer. The craft of playwriting demands that a dramatist consciously know and control every factor in the process. A playwright's working habits and the mental methods he employs determine the style of his art work, the play.

PART II: *PRINCIPLES OF DRAMA*

Yes, my friends, have faith with me in Dionysian life and in the rebirth of tragedy. . . . Arm yourselves for hard strife, but have faith in the wonders of your god!

FRIEDRICH NIETZSCHE
The Birth of Tragedy

3: *STRUCTURE*

> *A tragedy, then, is the imitation of an action that is serious and also, as having magnitude, complete in itself; in language with pleasurable accessories, each kind brought in separately in the parts of the work; in a dramatic, not in a narrative form; with incidents arousing pity and fear, wherewith to accomplish its catharsis of such emotions.*
>
> ARISTOTLE
> Poetics

The total action of a man makes up his life. All human activity, to some degree, is action. It is not merely movement, but change. Action in life ranges from the simple, such as when a person blinks his eyes or scratches his neck, to the complex, such as a man's decision to kill or not to kill his enemy. The continuing actions, from birth to death, of each human being are infinite and perpetual. A person's actions are the most fascinating things about him. Although everyone tries to impose order on his life and attempts, insofar as he is volitional, to control his activity, no one totally succeeds in structuring his action. Art, however, is different from life. Whereas life consists of diverse action, *drama is structured action*.

THE STRUCTURE OF ACTION

To study the composition of drama is to study the architecture of action. Simply defined, action is human change. Movement, activity, and alteration are kinds of change, types of action. Each drama is a connected series of changes. Each is action organized according to some logical probability. Changes of physical position, external milieu, inner feeling, mental attitude, interpersonal relationship—all are examples of human activities which can serve as materials. Whenever a playwright combines a group of separate actions into some sort of whole, he establishes a summary action. This becomes the overall form of a play. In this special sense, action operates as both material and form in dramatic art. When a dramatist relates human activities with unity and probability, he makes a beautiful art object, a play. And the special beauty of any play is beauty of action.

Generations of critics notwithstanding, Aristotle intended no mystery when he wrote of "the imitation of an action." He merely stated the principle that art is *imitative* in that it is *artificial*, and artificial simply means *man-made*. That which is artificial is not natural. A natural object comes into being through its own powers of generation, life, and death, or by means of universal physical principles. An artificial object, on the other hand, must be created by a maker. Only an artist can give a piece of marble an unnatural form; only he can make it into an artificial object. So it is with a play. Action exists in life, but when a playwright imitates such action, he makes an artificial object, a play. The true artist never copies life photographically. His work is mostly mimetic, i.e., constructive or formulative. A mimetic action, then, is to some degree crucial to every play.

Action and imitation are not the only special general qualifications for the formulation of a play. There are various kinds, or forms, of action and various sizes of actions. In the most notable definition of tragedy, Aristotle mentioned the word *serious*. With it, he identified the kind of action best suited to tragedy. Another kind is comic, and a third is melodramatic, or the seriocomic. Each kind, although identified by one word, has appropriate powers in and of itself. These appear more fully in this chapter's next section which deals with the basic forms of drama. Also, because each action properly occupies a certain length of time and needs a certain amount of detail, a certain magnitude, or size, is necessary for its completion.

Hence, to speak of the structure of a play is to speak of its formal cause. This is the action, as qualified by imitation, powers, magnitude, and completeness. Although these factors may appear abstract at first, they will become more concrete in explanations which follow. All these principles, and those to come, apply to any play. They function as fully in epic or absurdist dramas as they do in realistic or well-made plays.

Action in drama is best understood in terms of form-matter relationships. All components and all connections in a play have to do with the formulation of parts, both quantitative and qualitative. Quantitative parts of a play can be seen and counted. Moving from the larger to the smaller, they are: acts, scenes, segments, beats, speeches, clauses, phrases, words, and sounds. Qualitative parts of a play, on the other hand, are only partially identifiable after a play is finished, but they are readily apparent as more or less separate entities during the writing. The qualitative parts are: plot, character, thought, diction, sounds, and spectacle.

Action in drama is change, as both process and deed. In addition to movement and activity, it implies alteration, mutation, transformation, expression, and function. It also inheres in feeling, suffering, passion, conflict, and combat. Dramatic action consists of a series of singular acts. They can be simple or complex. Whenever such human activities are given unity in a drama, the resultant action assumes a structure. Thus, structure in drama amounts to the logical, or causal, relationships of characters, circumstances, and events. But the logic of drama differs from that of such disciplines as philosophy or physics. It is more nearly related to credibility, everyday likelihood, and common sense; it is the logic of both life and imagination. The logic controlling any single play is unique to that play. Thus in drama, action is in some manner reasonably related to the given circumstances, i.e., specific characters in a limited situation.

Structured action is plot in drama. Since every play possesses some sort of organization, *all plays have plot*. Probably only an insane person could write a completely chaotic, plotless play. The organization of a play is its plot. Because plot is so widely misunderstood, this chapter's title is "Structure," but plot is its subject.

One of the most important distinctions that every writer ought to understand is this: *plot and story are not synonymous*. They are, however, intimately related. Plot is overall organization, the form, of a literary work. Story is one kind of plot; it is only one particular way to make form in drama. Suffering, discovery, reversal, story, tension, suspense, conflict, penetration, contrast—these are examples of the various factors that may contribute to plot. Though extremely useful, story is but one among the many ways the materials of a play may be arranged. Story, simply defined, is a sequence of events, and such a sequence can provide one kind of unity in drama.

The structure of action in any play comprises the form of that play. In mimetic plays, unified action controls the selection and arrangement of all other parts. A plot may, for example, consist of an extended image of suffering, a group of discoveries or revelations, a series of events, or a chain of crises. Action may gain unity in many different ways. But, whatever form a play might assume, its plot will have some specific kind of structured action, a structure featuring wholeness, emotionality, and magnitude. In any given drama, the plot is the unique structure of its action.

TRADITIONAL FORMS

The form of an art object is its shape, the order of its parts. Plot is the characteristic general form of drama. Every work of art, or individual drama, comes into being as a result of four causes. Form is only one among them. It is always dependent upon the other three—the materials, the artist, and the purpose. Form, in turn, controls them during its construction and after its establishment. Actually, as a playwright composes a play, he never fully separates form and material. A perfectly formless object cannot, after all, be conceived, nor can a form be imagined without some compositional matter. Thus, form in drama consists of materials, or parts. In any play, the arrangement of the quantitative parts (e.g., the scenes) and the qualitative parts (e.g., the thoughts) describes the dramatic form. No two plays ever have precisely the same form.

All these considerations explain form in general. To speak of form more specifically is to mention the types of form and the plotting of an individual play.

Three broad types of form have always been useful to playwrights. They are tragedy, comedy, and melodrama. Although each of these admits many sub-types and even change somewhat from age to age, they still furnish useful conceptions of dramatic organization. Because form is not a fixed external pattern, the characteristics mentioned in the following discussion are neither absolute nor always essential. Each play must develop uniquely. Its material parts connect with each other, organically and internally. Extrinsic rules of construction usually limit the creativity of the artist and stunt the growth of the play.

Each major form of drama—tragedy, comedy, and melodrama—implies a kind of structure distinguished by certain powers, or emotive qualities. When a playwright arranges dramatic materials in one of these three forms, the resultant play generates powers both unique and appropriate to that form. For example, a tragedy will possess a special seriousness; a comedy will contain humor; and a melodrama will feature a mixture of apprehension and relief. The intrinsic powers of each play's form amount to its central core of emotionality.

Tragedy is unmistakably serious. This quality has many implications for the formulation and writing of a play. A genuinely serious play is first of all one that is urgent and thoughtful, though not dull or pedantic. It features an action of extreme gravity in the existence of one or more persons, often a matter of life and death. It deals with incidents and personages of consequence. The characters encounter forces, opponents, problems, and decisions actually or potentially dangerous. Second, a serious play comes into being through the use of action arranged to move from relative happiness to disaster, materials selected to generate gravity in the whole, style controlled to express painful emotions, and purpose applied to demonstrate life's meaning and man's dignity. Third, tragedy needs certain kinds of qualitative parts. The plot usually employs a story with movement from harmony through disharmony to catastrophe. The situations are *fearful*; they involve one or more characters of some value who are threatened by worthy opponents or great forces. The chief character, or protagonist, usually exhibits stature, enacts more good deeds than bad, struggles volitionally for the sake of something more important than himself, and suffers more dreadfully than he deserves. Insofar as this is precisely accomplished, the protagonist is *pitiful,* in the special tragic sense. The thought involved is more ethical than expedient.

Since Aristotle, many critics have echoed that the unique emotive powers of tragedy are pity and fear. The noun forms of these two words can be misleading. If a playwright sets out to arouse pity and fear, he may first think of an audience. But when a playwright worries primarily about audience reaction, the play suffers. The job of the dramatist is to create an object, not persuade a crowd. Although he can consider the audience, he should focus his attention on the play. He must establish the potentials of emotion in the play before it can ever infect an audience. Therefore, the pity and the fear should be thought of as qualities in the play and not as conditions of an audience. For a playwright, the adjective forms are more useful—a *fearful* action or situation, and a *pitiful* protagonist or set of characters.

Many extant tragedies could serve to illustrate all the characteristics mentioned above. But to study all the tragedies ever written would be to discover that each uses the principles differently and none holds exactly the same powers. *Oedipus the King* by Sophocles and Shakespeare's *Hamlet* well exemplify traditional tragedy in their respective representations of an admirable protagonist struggling with purpose against both human and cosmic forces. But even though Greek and Shakespearean tragedy have some overall resemblances, they also have many differences, such as style of diction and use of sub-story. From the eighteenth century to the present, however, tragedy has become ever more disparate. Tragedies about moral order are now less frequent; most contemporary tragedies deal with a protagonist's struggle against social and psychological forces. Widely divergent examples of tragedy abound, from Henrik Ibsen's *Hedda Gabler* to *A Streetcar Named Desire* by Tennessee Williams, and from August Strindberg's *The Father* to *Mother Courage* by Bertolt Brecht. In all cases, playwrights use the tragic form to show human action at its most intense and to examine the nature of man and the meaning of existence.

Comedy, the second broad type of recurrent dramatic form, deals not with the serious but with the ludicrous. It is not the antithesis of tragedy, but its complement. Representing another side of the nature of mankind, it thus employs a different kind of human action. The core actions in comedies explore the deviations of social man. Comedy upholds the normal and the sane, and it does so by exposing the anormal. Indeed, the ugly, sometimes as the grotesque and sometimes as the odd, is the very subject of comedy. The contrast between the normal and the eccentric in human nature and conduct is always crucial to comedy. Excesses, deficiencies, deviations, mistakes, and misunderstandings insofar as they are anormal are ugly and therefore apt for ridicule.

Three general principles of comedy are essential to its structure. First, *a mood of laughter*—sweet or bitter, pleasant or unpleasant—should be maintained throughout the comic action. Second, a humorous play comes into being through the use of action formulated to move from relative unhappiness, usually an amusing predicament, to happiness, or a pleasing resolution. Its materials generate laughter in the whole, and its style expresses wit. Its purpose may be to ridicule, correct, mock, or satirize. With good reason, writers often repeat the truism that tragedy requires an emotional view of life, while comedy demands an intellectual one. Third, comedy also needs special kinds of qualitative parts. The plot usually employs a story with movement from harmony through entanglement to unraveling. The situations are laughable; they involve one or more normal characters in conflict against, embroiled with, or standing in contrast to anormal characters or circumstances. The protagonist can be an eccentric facing a relatively normal world or a normal character encountering confusion. Insofar as one type or the other is established, the protagonist or what surrounds him is *ridiculous,* ludicrous in the special comic sense. The ridiculous may be defined as a mistake or deformity not productive of pain or harm to others. Further, the conscious thought articulated in comedy is more likely to be witty and satirical than moral or expedient.

The powers of comedy are best explained not by using the words ridicule and laughter, but by identifying a ludicrous or laughable action or character. For comedy, a playwright's job is to create such characters and direct them into such actions. Only incidentally should he consider the potential audience reactions. Playwriting is a matter of constructing an object with appropriate *beauty,* i.e., a verbal whole having appropriate unity, magnitude, and emotive powers; playwriting is not the practice of audience suasion. Similarly to tragedy, a comedy must internally establish and hold its own requisite powers.

Because comedy frequently depends upon the exposure of deviations from societal norms, it does not wear as well as tragedy from age to age, or culture to culture. Hence, not so many comedies continue to be performed after their first few productions. Some rare comedies contain relatively common or universal norms and abberations. Plays such as *Lysistrata, The Miser*, and *The Taming of the Shrew* have found production in nearly every age, from the time they were written to the present. But even with these, their success largely depends upon contemporaneous and inventive interpretation by a skilled director and an imaginative company.

Contemporary comedies tend to amuse contemporary audiences more readily than older comic plays. Perhaps for this reason, modern tragedy and comedy do not usually receive the same treatment from critics. *Tragedy* has become a value term, but *comedy* has not. Many people will permit the use of the term *tragedy* only in relation to a very limited kind of drama, a serious play of the highest and most formal sort. This practice does not extend to *comedy*. These people are likely to admit a far wider range of comic literature without much controversy. Reviewers, critics, and academicians often argue whether or not some recent tragedy, such as *Death of a Salesman*, is really tragic. But they seldom, if ever, discuss whether or not a current comedy, such as Neil Simon's *Come Blow Your Horn*, is funny. The well written comedies are obviously amusing. Also, because fine tragedies tend to endure longer than fine comedies, perhaps it can be understood why some people think of tragedy as a higher form than comedy. One is not better than the other; they are simply different.

Comedy as an overall form uses certain broad kinds of materials, but there are numerous comic sub-forms. Superficially, comedy differs from tragedy in variety. There are many sub-forms of tragedy, too, but they do not have widely used names. Some of the most common types of comedy are farce, satire, burlesque, caricature, and parody. Comic sub-forms sometimes possess names identifying the particular qualitative part which is most focal or most exaggerated: situation comedy, character comedy, comedy of ideas, comedy of manners, and social comedy.

Melodrama, the third form, is the least revered by critics, and yet it is the most often employed by twentieth-century playwrights. Although this kind of play is more easily constructed than a tragedy or a comedy, it can be equally skillful, beautiful, and valuable as an art object. A poorly written tragedy is boringly sentimental; a sickly comedy is thin and dull; but even a weak melodrama can command rapt attention from an audience. The proof of this is that two modern entertainment industries depend upon melodrama as their basic material—television and cinema. But from Euripides' *Electra* to *Look Back in Anger* by John Osborne, playwrights have written melodramas of high quality for the stage.

In the nineteenth century, the word *melodrama* came to be widely used to identify the third form. In other times, people have called it tragicomedy, *drame*, and romantic drama. To many, melodrama suggests the exaggeration and sentimentality of nineteenth-century versions of the form.

Because of the word's negative connotations, good melodramas in today's theatre are likely to be called "dramas," or "serious plays." The form, however, remains generally the same.

Melodrama, like tragedy, utilizes a serious action. Most often the seriousness arises from an obvious threat by an unsympathetic character to the well-being of one or more sympathetic ones. Although the seriousness is genuine enough, usually it is only temporary, and the sympathetic characters are happy at the end. But an ending that gives reward or happiness to the hero is not enough; the best melodramas also provide punishment or unhappiness for the villain. In melodrama, good and evil tend to be more clearly distinguished than in tragedy and comedy.

The emotive powers appropriate to melodrama have to do with fear in relation to the good characters and with hate in connection with the evil ones. Such a temporarily serious play comes into being through the use of an action formulated to move from happiness to unhappiness and back to happiness, materials selected to generate suspense in the whole, style controlled to express dislike and terror, and purpose applied to demonstrate life's potential for good and man's inventive vitality.

Melodrama, like tragedy and comedy, needs specific kinds of qualitative parts. The appropriate plot employs a story with movement from placidity to threat to conflict to victory. The situations are fearful and hateful; they involve good characters under attack by evil ones. The characters of melodrama are likely to be static because they have made their fundamental moral choices before the action starts. They do not change during the action, as they do in tragedy. In melodramatic characters, good and evil are unalterable codes. Thus, a melodrama usually features an obvious hero or heroine and an equally obvious antagonist or villain. Because an antagonist initiates the threat, normally he possesses more volition. Since a protagonist endeavors mostly to avoid disaster, the thoughts expressed in melodrama are usually more expedient than ethical.

Tragedy, comedy, and melodrama are the three most well-known and often employed forms of drama, but they are not the only forms. Each playwright, to some degree, develops an unique form every time he constructs a play. There are now many plays, and there will inevitably be many more, which have completely different forms of organization. R. S. Crane, a leading contemporary scholar, demonstrated that both Aristotle and the best modern critics have recognized the potential in drama for many new species. The most intelligent critics today realize that playwrights are constantly perfecting variations on the traditional forms and devising new ones. But playwrights must not be fooled by the journalists into thinking that such terms as "absurdist," "environmental," "total theatre," "theatre of cruelty," and the like identify forms of drama. Most such labels refer to theatrical or dramatic styles, albeit fascinating ones. But what about the form of such plays as *The Good Woman of Setzuan* by Bertolt Brecht, *The Chinese Wall* by Max Frisch, *Pantagleize* by Michel de Ghelderode,

Waiting for Godot by Samuel Beckett, *The Homecoming* by Harold Pinter, *Serjeant Musgrave's Dance* by John Arden, *Marat/Sade* by Peter Weiss, and *The Empire Builders* by Boris Vian? Although each has certain features of one or more of the three traditional forms, each is also unique in structure. And some of these plays represent new forms.

This book is not aimed at critical analyses, and therefore it will not explain every form in use today. But at least one other widely used general form should be identified, the didactic form.

Didactic drama differs from the three traditional forms in many respects. First, it is a totally different species of drama. Just as tragedy, comedy, and melodrama are the best known forms of *the mimetic species* of drama, plays constructed for purposes of persuasion represent *the didactic species.* The basic difference between the two has to do with thought as a qualitative part in the organization of a play. In mimetic drama, thought is material to character and plot. In didactic drama, thought not only serves that function, but also and more importantly thought assumes the position of chief organizing element. All the other parts of a didactic drama are selected and put together in such a manner so as best to propound a thought. By writing a mimetic drama, a playwright creates an object from which an audience *may* or *may not* learn. When he writes a didactic drama, he makes an instrument *to compel* the audience to learn. Mimetic drama at its best gives an audience an intense experience; didactic drama at its best stirs an audience emotionally in order to lead its individual members through a pattern of concern, realization, decision, and action in their own lives.

From Euripides to Jean-Paul Sartre, dramatists have written didactic drama, but many critics have usually confused it with one of the other forms. Their failure to recognize didactic drama appears to result from their prejudice against drama used as an instructional weapon or a propaganda device. Although the most obvious or the weakest didactic dramas, have been rightly identified, the best ones by outstanding playwrights have been labeled "problem play," "play of ideas," or with some other hedging term. Despite such mistaken critical attitudes, playwrights have always used the didactic form whenever they chose. The following well-known playwrights represent the number and range of those who have employed the form: Euripides, Aristophanes, Seneca, the fifteenth and sixteenth-century playwrights, Calderón, Shakespeare, Lessing, Schiller, Colley Cibber, Richard Steele, Edward Moore, Heijermans, Brieux, Hauptmann, Galsworthy, Ibsen, and Shaw. Writing didactic drama has long been a tradition for Americans too; Mercy Otis Warren, Hugh Henry Brackenridge, Colonel Robert Munford, Edward Sheldon, John Howard Lawson, Clifford Odets, Lillian Hellman, Albert Maltz, and James Baldwin are only a few. Throughout the Communist world, didactic drama is the dominant form. But the works of Bertolt Brecht prove that, whatever the propaganda value of didactic drama may be, such drama can be great art.

The structural nature of didactic drama not only relates to poetic principles, such as the mimetic forms, but also to rhetorical principles of persuasion. The purpose of didactic drama is a function (persuasion), rather than the creation of a thing (of beauty) with internal emotive powers. Nevertheless, as Shaw pointed out, every play must entertain. The twin goals, then, of every didactic play, while it is being performed, are entertainment and persuasion. Frequently used intentions of didactic plays are: to enlighten, to convince, and to provoke. The goal of every didactic piece, however, is action in the audience. The formal cause of all didactic plays is thought. It appears especially in argument, proof and disproof, expression of passion, amplification and diminution, and qualified language. Although the structure of most didactic plays includes some mimetic story, their form also adheres to rhetorical disposition—exordium, argument, proof, and peroration. In didactic drama, the causal sequence often gives way to a dialectic order of effective disposition. The didactic play does more than state a problem, it usually advances a solution. The sub-forms of didactic drama correspond to the various kinds of speeches in rhetoric. For example, political oratory exhorts the audience about the expediency of future action; forensic oratory attacks or defends somebody or something with relation to the past in order to establish justice; and ceremonial oratory, with concern for the present, praises or censures a person, an institution, or a system. Modern examples of these three sub-forms are respectively: *Viet Rock* by Megan Terry, *Blues for Mister Charlie* by James Baldwin, and *Marat/Sade* by Peter Weiss. *The Chinese Wall* by Max Frisch spans all three sub-forms, as does Brecht's *The Good Woman of Setzuan.*

The efficient cause of didactic drama, as reflected in the style of any such play, depends greatly on the attributes of clarity and appropriateness. Also, didactic style calls for simple characterizations and easily understood ideas.

The materials of didactic drama include rhetorical proofs, the mimetic qualitative parts as subsumed to thought, and appropriate factual items. The three rhetorical proofs are ethical proof (furnished by the character of the speaker), emotional proof (emotive powers for stirring an audience response), and logical proof (the connected arrangement of the argument).

The essence of didactic drama as a fourth form can be summarized like this: The chief material is persuasive language; the formative control is conceptual thought; the dramatist as effectuator speaks through the characters as performed by actors; and the purpose is the persuasion of an audience to attitudes or action. Didactic drama is not lesser drama but drama with a different purpose. An individual didactic play can attain greatness in proportion to the stature of its ideas, the functionality of its form, and the expressiveness of its materials.

A playwright should remember several things about overall form. Only the four most widely used forms—the tragic, the comic, the melodramatic, and the didactic—have been explained here, but potentially each has an infinite variety of sub-forms. Also, other minor forms exist. Although the playwright need not bother about what label should be applied to his plays, he can write a play by knowing all he can about the different forms and by consciously deciding, early in the process of writing, which form will best apply to the particular play he is constructing.

Before the conclusion of this section about form in drama, two other terms deserve attention, *catharsis* and *theatricality*. In fact, one of the purposes of this entire book is to help a playwright understand significant terms, thus enabling him to command structural principles with more authority.

Since Aristotle mentioned catharsis, or *purgation*, in his definition of tragedy (quoted at the opening of this chapter), argument has raged about what he meant. For the playwright, the meaning of catharsis is simple. *Catharsis means ending.* It does not refer to the purgation of audience emotions, although such a purgation may incidentally occur. If in a tragedy the situation is fearful and the protagonist is pitiful, then the author must see to it that those emotions of fear and pity which the play contains must be ended *in the play*. An easily understood example of catharsis is Hamlet's death. He follows a pattern of action throughout the play that leads him from one fearful situation to another; he is truly pitiful in the tragic sense. When he dies, he no longer stands in a fearful situation, and he is no longer pitiful, except retrospectively. Hence, the catharsis in *Hamlet* can be simply defined as Hamlet's death. Catharsis is coherent conclusion.

Theatricality refers to all qualities or occurrences *during a performance* of a play that arouse audience interest. Unlike catharsis, it is a term that has audience orientation. Some critics try to assign theatricality exclusively to the form of melodrama, or to some particular style of production. Just as any written drama will be to some degree dramatic, depending upon the amount and quality of its action, so every performed drama will be more or less theatrical, depending upon the quantity and variety of its visual and auditory stimuli. A play could be intensely dramatic without being equally theatrical, or vice versa. But the best plays possess outstanding degrees of both qualities. Theatricality can be defined as a play's capability, as a staged performance, of interesting and entertaining an audience.

Form, reflecting a playwright's vision of life and art, is something he consciously selects and thoughtfully carries out. A playwright decides to write a tragedy, a comedy, a melodrama, or a didactic drama partly because he consciously chooses one form or another as the best one for the play in process. Since form is the arrangement of all materials, a playwright should learn and use such principles of organization.

STORY

Although story is not the same as plot, they are crucially related. Plot is the overall organization of an action in a drama. Many factors—such as suffering, discovery, and reversal—may contribute to a cohesive plot. Since an action involves many individual activities performed by several characters, story is usually one of the most effective ways to structure an action and thus render a plot.

A story, in simple definition, is a sequence of events. If such a definition is interpreted too broadly, however, story becomes synonymous with plot. Actually, a story is a sequence of certain kinds of events, standing in a special relationship to each other. With an oversimplified idea of story, one might reasonably argue that every play has a story. But some plays have no story, or only a vestige of one. Documentary plays, such as some written by Peter Weiss and Rolf Hochhuth or all the Living Newspapers of the 1930's, have minimal stories, if any. Several contemporary playwrights sometimes compose plots with only vestiges of story—for example, Samuel Beckett, Eugène Ionesco, and Jean Genêt. Even some classic playwrights composed plays with minimal stories; an instance is *The Suppliant Women* by Aeschylus.

The definition of story as a sequence of events, however, contains several implications that lead to the detailed description of story elements which follows. Sequence means a continuous and connected series, a succession of repetitions, or a set of ordered elements. It implies order, continuity, and progression. An event refers to an occurrence of importance that has an antecedent cause, a consequent result, or both. An incident is an event of lesser importance but still of consequence. Event, occurrence, incident, and happening—all imply the instance of observable action. All refer to a rapid or definite change in the relationships of one or more characters to other characters or to things. The composition of a story as a sequence of events, then, is no simple affair.

Three other terms often associated with story are circumstance, episode, and situation. A circumstance is a specific detail attending an event as a part of its motivational setting. An episode is an event of more than usual importance and time span, one distinct or removed from others. A situation is a set of relationships within one character, between characters, or between characters and things. A situation refers to the sum of all stimuli that affect any given character within a certain time interval. It is a combination of circumstances.

Story in drama usually comprises part of a total narrative. The narrative consists of all the situations and events germane to the play, but it includes many that are not actually shown on the stage. All the events previous to the beginning of the play plus those enacted during the play make up the narrative. Only the sequential events during the play's time span form the story. The total narrative of *Oedipus the King*, for example, begins with the pregnancy of Jocasta by Laius and ends at the close of the play, or later if the entire trilogy is taken into account. The following illustration graphically represents the difference between narrative and story:

A B C

A is the beginning of the narrative; *B* the start of the story and the opening of the play; and *C* the ending of the narrative, story, and play. The important events in the narrative leading up to the play enter the play only as exposition.

The beginning of a story is the point of attack. It is the occurrence in the overall narrative selected by the playwright to serve as the initiator of the specific action that enlivens the play. Most plays contain a final climax, i.e., a significant decisive event, and a play is more or less climactic—or in tragedy, catastrophic—depending upon how close in time the point of attack is to the climax. Because the point of attack is late in *Oedipus* and comparatively early in *Hamlet*, the former is more climactic, or catastrophic, than the latter.

Amazingly, many people associated with drama do not know the elements of a story or do not understand sequential principles. Although most of the terms are commonly discussed by teachers, scholars, and critics, the functional knowledge about story has, for the most part, been handed from writer to writer. A surprising number of playwrights and fiction writers, however, apparently never learned much about story and thus often handle its elements in an uncertain, accidental way.

The story elements apply to all stories, but no two original stories use them in exactly the same way. The list of story elements does not make a formula. They enforce no pattern. A playwright may use them or not; he may even use some of them and not others. Most plays contain one or more of these elements, but only those having most of the elements can be said to have a complete story. Undoubtedly, a writer who knows the components of a story has the advantage of being able to handle them consciously. With an understanding of them, he will be a better craftsman. He can use them in any form of play, for any group of characters, and with any theatrical style. Story is not the exclusive possession of the writers of well-made plays. Ibsen, of course, used the story elements, but so did Brecht and Ionesco, Frisch and Pinter. It is also interesting to note that these elements operate in novels and short stories as well as in dramas.

The following discussion of the ten most basic story elements contains explanations of each. It also illustrates each with examples from Shakespeare's *Hamlet*, a tragedy; *Arms and the Man* by George Bernard Shaw, a comedy; and a typical Western melodrama. A writer can become more acquainted with them by identifying them in other works too.

Balance is the first element of a story. Balance implies a special situation, i.e., a set of relationships, that can exist at the beginning of a play. It means more, however, than mere happy circumstances. For the best kind of story, the opening situation should contain the possibilities for all the major lines of action in the remainder of the play. Furthermore, balance implies *stress.* A balanced situation should reveal the strained equilibrium between two contrasting or opposing forces. It should contain implications of potential upset, disharmony, or conflict. The stability at the opening of a play should be dynamic, not static. The balance at the beginning of *Hamlet* is like the deadly stillness before a storm. The guards are apprehensive; Hamlet is in mourning; the others in the court are attempting to establish permanent balance in the kingdom. In *Arms and the Man*, Shaw opened with a balanced situation in the Bulgarian home of Major Petkoff. When the play begins Catherine, the mother, and Raina, the beautiful daughter, happily share the good news about the Bulgarian victory over the Serbs in a battle that day. But they note that the battle occurred not far from their home. In the Western melodrama that will serve as an illustration, balance exists on a Wyoming ranch in the 1870's. A rancher, his wife, his daughter, and an adopted son are working pleasantly, choosing to ignore the news that an Indian raiding party has been marauding in their region. The rancher's brother, a visiting drifter, is off hunting in the nearby mountains.

The second element in a story is *the disturbance.* It is an initiating event that upsets the balanced situation and starts the action. Most often, the specific factor of disturbance is a person who appears and destroys established relationships. As a result, the world of the play becomes disordered and the characters agitated. The forces in the play reach a state of imbalance. The Ghost in *Hamlet* acts as the disturbance, not only perturbing Horatio and Marcellus, but also goading Hamlet into the action of trying to discover the true source of evil in Denmark. The upsetting factor in *Arms and the Man* is Bluntschli. Since he is a Swiss fighting as a mercenary with the Serbian army, he is fleeing from the victorious Bulgarians. He climbs the water pipe of the Petkoff home and enters Raina's bedroom. Raina is alone, and his intrusion discomposes her life and all the relationships within her world. For the Western melodrama, the Sioux war party, as the disturbance, attacks the ranch, kills the rancher and his wife, and kidnaps their daughter, a teen-age girl. The adopted son succeeds in hiding, and he witnesses the carnage and abduction. The girl's uncle, from a distant mountain, sees the smoke rising above the ranch. These three disturbances represent the wide variety of possible initiating elements. The Ghost is a minor character; Bluntschli turns out to be the play's protagonist; the Indians, a group antagonist, provide surprise and represent evil. Frequently, short melodramas skip establishing a balanced situation at the opening and begin with a disturbance, usually as an enacted crime, upsetting only implied order. A situation of relative balance and a disturbance which causes imbalance or unhappiness together comprise the formal *begin-*

ning of a story. The strongest sort of beginning involves balance of a highly desirable sort and a disturbance that depends little or not at all on antecedent events.

The protagonist, or central character, is the third story element. A volitional character who causes incidents to occur makes the most active protagonist. Oedipus, for example, forces the action throughout Sophocles' play. Some stories, however, utilize a protagonist who is central without being volitional. In this sort of story, the main character is victimized by opposing characters or forces. The Captain is largely a victim in *The Father* by Strindberg. Occasionally the protagonist is a group, as are the Silesian weavers in Gerhart Hauptmann's *The Weavers* or the village peasants in *Fuente Ovejuna* by Lope de Vega. In any case, the protagonist should be focal in the story by causing or receiving the most action. Hamlet, of course, acts as protagonist in Shakespeare's tragedy, as does Bluntschli in *Arms and the Man.* The Uncle in the Western melodrama comes down from the mountains to view the burnt-out ranch and becomes the chief agent for the action. The protagonist, then, is usually the character most affected when the disturbance causes imbalance, even when he himself acts as the upsetting factor. And it is usually the protagonist who sets about to restore order in the situation.

When the protagonist begins the activity of reestablishing balance, he normally has a *plan.* This fourth story element also appears in a variety of guises. The plan may be conscious or subconscious, carefully thought-out or eclectic. For a victim, the plan may be nothing more than writhing under the oppressive force. Most often the plan finds its way into one or more speeches by the protagonist soon after the disturbance occurs. Sometimes the plan involves a *goal* or *stake.* Stake, when used, is at best a specific object—person, place, or thing—desired by the protagonist and ordinarily wanted also by his opposition. In a triangular love story with two boys and a girl, the girl is the stake. In *Harpers Ferry* by Barrie Stavis, the town is the stake for John Brown as well as for the defending forces. Another advantage in using plan, goal, and stake is that the characters can credibly reveal their motivations and objectives. Hamlet's plan is complex and constantly changing. It begins with three key speeches in Act I, Scene 5. The first is a soliloquy immediately following his conversation with the Ghost; the second is a speech to his friends; and the third speech comes at the end of the scene. He swears to remember; and he realizes that "the time is out of joint" and that he must "set it right." The next and most concrete part of his plan appears in Act II, Scene 2, when Hamlet welcomes the Players to Denmark; with them he plans the play to be performed before the King. In *Arms and the Man,* Bluntschli charms Raina into helping him plan his escape from the pursuing Bulgarian Army. In the Western melodrama, the Uncle swears vengeance on the Indians for their murders. He discovers that his brother's adopted son is still alive, and together they plan to track down the Indians and get back the girl. She is the stake. The success of any plan as a story element depends variously on the action required, the worthiness of the goal, and the importance and concreteness of the stake.

Obstacles are the fifth story element. An obstacle is any factor in a story that opposes or impedes the progress of the protagonist as he attempts to restore balance by carrying out his plan. Tension arises whenever any character confronts an obstacle. The best obstacles, the ones which incite the most action, impose three conditions on the protagonist: An obstacle, when it becomes apparent, should aim a *threat* toward the protagonist. Being properly as strong or stronger than he, it will cause *conflict* when the protagonist attempts to remove it. And it will, thus, lead to a minor or a major *climax* in the story. Obstacles are typically of four kinds. First, they can be physical obstructions, such as a mountain to climb, a distance to traverse, or an enemy to be found. Second, obstacles are frequently antagonists—other characters in the play who oppose the protagonist. Often one chief antagonist leads the opposition. Third, obstacles can occur within the personality of the protagonist himself. He may have intellectual, emotional, or psychological problems to overcome before being able to achieve his goals. Fourth, obstacles can be mystic forces. These enter most stories as accident or chance, but they can be personified as gods or expressed as moral and ethical codes. The plays that have the best stories employ all four types. The major obstacles facing Hamlet are: (1) a time and place for him to confront the King; (2) Claudius as chief antagonist; (3) Hamlet's own reflective nature; and (4) accidents, such as when he kills Polonius by mistake. Throughout *Arms and the Man*, Bluntschli encounters such obstacles as: (1) finding a hiding place; (2) Sergis and others as antagonists; (3) his own realistic but free-wheeling nature, and (4) the romantic conceptions of war and love. In the Western melodrama, the obstacles facing the Uncle are identifying the Indians' trail, crossing a flooded river, pushing through a snowy mountain pass, and getting the girl away from the Indians. If an obstacle is clear and possesses sufficient strength, the story will contain a proportionate amount of suspense. Of special importance is the fact that properly conceived and presented obstacles force decisions on all the major characters. They will have to decide what to do and whether or not to do it. Such decisions produce dramatic action.

Stories of length usually contain *complications*. Although complications come sixth in this list of elements, they can rightly enter a story at any time. The initial disturbance is, for example, a specialized complication. A complication is any factor entering the world of the play and causing a change in the course of the action. The best complications are unexpected but credible. They can be positive or negative factors for any of the conflicting forces in the story. They can, for instance, aid or hinder the protagonist or the antagonist. Typically, complications are characters, circumstances, events, mistakes, misunderstandings, and best of all—discoveries. Most often complications present new obstacles to the protagonist. *Hamlet* contains many complications, such as the entrance of the Players, Ophelia's suicide, and Gertrude's drinking of the poison. *Arms and the Man* also abounds with them. Louka, a maid, complicates the story by chasing and capturing Sergis, Bluntschli's main rival for Raina. Another important one is the arrival of Bluntschli's mail. It informs him that he has inherited his father's hotels in Switzerland. As a result, he is able to convince Raina's parents that he is a worthy suitor for her hand in marriage. Often in melodrama, a complication enters the story unexpectedly and then becomes an obstacle. The Western melodrama well illustrates this. As the Uncle rides after the Indians, he stops at a tavern in a small crossroad town. Inside waiting for him are three old enemies, gunfighters who have been trailing him for months. Of course, the Uncle must shoot them before continuing his pursuit of the Indians. Complications are significant factors in the maintenance of tension and activity in a story. Their major contributions are surprise and story extension.

The seventh story element may or may not be employed in a given play. It is *sub-story.* To use the word *sub-plot* is misleading as well as incorrect. Since plot is the total, inclusive organization of all materials and activities in a play, there can be no such thing as a sub-plot. Sub-stories, however, are not merely possible; in long stories, they are often essential. Simply constructed plays, such as *The Homecoming* by Harold Pinter, contain no sub-story; but complex plays, such as Brecht's *The Caucasian Chalk Circle*, have one or more. Being longer, novels are more likely to contain sub-stories than plays. A sub-story usually includes all, or most of, the elements of main stories, but sub-stories, being subordinate, do not require so much detail. For a sub-story to contribute successfully to a main story, it should involve some of the same characters, and its climax should come before or during the major climax in the main story. Also, the results of the various segments of the sub-story should reflect, contrast with, or affect the main story. Shakespeare and other Elizabethan playwrights were masters of sub-story. In *Hamlet*, the secondary story of the House of Polonius, for example, complements and affects the primary story of the House of Hamlet. In *Arms and the Man*, the sub-story entails the love triangle of Sergis, Louka, and Nicola (another servant). It ties in with the major triangle of Bluntschli, Raina, and Sergis, because Sergis is involved in both. In the Western melodrama, the sub-story depicts the boy's transformation into a man as he accompanies his Uncle in hunting the kidnapped girl. Secondary stories can be substantial or minimal. But they should not be confused with exposition, the narrative material leading up to the action of the play. Sub-story is an effective tool for contrast and complication.

Crisis is the eighth story element. Along with climax, it is more essential to a drama than any of the other elements. All good plays contain at least one crisis of some sort. Crisis can appear in many guises, and it can operate at numerous levels. Simply explained, crisis is a turn in the action. More complexly, crisis is a period of time in a story during which two forces are in active conflict and throughout which the outcome is uncertain. In relation to the preceding story elements, a crisis occurs whenever the protagonist confronts an obstacle. This meeting of opposed forces usually engenders conflict. One agent normally is performing an action while another is trying to obstruct that action. The action may be an attempt to reach a goal or to capture the stake. Because the outcome of the crisis remains undetermined until the climax, crisis naturally arouses suspense. It also necessitates decisions. The protagonist and his opponent must decide whether or not to fight, and how to fight or avoid fighting. Thus, crisis always produces dramatic action. It forces change. The conflict which crisis contains is significant to most plays. Some theorists, e.g., Ferdinand Brunetière and John Howard Lawson, even consider conflict to be the chief component of drama. Although conflict makes the most dynamic kind of crisis, it is not always essential. But change, or action, is always necessary. Crises involve some combination of physical, verbal, emotional, and intellectual activity on the part of one or more characters. For example, a crisis could be a physical fight, a verbal argument, or an introspective search. Since crisis always requires a period of time, certain scenes in plays can be identified as crisis scenes. The first of the two major crises in *Hamlet* is the scene in which the Players enact "The Murder of Gonzago" and by which Hamlet hopes to "mousetrap" the King. The second is the dueling scene near the end. The latter contains intense conflict, but the former does not. Both encompass great activity, make extreme action, and lead to the two major climaxes. The major crisis scene in *Arms and the Man* is the last segment of the third act. It involves the revolving conflict between Bluntschli, Raina, and Sergis about whom Raina will marry. In the Western melodrama, the major crisis is, predictably, a physical fight between the Uncle and the Indian chief—with knives on the edge of a cliff. The crises of most plays function as the chief periods of concentrated activity and violent change.

Climax, the ninth element, always follows crisis. Their relationship is, at best, causal and necessary. Climax is a high point of interest for the characters, a single moment following a crisis. It is the instant when conflict is settled. Usually, it involves discovery or realization for the characters, and it can be a moment of reversal in the story. A climax cannot happen without a crisis, some specific rising action, building up to it. And every crisis results in a climax. The climax, however, may be immediate or postponed. A writer can extend a main story and interweave it with substories by interrupting a crisis before it reaches the necessary climax and then by letting a later climax end multiple crises. Shakespeare employed this technique in many of his plays. In *The Taming of the Shrew*, for example, the final climax when Katharina appears at Petruchio's call settles several crises. Climaxes, like crises, can be major or minor in impact. In most stories, the final major climax is the moment when some sort of balance, or order, is reestablished. One of the two forces in conflict during the action wins, or sometimes they reach an absolute stalemate. Or the stake gets into the "right" hands. Finally, a climax may be a moment of decision. A character's deliberation, as crisis, spans a period of time; his decision, as climax, is one moment. The first of the two major climaxes in *Hamlet* occurs during the performance by the Players; Claudius stands and cries: "Give me some light! Away!" The second major climax is the final one in the play; it is Hamlet's moment of death. The major climax in *Arms and the Man* happens at the end when Raina indicates she will accept Bluntschli's proposal. The Western melodrama's climax comes when the Uncle stabs the Indian and shoves him off the cliff. A climax is a high point in a play, and the final major climax is the highest point in a dramatic story because the entire action is so structured to culminate in one moment during which the outcome of the whole is suddenly certain and apparent.

The tenth and final story element is the *resolution*. Whereas most of the elements in a story are kinds of characters, actions, scenes, or events, resolution—like balance—is essentially a situation. It may depend upon activity or explanation as the means for its expression, but resolution is a set of circumstances resulting from the certainty of the climax. In composing a resolution, a playwright's job consists of making the reestablishment of some kind of balance clear in every necessary detail. Because the protagonist wins or loses, gets or misses the stake, reaches or does not reach his goal, all the other characters are affected. During the resolution, the world of the play settles into some relative state of balance, or perhaps permanent imbalance.

Just as no significant antecedents precede the opening of a play, at best no essential consequences follow the ending. The *ending* of a story amounts to a combination of the final climax and the ensuing resolution. Another popular term for this combination of final climax and part of the resolution is *denouement*. A denouement is the outcome of a series of events, the final unraveling and settlement of the complications and conflicts.

Because a skilled playwright will indicate during the final crisis just how all the major and minor characters are involved, the best sort of climax brings all noteworthy involvements to a close. The climax fixes the fate, insofar as the play's story is concerned, of every character. The resolution, then, simply shows the resultant *situation*. The resolution in *Hamlet* is all that occurs after the death of Hamlet. Shakespeare used only nine speeches for it. Horatio says goodbye to Hamlet. Fortinbras enters, learns what has happened, and claims the throne of Denmark. He then venerates Hamlet and orders a funeral procession. Shaw, in *Arms and the Man*, employed only one speech after the moment of climax to resolve the play. After Raina accepts him, Bluntschli simply tells the characters to hold everything until he returns in a fortnight. In the Western melodrama, the resolution is longer. After the Uncle kills the chief, the other Indians give him the girl as a trophy; however, she runs away because she is ashamed about having been the victim of the chief's attentions. The Uncle sends the boy after the girl. He catches her, calms her down, and the three of them start the long trip back to the ranch. They ride off into the sunset! A resolution should hint, indicate, or explain the new balance achieved at the close. It shows the results of the entire action, and it culminates in an ending.

The ten elements—balance, disturbance, protagonist, plan, obstacles, complications, sub-stories, crisis, climax, and resolution—are the means for the construction of a story. Although the playwright can consciously apply them to his chosen materials, they are not formulary. Every story can be quite unique. And not every play needs a story. But when a playwright wishes to combine story elements with actions, situations, characters, and events, the variety of potential stories is infinite. The list of elements, however, does not automatically make a story. Each element must have a specific representation, and all the elements as represented must be appropriately combined and then divided between various scenes and acts. This work is best accomplished in the rough scenario stage of the writing process. A rough scenario might well contain some such list as the following:

ACT I

Balance: a description of a situation indicating potential upset
Disturbance: an event
Protagonist: a character
Plan: an explanation of goal or stake and how the protagonist will proceed
Obstacle 1: one identified
Crisis: a scene described, including conflict
Complication: interrupting the crisis

ACT II

Protagonist: in a situation
Obstacle 2: one identified
Crisis: a scene described, including conflict
Complication: a new force enters and changes the conflict, perhaps temporarily defeating the protagonist
Obstacle 3: one identified
Crisis: the scene develops with heightened conflict
Climax: a discovery or reversal occurs, and the crisis ends with a promise of further conflict

ACT III

Protagonist: in another situation
Obstacle 4: one identified
Crisis: a scene described, including conflict
Complication: a new force changes the balance of power in the struggle
Obstacles 1 and 2: reentry
Crisis: the major conflict scene
Climax: the protagonist wins
Resolution: he gets what he wants

This list merely exemplifies how some particular story might appear in a rough scenario. Just as this story illustration leaves out any sub-story, so any element could be omitted, except crisis and climax. A fully developed story is not absolutely necessary for every play, but most sets of dramatic materials are better organized for having one. Story is one kind of structure; one method of making plot; one way a playwright can organize actions, characters, and thoughts.

UNITY

Unity is one attribute that brings beauty, comprehensibility, and effectiveness to any work of art. Each of the art forms has its own proper kind of unity, and every one ought to possess the quality in some form. If one were to pull 20,000 words out of a hat, one at a time, then put a period after every tenth one, and arrange these "sentences" in groups of five under various characters' names, the result would probably be a play without unity. But all plays made by an artist and understandable to an audience have unity. Some, however, have better unity than others. To unify a play means to organize the parts, on the basis of some plan or logic. Thus, plot, as an overall order, comes into being. The unity in any work of art depends upon the kind of parts used, the purpose to which they are arranged, and the manner in which the artist works. In drama, the parts to be unified into a plot are characters and all their doings, thoughts, and sayings. *Unity of action is dramatic unity; it is a quality of plot.*

Before examining the numerous means for achieving unity in drama, another question must be answered. What about unity of time and unity of place? Some critics and teachers still demand the three "classic" unities —time, place, and action—of unsuspecting playwrights. But the unholy three are not even classic. The Greeks, anyway, concerned themselves only with dramatic unity itself. Although they were obviously aware that each play should be developed to a certain length in relation to its proper magnitude, none of the great tragedians evidently worried about unity of time. A long play need not absolutely observe the unity of time, and thus avoid time-lapses. Neither did the Greeks hesitate to change locations, thus breaking unity of place. Although Aristotle discussed unity of action, he actually was concerned with unity of material parts making up an action. He hardly mentioned unity of time and place. A series of pedantic writers and critics of a much later time succeeded in establishing the three unities as necessary. Few playwrights have accepted all three of them. Unity of time and unity of place should be used only when it suits the play and the playwright. They are not necessities. Unity of action is not a rule for writing; it is simply a desirable quality in a play.

Since drama is a time art, it naturally needs a beginning, a middle, and an end. These three elements are not so obvious as they may first seem. They mean more than just starting a play, extending it with words, and stopping it. First, a functional beginning and end imply wholeness and completeness; a middle emphasizes full development. A beginning is an event, arising out of the given circumstances of a situation, that has no significant antecedent but does have natural consequences. A middle is one or more events or activities having both antecedent causes and consequent results. An end, precipitating a more or less balanced situation, is one item of action having antecedents but no significant consequences. Someplace in every narrative there is a proper beginning for a play, and a proper ending. The writer needs to discover the best points of attack and conclusion. A play cannot just start anywhere. Also, if a play has one or more sub-stories, each story within the whole will need a beginning, middle, and end. Sub-stories, however, do not need the same extension as main stories.

Beginning, middle, and end in drama also imply dramatic probability, especially probability of action. Simply defined, probability means credibility and acceptability. Probability and unity are twin qualities that help to bring beauty to an art object. In a play, when one event causes consequences—incidents, emotions, thoughts, or speeches—the causal relationship creates unity. A chain of such antecedents and consequences contains unity and makes probability. The playwright depicts not what might happen in a given situation, but among all the possibilities, he selects the most probable, or the clearly necessary. At the beginning of a play anything is possible. As lights first illuminate a stage, Oedipus or little green men from Mars could walk on. After the first moment, however, the possibilities of what can credibly happen are progressively, minute by minute, more limited. Once the beginning indicates a specific situation and a group of characters, the realm of the possible becomes the realm of the probable. In the middle of a play, the characters follow one or more lines of probability. Before or after they perform an action, it should be probable. The end of a play is limited to the necessary. Because the characters have done and said certain things, one resolution is necessary. Thus, the possible, the probable, and the necessary make probability in drama, and they establish the quality of unity.

At the beginning of *Hamlet*, it is possible that Hamlet will refuse to speak with the Ghost or run away at the first sight of it. But once he talks with it, he probably will try to make sure Claudius is guilty. Hamlet, therefore, arranges the play to trap the King. Once Hamlet knows Claudius is the murderer and once Claudius recognizes Hamlet's knowledge, the two courses of probability are that either man may destroy the other. It becomes necessary, then, in the play for them to meet—a crisis—and for them to fight. And finally, the climax, at which point one or the other wins, is necessary as a resolution to the crisis. Because of all the foregoing circumstances in the play—the poisoned rapier included—Hamlet must necessarily die. Although this is an oversimplified rendering of the possible, probable, and necessary in *Hamlet*, it serves to illustrate how unity and probability are allied.

Probability is, however, only one of the possible ways a playwright may create unity of plot in a drama. Causality is another means, one closely related to probability. Story is another; it is a particular pattern of causality in events. Also, what Kenneth Burke called the pattern of arousal and fulfillment of expectations can lend unity to a fictional form. Furthermore, other qualitative parts may affect plot to such a degree that they gain control of it. This discussion has focused on unity of action so far, but unity by means of character and thought is also possible. A play's organization can depend primarily on character change. This is one of the unifying factors, for example, in Ionesco's *Rhinoceros*; every character in the play, except one, changes into a rhinoceros. The third principal type of unity is that of thought. An idea, or a thought complex, may control a play's structure. When thought acts as the primary unifying factor, the play is didactic—e.g., Brecht's *The Good Woman of Setzuan*. In some plays, all three kinds of unity—action, character, and thought—operate in a nearly equal manner. *King Lear* is one such play, and *Becket* by Jean Anouilh is another.

Suffering, discovery, and reversal are also significant qualitative elements of plot. These, too, are related to beginning, middle, and end. Suffering can be defined as anything that goes on inside a character. It can be tragic, comic, or intermediate. It is not only the basic material for every characterization, but also it is the condition of each and the motive for the activities of each. The most intense suffering leads to, or forces, discovery. Discovery means change from ignorance to knowledge. There are many kinds of discovery. A character can discover a physical object, another person, information about others, and information about himself. It includes finding, detecting, realizing, eliciting, identifying, and recognizing. Any significant discovery forces change in conditions, relationships, activity, or all three. Thus, discovery is a major means for action in drama. The climaxes of both *Hamlet* and *Death of a Salesman* hinge on discovery. During the enactment of the play-within-the-play, Hamlet discovers that Claudius is truly guilty, and Claudius simultaneously discovers that Hamlet has found him out. This is a double discovery. In *Salesman*, the climax occurs when Willy discovers that his son, Biff, truly loves him. Additionally, false discovery can usefully complicate a plot. In *Oedipus the King*, Oedipus mistakenly discovers that Creon and Tiresias are plotting against him. Such a false discovery requires a later discovery of the truth. The best kind of discovery forces reversal, or peripety. A reversal is a violent change within a play from one state of things to a nearly opposite state. The situation—including both relationships and activities—completely turns around. In *Hamlet* for example, when Hamlet and the King make their double discovery, a reversal occurs. Up to that moment Hamlet has been the volitional pursuer of Claudius, but from the moment of reversal to the final crisis, Claudius is the pursuer of Hamlet. Thus, in the best kind of reversal, agent becomes object, and object becomes agent. After the initiating disturbance in a play, suffering most likely will precede discovery, and discovery precedes or produces reversal.

Beginning, middle, and end apply to the form of drama in many other ways. Francis Fergusson in his book *The Idea of a Theater* identified another related formative pattern. Kenneth Burke also treated it in his books *A Grammar of Motives* and *Philosophy of Literary Form.* Although Fergusson called the pattern "tragic rhythm," its elements apply to other kinds of drama as well. The rhythm of action in drama often goes from purpose through passion to perception. At the beginning, a protagonist initiates an action; in the middle he suffers while carrying it out; and at the end he, or some other character, has increased insight as a result of the action.

Thus as formative parts of plot, beginning-middle-end imply other qualitative means of organization: (1) possible-probable-necessary, (2) suffering-discovery-reversal, (3) purpose-passion-perception, and when related to story (4) disturbance-crisis-climax.

Although strict unity of time is not an absolute rule for the writing of drama, a playwright should consider his play chronologically. How time is handled makes a significant effect on the form—action, unity, and magnitude—of any play. Time can affect a play in various ways. Chiefly, time applies to drama as location and as period. First, a play occurs at some location in time, a relatively specific point in the infinity of time. It can happen in the past, the present, the future, or all three; or it can happen at an unspecified time, or even in non-time. Shakespeare placed *Julius Caesar* in the past; Williams placed *A Streetcar Named Desire* in the present; Capek placed *R. U. R.* in the future; Kaiser placed *From Morn to Midnight* in unspecified time; and Beckett placed *Waiting for Godot* in non-time. Location in time can also be affected by the "when" of production, i.e., the performance date or the director's interpretation of time.

The second consideration about time in a play has to do with period, or time span. A play has a "real time" or "performance time"; this is the time consumed by the actual performance. But within the play there is some period depicted too. This segment of time can be handled in a multitude of ways. Sequential time means straightforward and causal progress through time, perhaps with some leaps between scenes or acts. *Macbeth* spans such a period. Diffuse sequence means interrupted progress in time; the focal period may be interrupted by flashbacks. *Death of a Salesman* contains a diffuse sequence. Circular time, as Richard Schechner has pointed out, means that time passes and events occur; but there is not much causal relationship; and the series of events in that time passage repeats itself. Ionesco's *The Lesson* uses such a circular time period. Episodic time means a series of short, relatively unrelated periods. Brecht's *The Private Life of the Master Race* proceeds in such a manner. Denied time means non-time and non-period as chronological circumstances of the play. Strindberg worked with such conceptions of time in his dream plays; Pirandello depicted time, and other "certainties" in life, as relative. In *The Bald Soprano*, Ionesco attempted to deny time. Time sequence, then, is a crucial matter in formulating a play, and it is a factor with which playwrights constantly experiment. A playwright's decisions about time in his drama are decisions about form.

PREPARATION, SUSPENSE, AND SURPRISE

Planning an overall organization for a play is somewhat abstract and relatively easy. Selecting specific events and establishing relationships to carry out the plan is difficult. Composing a story, giving credibility to characters, and weaving thoughts into a play is harder. Establishing interrelated causal chains is even more complex. And among the most demanding aspects of playwriting is the task of inserting items of preparation. All the kinds of specific details that give a play apparent probability can be grouped under the heading of preparation. Items of preparation make actions and causal relationships seem inevitable. They make the characters believable and intelligible. Lastly, they enhance the play's emotional effects.

Although preparatory details can appear in a number of guises, they are best understood as exposition, plants, and pointers. Each of these has various types and serves somewhat different functions. Taken together, they make overall credibility, permit surprise, and stimulate suspense. They can serve necessary functions in any form of drama or any style of play.

Exposition is any information in the play about circumstances that precede the beginning, occur offstage, or happen between scenes. It can be sub-divided into exposition about the distant past or exposition about the recent past. Whenever one character explains to another any circumstance from the distant past, the present action may be enhanced, but the playwright must take care only to include essential information. No less important are the items from the recent past; these may range from a major discovery to an entrance motivation. It is important, for example, to explain the causes of the conditions at the opening of a play, and often it is important, though of less significance, to reveal why a character enters a particular scene. Exposition may occupy a relatively small proportion of the script, as in *The Birthday Party* by Harold Pinter, or it may take up a great amount of the dialogue, as in *Slow Dance on the Killing Ground* by William Hanley. Oddly, most beginning playwrights pack their scripts with too much exposition. Exposition should be minimal, but sufficient to the needs of the action. Always, it best enters a play subtly and spreads over more than one scene. Obvious expositional devices impair the excellence of any play, and most contemporary playwrights avoid them. Some typical rusty devices are: the narrator, the servant, the telephone, and the foil (a minor character who acts as a contrasting confidant to a major one). Exposition best appears during a conversation about something else, as brief information precipitating a major discovery.

A *plant* is a more specific device. Usually, a plant is an item of information, one that the playwright inserts early in the play and that turns out to be significant later. Often it is an item of exposition, but not always. As one form of preparation, plants provide evidence for subsequent deeds and speeches. Plants assume importance for the characters, and for the audience, in retrospect. Their initial impact is slight, but eventually it is great. They have many uses. Plants should be used to establish character traits before those traits occur in action. They may indicate relationships, provide evidential information, or reveal attitudes. They make possible both surprise and accident. When a surprising event takes place, it may be startling, but it must be credible. Plants establish the basis for such credibility.

There are at least eight chief devices for planting. (1) An attitudinal speech from, or about, a character prepares for later action. Brecht's *Mother Courage* is full of such speeches, and they prepare for the title character's stoic state at the play's end. (2) A minor crisis frequently sets the possibilities for a later major crisis. In scene 7 of *The Chinese Wall* by Max Frisch, the Contemporary and Mee Lan carry on a mild conflict and then fall in love; thus, their love, their individual human contact projected as an answer to war, is possible in the final scene. (3) A piece of business, i.e. a physical action at first seemingly unimportant, often gains significance later in the play. Although the device appears in most detective stories, it is also used in many kinds of plays. In Lillian Hellman's *The Little Foxes*, Horace's activity of taking medicine in Act II establishes his weak heart condition and makes his death in Act III credible. (4) A suggestive or explanatory speech not having much apparent importance can turn out to be crucial. In *Ondine* by Jean Giraudoux, the King of the Sea explains that if Ondine, a water sprite, marries Hans, a human, their relationship must be perfect. If Hans is unfaithful to her, he will die, and she will forget him. That is exactly what later happens. (5) Minor characters sometimes function as, or present, plants. The confidant, such as Horatio in *Hamlet*, and the *raisonneur*, such as the Ragpicker in Giraudoux's *The Madwoman of Chaillot*, make obvious plants. Charley enters *Death of a Salesman* primarily for the sake of plants and contrast. (6) Physical items of spectacle—a setting, a prop, a costume, or even a sound—occasionally operate as plants. The locale of a theatre acts as a credibility plant in *Six Characters in Search of an Author*. The handling of a key early in *Dial "M" for Murder* permits the discovery of the villain at the end. Harpagon's costume in *The Miser* helps make his actions and speeches believable. The siren-like sound effect in *The Empire Builders* by Boris Vian is a major preparatory item for all the action in the play. (7) Relationships, especially those established early, can function as plants for later action. The suspicious and suspended relationship between Stanley and Blanche in *A Streetcar Named Desire* provides the logic for Stanley's final action of committing Blanche to a mental institution. (8) A minor incident often serves as a plant for a major event. In *The Taming of the Shrew*, Kate's early ill treatment of Bianca sets up her later violent behavior with Petruchio.

A *pointer* is also a specific device of preparation. Whereas a plant stimulates a backward view, a pointer impels the characters, or the audience, to look ahead. A pointer is any specific item in a play that indicates something of interest will occur later. Pointers arouse questions and anticipations. One great speech full of pointers is the Watchman's opening speech in *Agamemnon* by Aeschylus. Although in a well written play nearly everything before the final climax stimulates forward interest, pointers are the special items that the playwright inserts to heighten interest, expectation, concern, or dread.

In general, most of the eight kinds of plants can also function as pointers. But more specifically, pointers frequently take on one of the following particular shapes: (1) a statement that some event will take place; (2) a question about the future; (3) a prop, scenic item, or piece of business suggesting something to come; (4) an assertion opposed to the obvious course of activity. Additionally, the existence of these general things in a play point to the future: a brief conflict leading to a future major conflict, emotional behavior, antagonistic attitudes, and any kind of delay. All the devices of planting and pointing in a drama amount to the overall foreshadowing.

Suspense in a play for the characters—incidentally for the audience—is obviously related to the kind of preparation that produces expectations in them about the future. Thus, pointers nearly always arouse *suspense*. But most skilled playwrights employ *the hint-wait pattern* as a technical device. They either check to make sure their play contains several such patterns, or they insert them according to the needs of that particular play. This pattern is simple to learn and to employ, and it is usually effective. The pattern consists of a hint, a wait, and a fulfillment. Some character in a play indicates that something is likely to happen; other activity forces a wait; and then the expected event does take place—in a slightly different way than anticipated. In any given play, when the pattern first occurs, the resultant suspense will be small, but each succeeding time the pattern comes up more suspense proportionately arises. The hint-wait pattern occurs many times in *Macbeth*; in fact, the major formal relationship between Acts IV and V depends upon multiple usage of the pattern.

Suspense automatically occurs during all crises. A crisis requires these steps: identification of two opposed forces, an indication that they will fight and that one or the other will win (hint), the occurence of the conflict (a suspenseful wait while the fight goes on), and a climax (fulfillment as resolution of conflict). Hundreds of motion picture Westerns have employed this type of crisis order as their total plot; nevertheless, any playwright can use crisis and conflict to good advantage to make both plot and suspense. In *Who's Afraid of Virginia Woolf?*, for example, Edward Albee made them central.

Finally, suspense also proceeds from deliberation and decision. When a character faces a problem, he usually deliberates about solutions or alternatives. This is another form of crisis, another kind of conflict. Whether he expediently wonders how to do something or ethically reflects about whether or not to carry out an activity, suspense arises. The fulfillment in such cases is the decision following deliberation. Decision is also, and even more significantly, action and climax. It is action because it demands a change in state or activity, and it is climax in that it resolves deliberative crisis. The overall form of Brecht's *The Good Woman of Setzuan* follows this pattern, and it recurs in the individual segments of the play. In this special sense, suspense can become action.

Many theorists have recognized the frequent incidence of *surprise* in drama, and most playwrights try to get it into their works. But too often they fail to distinguish between simple surprise, which requires little preparation, and credible surprise which demands a careful establishment of probability. On the plot level, surprise is at best an unexpected event that is fully believable during and after its occurence. Hence, surprise depends upon antecedent plants for its probability. *Rhinoceros* by Ionesco contains many surprises, but most of them are well grounded in preparatory devices and thus are believable within the limits of the milieu and logic of the play. Surprise can also proceed from dual lines of probability. In a series of events, one line of probability is obvious and leads to an apparent outcome; the second line of probability is hidden, or seemingly unimportant, and leads to an unexpected outcome. When the second line suddenly comes forward, surprise results. In this manner, dual probability produces surprise. The other qualitative parts of drama can also produce surprise. A character with a surprising trait, an unexpected thought, a fresh combination of words, a startling series of sounds, a stunning item of spectacle—all such things can produce surprise.

Additionally, surprise can come from *chance* or accident in a play. Although a play is a network of probability, chance always takes an important part in the action. A playwright should identify all accidents in his play and attempt to surround them with probability. It is accidental, for example, that Fortinbras returns to Denmark exactly at the end of the action in *Hamlet*, but because of several references to him and his one earlier appearance, it is acceptable that he enter the scene at precisely the right moment to conclude the play.

Beginning playwrights, and unfortunately some who should know better, tend to make mistakes with preparation, exposition, plants, pointers, suspense, surprise, and chance. Most often errors come from over-preparing the obvious and failing to establish probability for the unusual. Exposition is best kept to an absolute minimum, and then presented straightforwardly during the course of interesting action. A need for the information should arise before the exposition appears. Planting errors are usually the result of too few plants rather than too many. The beginner often lets characters discuss an event after it has happened rather than pointing to it before it happens. Most plays could have better suspense if their authors would more consciously employ the hint-wait pattern. With surprise and accident, the common flaws have to do with setting up lines of probability. The work of investing a play with sufficient items of preparation is complex. It is best done during the composition of the full scenario. Proper preparation creates the qualities of unity, probability, and economy. Hence, structural preparation is crucial to plot and necessary for beauty in drama.

MAGNITUDE AND DEVELOPMENT

When a writer catches a germinal idea and then begins to expand it, questions about scope bombard his mind. What should be the play's size? Should it be a short play or a long one? What will be the proper development of the material? How many crises, characters, and sets does it demand? All such questions are structural considerations regarding the plot, but more particularly they are factors having to do with the play's magnitude.

Magnitude is a sign of unity and a condition of beauty. It is the proper development of an action to achieve internal completeness and total integration of all parts. Length and quality are the first two determinants of magnitude. Length is a result of the admitted quantity of material for each of the parts of the drama. It depends upon the number of situations, events, complications, and sub-stories; characters; expressed thoughts; words per average speech; sounds per average word; and settings. On the level of plot, length depends upon the number of events, obstacles, crises, and climaxes included; and upon the amount of preparation, suspense, and surprise developed. A playwright can best decide the length of a play by making a series of choices concerning quantities. Decisions about quantity and length, however, should not be merely arbitrary nor rationalized as inspiration. All should relate to the basic action at the core of the play. Every choice about quantity should be the reasonable consequence of a recognition of the needs of the specific action to be represented.

In addition to length and quantity, organic wholeness is another determinant of magnitude. Only if every item in a play is bound to another, in some logical relationship, can the whole be complete and economic. Incidents, characters, or speeches that appear in a play merely for their own sake increase the play's magnitude to the detriment of the action. When the play is "finished," if there is anything omitted that should be there, or if anything can be cut without harm to other elements, then the magnitude is wrong. For proper magnitude, the parts should have an ordered arrangement, and each part should in some way be related to one or more complementary parts. Economy in drama means that every item in a play, from a single physical action to a major climax, must serve more than one function and must be interrelated with some other item. Usually, the best kind of organic relationship in drama, since it deals with human action, is that which is causal.

Magnitude also depends upon the amount and quality of contrast in a work. Contrast is the variety and diversity of adjacent parts. Like all other works of art, a drama needs contrast. It should be apparent in groups of events, characters, ideas, speeches, sounds, and physical actions. A playwright needs to look for practical and apparent variety as he surveys the potential materials for his play. He should choose events, circumstances, and characters that obviously differ from each other. Contrast, then, affects the magnitude of a play because to demonstrate differences takes time and space. The length of a story, for example, depends partly upon the variety and complexity of its situations and incidents.

Plays with exemplary magnitude do not abound in any one period of drama, but many outstanding examples exist. Magnitude as appropriate length and quantity is especially fine in Sophocles' *Electra* and in *Look Back in Anger* by John Osborne. The over-extended length and quantity of materials in *Rhinoceros* by Ionesco and *The Iceman Cometh* by Eugene O'Neill, both commanding plays in other respects, prevent them from achieving the highest excellence. The organic wholeness of Euripides' *Hippolytus* and Jean-Paul Sartre's *No Exit* is apparent, but it is less than it should be in Thornton Wilder's fascinating play *The Skin of Our Teeth* and in García Lorca's brilliantly worded *Blood Wedding*. Contrast and variety are special virtues of magnitude in Shakespeare's *King Lear* and in Michel de Ghelderode's *Pantagleize*, and they are noticeably lacking in Seneca's *Medea* and T. S. Eliot's *The Cocktail Party*.

The best magnitude for any given play has to do with its comprehensibility as a whole. If the development of a play is appropriate, the whole will appear as a balanced and ordered composition of parts. Length and quantity, organic wholeness and causal relationships, contrast and variety —these are the chief determinants of magnitude. In summary, beauty of form in drama depends upon the qualities of unity, probability, and magnitude.

EXPLORATIONS OF NEW STRUCTURES

Every work of art involves an artist's actual formulation of some sort of concrete life image. Such a formulation requires the use of details, materials, and parts. Thus, in art, form is the organization of parts into a whole. And to create organization, every artist employs, however consciously, certain principles of order and arrangement. As he particularizes such principles in a specific work, that work assumes a structure, a set of relationships that join the parts to form the whole. Hence, in any art work, form is particularized structure. And a drama's structure is its form, its organization, its *plot*.

There is, however, no single kind of plot appropriate for all dramas. Each play possesses, at least in some degree, an unique plot, a particular structure, an individual form. "The best kind of plot" as an universal formula simply does not exist.

This discussion of structure, or plot, in drama would be misleading if it did not take into account particular versions of form that significant twentieth-century playwrights have employed and that influential theorists have promoted. But before exploring some contemporary varieties of dramatic structure, one should realize that the basic principles discussed earlier in this chapter apply as much to current as to older dramaturgy. The major differences between traditional drama (if there is such a thing) and most of the innovative new plays have to do with how the principles are used rather than discovery of absolutely new principles. It is not true that traditional principles are entirely absent in any contemporary work. Contemporaries—such as Bertolt Brecht, Eugène Ionesco, Peter Weiss, and Jean Genêt—have simply employed them in new ways within their unique constructions. And, after all, Samuel Beckett's *Waiting for Godot* is structurally no more different from *Oedipus the King* than is Shakespeare's *Antony and Cleopatra*, Goethe's *Faust*, Alfred Jarry's *Ubi Roi*, Leo Tolstoy's *The Power of Darkness*, or August Strindberg's *The Ghost Sonata*.

That twentieth-century dramatists have created a marvelous variety of structural systems in specific works is undeniable. To comprehend them, one needs both to examine the distinction of specific works and to identify the distinguishing features of various groups of works. But since the purpose of this book is more theoretical than critical, this discussion concentrates on structural features common in groups of plays and offers illustrative examples. The expanse of organizational principles cannot, however, be treated merely with a single set of simple terms. Hence, this investigation approaches form in three ways: as dramatic species, as organizational movement, and as graphic arrangement.

Two broad species of *form* are the *mimetic* and the *didactic*. The primary purpose of a mimetic drama is being; a writer constructs a mimetic play as an aesthetically complete object. The central purpose of didactic drama is persuading; a writer constructs such a play to inculcate ideas. In the former, meaning is implicit, and in the latter, meaning is explicit. A writer creates mimetically when he constructs a play primarily to reflect the human condition, and a writer who creates didactically devises a play mainly to affect human attitudes and change the human condition. Form in mimetic plays may be said to be centripetal, organized so that the chief parts (human feelings and events) cohere for the sake of the whole and so that the structural force acts inwardly toward an axis. And the axis in mimetic drama is usually an action. Thus, mimetic drama is primarily afferent. On the other hand, form in didactic plays can be called centrifugal, organized so that the parts adhere for the sake of a process (persuasion) and so that the structural force impels the chief parts (argumentative thoughts) outwardly, away from a center. And the center in didactic drama is usually, and at best, a metaphysical complex of ideas. Thus, didactic drama is primarily efferent.

Examples of the didactic species are: *Mother Courage* by Bertolt Brecht, *The Devil and the Good Lord* by Jean-Paul Sartre, and *The Deputy* by Rolf Hochhuth. Other important playwrights who have written dianoetic (thought-controlled) plays include Peter Weiss, Gunter Gräss, LeRoi Jones, Max Frisch, and Megan Terry. Examples of the mimetic species are: *Cat on a Hot Tin Roof* by Tennessee Williams, *The Caretaker* by Harold Pinter, and *Rhinoceros* by Eugène Ionesco. Many significant contemporaries have written mostly mimetic plays—for example, Samuel Beckett, Fernando Arrabal, Arthur Adamov, Jean Genêt, Arthur Miller, and Edward Albee. A number of dramatists have written both didactic plays and mimetic ones, for instance: Albert Camus, Friedrich Duerrenmatt, Slawomir Mrozek, John Osborne, and Michel de Ghelderode.

The most influential theorist who has written extensively about didactic form is Bertolt Brecht. *Brecht on Theatre*, edited and translated by John Willett, is the best volume in English for a study of Brecht's principles. A few of his key ideas about the structure of what he calls "epic" drama are: (1) that didactic drama should appeal primarily to spectators' reason, (2) that it should progress through narrative more than through story, and (3) that scenes can well be episodic since a central idea holds them together. He also wrote about such principles as montage, curved development, man as a process, and alienation effect.

Although many writers have written theoretically about mimetic drama, the essays of Eugène Ionesco have the most to say about recent developments in mimetic form. *Notes and Counter Notes* is the best collection of Ionesco's theoretical pieces in English. Some of his ideas about form in mimetic drama are: (1) a play is a construction of a series of conscious states, or of conditions, with mounting tension until the states become knit together and finally are unraveled or else culminate in absolute confusion; (2) the heart of drama is division and antagonism, crisis and the threat of death; (3) a play is a set of emotional materials, including moods and impulses; and (4) a dramatist discovers form as unity by satisfying his inner emotive needs rather than by imposing some predetermined, superficial order. Ionesco has also discussed the creation of dramatic microcosms, the use of symbols, the making of myths, and the employment of enigmas.

To think of contemporary dramas as mimetically or didactically organized wholes is but one way to discern the recent developments of dramatic form. Another approach to form is to consider the *organizational movement* of dramas. And in simple terms, one can perceive two basic kinds of organizational movement—*horizontal* and *vertical.*

A play that moves horizontally usually has a structure that is causal. One event causes another, and thus they form a connected series of antecedents and consequences. Connections in horizontal plays are conceived imaginatively and executed logically and objectively. The characters' motivations are important in order for the causality to be probable and clear. A story is most often the control of a horizontal play. The emphasis in such a play's action is on progression, extension, and distance. The activities, taken together, tend to be egressive. The movement from one action to another is connected, continual, consecutive, and sustained. And the peak of interest is climax.

In contrast, a play that moves vertically usually has a structure that is far less causal, and many such plays are quite adventitious, or non-causal. One event occurs for its own sake, rather than as an antecedent to a succeeding one or as a consequence of a preceding one. The events are sequential in that they follow each other in performance, but they do not make up a causally connected series. Connections in vertical plays are conceived imaginatively and executed imaginatively and subjectively. Some plays of this sort penetrate character motivations deeply, but not for the sake of identifying causality; they do so as contemplation and for intensity. Other vertical plays avoid motivations altogether. Story is seldom of major importance in vertical plays. When critics call such plays plotless, they ordinarily mean storyless, since all plays have plot as structure. The suspense in horizontal dramas usually comes from conflict, but in vertical ones it most often arises from tension. Conflict is a clash of forces, and tension is stress, anxiety, dread, or anguish within characters. The emphasis in a vertical play's action is on convergence, not progression; penetration, not extension; and depth, not distance. Direction is important in a horizontal play, and deviation in a vertical one. The non-story play stresses being; the story play emphasizes becoming. The activities in a vertical play are usually fewer and, taken together, are introgressive. The movement from one action to another is likely to be disconnected, transformational, and intermittent. It features interval, not connection. And the peak of interest is more often convulsion, convolution, or pause rather than climax (as denouement). A horizontal play usually has a beginning, middle, and end as a causal series; a vertical play often has simply a start, a center, and a stop as a broken, or random, sequence.

There is no implication here that one sort of structural movement is necessarily better than the other. Great playwrights have composed plots of both sorts. Chekhov and Pirandello wrote mostly vertical plays; Ibsen and Shaw mostly horizontal ones; and Strindberg wrote both sorts. Some contemporary examples of well written horizontal plays are: *The Crucible* by Arthur Miller, *The Just Assassins* by Albert Camus, and *The Visit* by Friedrich Duerrenmatt. Others who have written such plays include Tennessee Williams, John Osborne, Bertolt Brecht, Ugo Betti, and Jean Anouilh. Some of the significant vertical plays are: *Endgame* by Samuel Beckett, *The Blacks* by Jean Genêt, and *America Hurrah* by Jean-Claude van Itallie. Others who have written such plays are Eugène Ionesco, Peter Weiss, Michel de Ghelderode, and Fernando Arrabal.

And where can one find discussions of these two sorts of dramatic form? Three of the best sources for the structural theory of the horizontal form are: (1) the essays of Bertolt Brecht, (2) the Introduction to Arthur Miller's *Collected Plays*, and (3) *Tragedy and the Theory of Drama* by Elder Olson. For explorations of the vertical form, one can read: (1) the non-fictional works of Beckett, Ionesco, and Genêt; (2) "The Ostend Interviews" of Michel de Ghelderode; and (3) the articles and interviews in *The Drama Review* (formerly *Tulane Drama Review*) that deal with the "vertical" playwrights listed above.

There are many other interesting or productive ways to view dramatic forms in all their contemporaneous variety, but only one more of these will be useful here. One can consider plays as *graphic arrangements*. And again, a division of dramas into two groups provides an overview of the two basic sorts of graphic form. One is *linear* and the other *configurative*.

Linear form in drama is characteristic of those works that have single or parallel lines of successive events. The characters appear in psychological perspective; they are recognizably life-like and causally related to the action. Situations are important, and the movement is from one to another with increasing complications leading to final resolution and an ending situation. The whole is organized as rational reality; the structure is concrete. The arrangement of parts represents the arrangements in actual life. Most plays of horizontal movement can be graphically diagrammed as linear.

Configurative form in drama is characteristic of those works that have curved patterns of activity, episodic lines of action, and asymmetrical or random arrangements. The characters are more fragmentary, distorted, or simultaneous—as in cubistic paintings, for example—than are those in linear drama. Their motivations are often missing; they appear to be fantastical; and they are seldom causally related to the action. Conditions are more important than situations; often a configurative play is simply a presentation of only one life condition as seen through a distorting lens of imagination. Such plays concentrate on stasis or circularity. The connections between people and other people, or between events and other events, are often more surreal than real; the relationships depend upon imaginative association rather than on causal progression. A configurative play is likely to be variegated and rhapsodic. Exposition and preparation are generally absent, and rhythm or pattern usually replaces story. Transitions are likely to be abrupt, rather than smooth as in linear drama. The whole of a configurative structure is organized as a vision or a dream in order to penetrate to the reality of existence beneath the level of sensory reality. Such a structure is abstract. The arrangement of parts presents the arrangements of the imagination. Most plays of vertical movement can be graphically diagrammed as configurative.

Although no one play possesses all the characteristics of one of these two forms and absolutely excludes any feature of the other, some plays can serve as examples of each form. Some linear plays are: *The Plough and the Stars* by Sean O'Casey, *Long Day's Journey into Night* by Eugene O'Neill, and *Waltz of the Toreadors* by Jean Anouilh. And some well wrought examples of configurative plays are: *Marat/Sade* by Peter Weiss, *Serjeant Musgrave's Dance* by John Arden, and *Comings and Goings* by Megan Terry.

Among the many books that discuss dramatic form, there are several that would perhaps epitomize the linear or the configurative. John Howard Lawson's *Theory and Technique of Playwriting* offers an informative discussion of dramatic structure as linear. It reveals the important considerations about constructing a line of action involving a volitional character who meets obstacles, enters into conflict with them, and eventually wins or loses. Marian Gallaway also wrote a treatment of linear form in her book, *Constructing a Play*. Richard Schechner has written cogently about configurative structure—what he calls "open" form—in his book, *Public Domain*, a collection of his articles from *The Drama Review*. In his article entitled "Approaches," Schechner wrote significantly of the employment of time as a control for the playwright or a cue to the critic. And in exploring various aesthetic approaches to drama, he also discussed rhythm and circularity of structure. Additionally, the works of Antonin Artaud, especially *The Theatre and Its Double*, are important for an understanding of configurative drama. Although Artaud included little about dramatic structure, per se, in that volume, he presented a vision of dramatic art that is imaginative and stimulating. Rather than spelling out the principles of configurative drama, Artaud suggested them. And his theories have, in fact, impelled many contemporary playwrights to innovate with abstract drama. Artaud's theories, however, have more to do with dramatic style than with dramatic structure. But they can, nevertheless, provide an attitudinal base for comprehending recent innovations in configurative or vertical dramas.

It is appropriate at this point to consider some other books that explore dramatic forms. Although many volumes of dramatic theory and criticism exist, only a few are likely to be of great help to the playwright or the person examining dramatic structures. Among the most useful older works are: *Poetics* by Aristotle, *Ars Poetica* by Horace, *Hamburg Dramaturgy* by Gotthold Lessing, *The Technique of Drama* by Gustav Freytag, *The Birth of Tragedy* by Friedrich Nietzsche, *The Law of the Drama* by Ferdinand Brunetière, the *Journals* of Friedrich Hebbel, and the many essays on drama by Maurice Maeterlinck. Among the most pertinent modern works are those, previously mentioned, by Brecht, Artaud, and Ionesco. Other revealing contemporary discussions of dramatic form are those by Eric Bentley, Elder Olson, and Richard Schechner. Kenneth Burke, in such works as *Grammar of Motives* and *Counter-Statement*, has investigated poetic forms of many sorts, the dramatic as well as the lyric and the narrative. Burke is one of the most significant structural theorists of the century. Jean-Paul Sartre and Albert Camus have devoted no full-length works to dramatic theory, but they have influenced contemporary dramaturgy with many of their ideas about literature and art. Others who have written interestingly, and have to some degree been influential, about dramatic structure or closely related matters are: Susanne Langer, Ernst Cassirer, George Santayana, Carl Jung, T. S. Eliot, Francis Fergusson, Gerald F. Else, Northrop Frye, Hubert Heffner, Harry Levin, Marshall McLuhan, and Susan Sontag. And many dramatists have written informatively about structure—for example, Ibsen and Strindberg, Arthur Miller and Friedrich Duerrenmatt. If one were to read only a few works in order to understand the multiplicity of dramatic forms, one might best choose those by Aristotle, Nietzsche, Brecht, Ionesco, Elder Olson, Kenneth Burke, and Richard Schechner.

To comprehend fully the possibilities of dramatic forms, plays themselves are even more revealing than theorists—provided that one first knows at least some of the basic principles and approaches. The following plays indicate the range of contemporary structural practice: *The Good Woman of Setzuan* by Bertolt Brecht, *A Streetcar Named Desire* by Tennessee Williams, *Waiting for Godot* by Samuel Beckett, *The Blacks* by Jean Genêt, *Rhinoceros* by Eugène Ionesco, *No Exit* by Jean-Paul Sartre, *Marat/Sade* by Peter Weiss, and *The Homecoming* by Harold Pinter. All of these, however, are well established plays with international reputations. Among lesser known works, one can also discover fascinating new structures, for example: *The Empire Builders* by Boris Vian, *Tango* by Slawomir Mrozek, *Viet Rock* by Megan Terry, *Fire!* by John Roc, *Anna Kleiber* by Alfonso

Sastre, *Pantagleize* by Michel de Ghelderode, *Chicago* by Sam Shepard, *Futz* by Rochelle Owens, *The Curve* by Tankred Dorst, and countless others.

One should think again and again of the facts that form in drama is the organization of parts into a whole and that the key to structure in any play is unity. The principles of form are the principles of unity. Structure is the dynamic connections between parts. And form, unity, organization, structure —all amount to plot in drama. Whether a play is mimetic or didactic, tragic or comic, horizontal or vertical, causal or random, linear or configurative, open or triangular, it will have some sort of unity, a structure, and therefore a plot. The principles described throughout this chapter do not alone dictate the structure of a play. But rather as an individual playwright chooses to employ some and avoid others, he devises an unique structure for the particular materials he wants to unify. The potential multiplicity of dramatic forms is probably infinite, and one of the exciting trends in contemporary theatre is the exploration of new combinations of structural principles.

This chapter contains many of the chief principles of form in drama. The preceding section indicated the variety of specific forms that playwrights now employ. There are, of course, other conceptions of form; the treatment here does not pretend to be exhaustive. But if a writer wishes to create a play, any kind of play, he will be better able to do so if he recognizes the potentials of the basic structural principles. And consequently, he will then have greater freedom for innovation.

Handling dramatic form requires consideration of action, structure and plot. These three comprise the formal cause, the essence, or the basic conception of any play. The nature and size of the whole, the order and magnitude of the unity, determine the actuality of a drama. And dramas are actual. They do not so much mimic life as intensify it.

All the formative principles here are flexible and functional. The variety of their potential uses is magnificently infinite. But none are rules. Any one of them can be ignored or avoided. But it is possible for them to make any individual play better, and if a playwright uses them regularly, his plays will be better.

To write a work of poetic art requires judgment and imagination. Although many writers enjoy claiming that mystic mental fancy is the epitome of their art, inspiration is not enough. An artist's imagination depends mostly upon his vision, discipline, and psychological habits. His judgment depends upon his perceptiveness and knowledge. It is knowledge, of course, that the foregoing considerations about plot can furnish. A great playwright will probably make great plays no matter what, but the more knowledge he possesses about his art the greater his plays will be.

The following modern plays serve the accompanying items as examples worthy of study: action, *Serjeant Musgrave's Dance* by John Arden; tragedy,

Pantagleize by Michel de Ghelderode; comedy, *Ring Round the Moon* by Jean Anouilh; melodrama, *The Homecoming* by Harold Pinter; didactic drama, *The Good Woman of Setzuan* by Bertolt Brecht; story, *The Crucible* by Arthur Miller; unity, *Waiting for Godot* by Samuel Beckett; probability, *A Streetcar Named* Desire by Tennessee Williams; preparation, *The Just Assassins* by Albert Camus; magnitude, *The Empire Builders* by Boris Vian; and contrast, *The Chinese Wall* by Max Frisch. All these playwrights are masters of forming the actions of plays.

For skilled playwrights, form is structured action. And structured action —in all its clarity, emotive power, and dramatic beauty—is an active thing. It achieves actuality during the writing and remains actual, forever, in the play itself. As plot, a structured action stands as organization to all the other parts—character, thought, diction, sounds, and spectacle—and is both dependent upon them and inseparable from them. It is, however, the most necessary part of a play. Drama depends on language, for example, only as one means for expressing the action. A structured action is the heart of any drama.

A dramatist, then, is a maker of plots. Sometimes he uses a story and sometimes not. But insofar as he creates a structured action of whatever kind, he is a myth-maker. As Ernst Cassirer has pointed out, form is not best measured in terms of meaning and truth. A drama as a formulated action contains its own intrinsic value, its own laws of generation, its own symbolic system. A play is not mock life; it produces a world of its own by enduring as an organized entity. Drama is apprehensible and comprehensible only when it takes on a definite form. It is, then, an organ of reality. A dramatist is a myth-maker because he connects essences with reality in a specialized (i.e., "beautiful") object. A playwright creates a structure of action as his conception, and he symbolically expresses human life as his myth.

4: *CHARACTER*

My souls (characters) are conglomerations of past and present stages of civilizations, bits from books and newspapers, scraps of humanity, rags and tatters of fine clothing, patched together as is the human soul.

AUGUST STRINDBERG
Foreword to Miss Julie

A play primarily consists of human action, and personages of some sort must enact or reveal that action. The agents in a play carry out its action through their individualized being, their words, and their deeds. In this special sense, character is the material of plot, and plot is the form of all the characterizations, what the words and deeds amount to as a whole. Insofar as drama explores the potential of human action, it also explores human character, and vice versa. Thus, drama is an art chiefly concerned with the relationship of human character to human action. And because in a play characters are identifiable and self-explanatory, each character must be to some degree different from all others. Thus, for a playwright, characterization in drama requires the presentation of characters in action, characters as differentiated one from another. Drama is character in action.

CHARACTERS ARE NOT HUMAN BEINGS

An action in a drama, as the preceding chapter demonstrated, is artificial, i.e., chosen and structured by its author. An action is related to human life but is not identical with it. Likewise, a writer composes characters; he must make them up. And however much they may resemble actual people, personages in a play are not human beings. Nevertheless, in order to formulate life-like characters, a writer should understand living persons and fully employ his knowledge of them. Hence, to comprehend the principles of characterization, a playwright needs first to consider some qualities of a human personality and their relationships to the qualities of a dramatic character.

A human being is an unique complex of physiological and psychological elements. As an individual, each human is distinctive, and as a feeling and behaving organism, each person possesses conation, the power of striving to extend life. Each has a personality, a totality of characteristics distinguishing that individual from all others, composed of specific personal and social traits, attitudes, and habits. The personality of every individual is manifest on the levels of instinct, emotion, and sentiment.

Instincts are inherited dispositions, natural tendencies impelling an individual toward certain patterns of behavior for attaining specific ends. They stimulate impulses that require attention and produce action. Some typical and basic instincts in human beings are: attraction and repulsion, domination and submission, flight and pursuit, destruction and construction, display and concealment, curiosity and aversion. The instincts are closely related to, but not necessarily identical with, the basic human drives: hunger and thirst, air and temperature, elimination, sex, absence of pain, sensory stimulation and activity, and sleep. Some of the goals, then, of instincts are biologic: food, shelter, warmth, light, reproduction, and protection of the young. Some of the ends are invented: mimicry, knowledge, well-being, happiness, perfection, wealth, fame, and power. Instinctual behavior patterns tend to arouse emotions, and emotions are seldom independent of instincts. When something thwarts instinctual impulses, fear, anger, or some similar emotion arises. When the individual achieves an instinctual goal, he has joy, satisfaction, or a related emotion. Failure gives rise to sorrow, despair, or emotions of that sort. Compared to emotions and sentiments, instincts are least affected by intelligence.

Emotion in a human being is a state of excitement productive of vivid feeling. Emotions are potentially more complex than instincts and occur more consciously in a person. When an individual fulfills an instinctual impulse without difficulty, emotion does not usually result. When such impulses are blocked, emotion always results. Instincts lead to habits; emotions lead to action and sentiment. Unrestrained emotion is unstable and disorderly, violent and recognizable. Habit is the opposite—predictable, calm, and unobtrusive (at least to their possessor). Intense emotion involves diffuse nervous disturbance. Although complex, emotions are more adaptable than instincts or habits. Thus, emotions can provide the force to impel an individual to cope with a situation or an obstacle. Emotion as tension cannot persist for long, or else it becomes pathological. Nevertheless in moments before action, emotion stimulates an individual to greater energy and higher behavior potentials. Emotions initiate most notable changes of character. Both instinct and intellect can still emotion, and yet emotion is indispensable to sentiment.

Every playwright would do well to be aware of the various basic views of emotion advanced by professional psychologists: (1) Some follow the James-Lange theory and consider emotions to be skeletal and visceral sensations. (2) Many behaviorists identify body changes, rather than nerve sensations, as emotion. (3) Instinctivists judge emotion to be an aspect of consciousness accompanying instinctive behavior patterns. To them, emotion represents selfhood in all action. (4) The psychoanalytic view assumes an id, a human core of psychic energy. From the id, they derive the subconscious libido and the conscious ego. The libido channels energy for psychic changes and for emotions related to primitive biological urges, such as sex. The ego, the conscious part of personality, mediates the demands of the id, superego, and external reality. The psychoanalytic view of emotion, then, focuses on sexual and egoistic motivations. (5) Another theory which might be called psychological-physiological, is that all emotions, as intense feelings, are distinctive products, consciously perceived, of certain parts of the nervous system or brain which in turn affect the entire organism. Those who favor this theory often subdivide consciousness into motor consciousness, sensory consciousness, and the like.

The playwright need not join any of the separate schools of psychology, because he can learn from all of them. But every writer should attempt to understand emotional motivations, states, and reactions. He ought to recognize, for example, that love and appetite are two basic emotional compounds. Love between a man and a woman usually has to do with sex, and sex impels a human into such activities as inducement and submission. Appetite connects with ego-emotion, and relates to such activities as compliance and dominance. In *The Emotions*, Robert Plutchik identifies eight primary emotional dimensions; the following list includes those and several connected emotions: (1) destruction: annoyance, anger, rage; (2) reproduction: serenity, pleasure, happiness, joy, ecstasy; (3) incorporation: acceptance, admission; (4) orientation: surprise, amazement, astonishment; (5) protection: timidity, apprehension, fear, panic, terror; (6) deprivation: pensiveness, gloominess, dejection, sorrow, grief; (7) rejection: tiresomeness, boredom, dislike, disgust, loathing; (8) exploration: set, attentiveness, expectancy, and anticipation.[1] Many emotions connect to the sentiments of love and hate. Other emotions commonly identified by psychologists are: anxiety, shame, awe, embarrassment, envy, and many more. Emotions usually arise from conflicts and maladjustments of, or between, various emotional motives, states, and activities. Humans can control their emotions more easily than their instincts, and their control can be subconscious or conscious, natural or cognitive.

A sentiment is a human psychical component that controls a person's emotions, behavior, and instincts. Sentiment means, in this context, a mental attitude or an intellectual feeling; it is not to be confused with sentimentality (a softly emotional quality of disposition) or sentimentalism (a weakly emotional tendency of character). Sentiment, as the governing part of personality, gives warnings to the individual and moves him to heed those warnings. Sentiments must utilize something other than emotion for motive force. It depends upon perception, knowledge, memory, and intelligence to produce self-control. It must stem from a higher system than instinct or emotion. By various definitions, the sentiments, taken together, contribute to the composition of the mind, the soul, or the superego. Repression is an extreme form of self-control; sentiments, thus, can become destructive or can exclude an item from consciousness. Conscience, or behavioral code, can be thought of as a broad repository of sentiment. It consists of opinion, beliefs, ideals, duties, and the like. These are established or accepted by the intellect. They usually result from an individual's environment and education. Values are correlative to person—person being psyche, or self, in every human being—and each person carries values to experience. Thus conscience differs from person to person, and it grows with experience and life contacts. Reason comes into play as the factor which sorts through the sentiments and aligns one or more of them with the rising instincts and emotions to cope with a given condition. It produces ideas, concepts, and knowledge. Hence, physical traits, instincts, drives, habits, emotions, desires, thoughts, and values function in a person's sentiments and there compose the unity of his character.

Personality is self. It is the form, or overall unity, of an individual's traits. It implies the quality of reality and the state of existence. It includes the complex of characteristics that distinguishes one person from all others, and it admits the behavioral potentials of the individual which transcend all his attitudes and actions. Since the human brain is man's most distinctive feature, man can live best, maximizing his humanness, by exercising his brain. As Jesse E. Gordon demonstrated in *Personality and Behavior*, man reaches the heights of behavioral potential by making optimum use of his intellectual functions.[2] Thus, he must strive to use his ability to incite

[1] Robert Plutchik, *The Emotions* (New York: Random House, Inc., 1962), p. 63.

[2] Jesse E. Gordon, *Personality and Behavior* (New York: The Macmillan Company, 1963), pp. 548–49.

and inhibit bodily processes and balance the demands of his biological instincts and his social environment. Personality is the totality of a human being's physiological and psychological traits, and therefore it is the epitome of whatever differentiates one human from every other human.

Characters in plays, stories, and novels must to some degree resemble human beings, but since a writer composes them, they are after all not human beings but man-made characters in plays, stories, and novels. Although a character in a play and a personality in everyday life possess some features in common, there are vital differences between them. A personality exists in and of itself; a dramatic character will eventually reach full actuality only when blended with the personality of an actor. A human personality lives in an unlimited milieu, the everyday world of extensive possibility for situation and activity; a character exists in the delimited milieu of a play, an imagined world of diminutive probability and necessity. A personality is an individual performing actions; a character is an agent of action. Infinite change is possible and likely for a human being. Each body cell is regenerated in a human about every seven years. The personality of a person alters at least somewhat each day and changes greatly during a longer period of time, because of rapidly stored new experiences and developing attitudes. Today's close friend may be tomorrow's acquaintance and next week's stranger. The components of a character, however, once established, never change. Although Hamlet goes through several changes in the course of the play, the script never changes from year to year. Although any reader's interpretation of Hamlet, or any other character, may change, a character himself is always the same within the limits of the sayings and doings of the play. Thus, *a human personality can be identified as infinite and always changing, while a dramatic character is finite and never changes.* A human possesses an open personality with a series of altering and alterable distinctive qualities. A character possesses a closed structure with a series of specific and fixed distinctive qualities. If a person is a friend of a number of people, each one will know the person by different traits. But everyone who comes in contact with a character can, potentially, know that character in exactly the same way, because the same traits are apparent to all. A human personality can assume nearly any set of traits it wishes, but a character can only possess those which a writer gives it. Hence, characters are not human beings. They can and usually should deliver the illusory appearance of personalities, but they are truly agents more or less causally related to a structured action.

CONTRAST AND DIFFERENTIATION

Since personality implies selfhood and individuality in a human being, a playwright can usefully approach the work of building characters by considering them as unique individual personages. To do so, he must develop ways to differentiate one character from another, or from all others, within a play. Although a play's action demands certain behavior from the characters, each of them materially helps to determine the plot. If the characters are to be more than functionaries, they must differ in the kind, number, and quality of traits they possess. A playwright, therefore, most easily characterizes the people in his play by choosing and assigning traits to each. Knowingly or not, writers ascribe the following six kinds of traits to their characters:

Biological
Physical
Dispositional
Motivational
Deliberative
Decisive

Biological traits are those which establish a character as some identifiable being—human or animal, male or female. Many children's plays have animal characters, and some adult plays do too. The animal characters in such plays, however, usually possess human traits. Chantecler is, for example, a rooster in Edmond Rostand's play of the same name. Characters in most plays are, of course, human. The posse in most movie Westerns is always made up of a group of men, most of whom have only their maleness as a characterization. In *Lysistrata* by Aristophanes, the biological traits of the various characters provide the story's basic conflict, women vs. men. The biological is the simplest, yet the most essential, level of characterization. With some minor characters, biological traits are the only ones desirable or necessary.

Physical traits provide a slightly higher level of characterization. They, too, are simple but usually necessary. Any specific physical quality, such as age, size, weight, coloring, and posture can be used. Features of the body and face are physical traits. Vocal quality can serve, as can habitual activity or manner of moving. Physical states—of health and illness, normality and abnormality—may characterize an agent. Even clothing or possessions can indicate individuality. All these are general kinds of physical traits. In *Dr. Faustus*, Christopher Marlowe used mainly physical beauty to characterize Helen of Troy. In *The Taming of the Shrew*, Katharina's voice is one of her significant traits. And in *The Miracle Worker* by William Gibson, the young Helen Keller's physical conditions of deafness, dumbness, and blindness precipitate the action of the play. Physical traits give characters visual distinctiveness, and they are significant aids to the actors who will play the characters. Sometimes a playwright gives no physical details about a character, but merely leaves those to the actor who automatically supplies them by his presence on stage.

Dispositional traits reflect the basic bent of a character's delimited personality. In every play, most characters have a prevailing mood, controlled by the individual's temperamental makeup. Disposition in characters consists of a customary mood and life-attitude as demonstrated in speech and activity. Some authors depend especially on focal disposition as the key to each of their characters. A simple example of dispositional traits appears in the children's story of "Snow White and the Seven Dwarfs"; each of the dwarfs has mostly a dispositional character, such as Happy, Sleepy, and Grumpy. Balzac and Dickens are two novelists who emphasized character dispositions, and Lorca and Pinter are two twentieth-century playwrights who have utilized such traits to a significant degree. In Dickens' *David Copperfield,* for example, Uriah Heep appears to be humble; his disposition is central to his characterization. All the characters in both *The Birthday Party* and *The Homecoming* by Pinter have central dispositional traits as major components of characterization. Most playwrights have found that singularity of overall disposition is preferrable to multiplicity of dispositional traits. A playwright will find that giving one basic mood or temperament to each character will provide each with optimum credibility, probability, and unity on that level. A character's mental attitudes and physical tendencies, as expressed and performed, will therefore be more believable.

Motivational traits are even more complex than the three foregoing types. Also, each major character is likely to possess a larger number of them. Whereas dispositional traits often appear in a play as a character's attitudes, motivational traits come forth as desires. An attitude can be passive, but a desire usually impels a character into action. Motivations occur on one or more of the three levels—instinct, emotion, and sentiment. These correspond to those same elements as described in the previous segment of this chapter in the discussion of human personalities. Instincts furnish basic drives and stimulate impulses to activity. Not many playwrights consciously use instinctual motives. Most of the great playwrights, however, use them frequently, and modern playwrights would do well to use them similarly. Many authors are evidently too lazy to take the time to study psychology. Every character should have natural and unavoidable wants on the level of instinct. On the emotional level, most playwrights adeptly handle motivational traits. A character wants something and is emotional about that want; thus, the character possesses a desire. On the level of sentiment, characters are more immediately aware of concrete objectives; they consider and choose ends for their actions. On this level, ethical and expedient thought is suggested, means arise, plans become important, and stake or objective comes clear. Instincts appear in characters as motivational traits that are subconscious needs; emotions appear as semi-conscious desires; and sentiments appear as conscious goals.

Understanding motivational traits means considering such significant psychological theories as the associative learning theory. In *The Psychology of Human Conflict*, Edwin R. Guthrie, a twentieth-century behaviorist, discussed the clashing motives within each individual. He maintained that human motives are actually stimuli to action. Associating motivation with a pattern of stimulus-response, and excitement-activity, Guthrie described the direction of learning and habit adjustment. He outlined the events in six steps: (1) stimuli upset the individual and produce excitement, heightened activity, and a drive to remove the stimuli; (2) this response blends with previously learned responses; (3) a series of such activities, when repeated, leads to habit; (4) the excitement generated by stimuli produces not only activity but also new stimuli; (5) every response connects to the initial drive; (6) ultimately after either short or prolonged activity some response removes the drive. Guthrie summarized his treatment of motive by referring to persistent stimuli and organic conditions that induce and maintain excitement.[3] This theory of motivation is, of course, only one among many which are potentially useful to the playwright, not generally but as specifically employable in a given play.

Motivational traits ordinarily appear in a play as spoken or implied reasons for activity. They can be as simple as one character's reason for entering a room or as complex as the group of motives inciting Hamlet to commit regicide. Many teachers, writers, and critics overemphasize the importance of motivational traits. They are right that each character's key motives should be apparent, but other kinds of traits are even more important. Clarity and multiplicity of motivational traits make for "depth of character," or more precisely these traits make a character more understandable and credible. When a character's drives and desires are clear, probability of character contributes to the overall and more crucial probability of the action and of the whole play. A playwright may decide, of course, that motivational traits are unnecessary for a given kind of play. Ionesco and Pinter, for example, have written plays in which ordinary motivational traits would have spoiled the suggestivity of the characters. About multiplicity of motives, August Strindberg wrote brilliantly in his foreword to *Miss Julie*. He suggested that although motives must be clear, they should also be multiple and, if possible, paradoxical. A character is usually more interesting if he acts as a result of several contrasting motives. When motives become conscious to a character and he begins to think about how to satisfy his desires or accomplish his goals, the next level of characterization becomes necessary.

Deliberative traits refer to the quality and quantity of a character's thoughts. The foregoing four types of traits can respectively show a char-

[3] Edwin R. Guthrie, *The Psychology of Human Conflict* (Boston: Beacon Press, 1962), pp. 89–104.

acter living, acting, feeling, and desiring. Deliberative traits have to do with a character thinking. The deliberative traits of all a play's characters, taken together, comprise a major portion of the thought in that drama. The transition from the characterization level of feelings, attitudes, and desires to the level of thinking is not always readily apparent. In few plays is there an absolute demarcation between emotion and reflective thought. Emotions themselves are thoughts of a simple sort. The deliberative traits, however, which a playwright can consciously give a character, functionally occur in the dialogue as obvious reflection. When a character recognizes, considers, evaluates, or weighs alternatives, he thinks. Thinking is an active process and produces dramatic action in itself. While reflecting, what a character does is to plan, to ponder, to remember, to determine, to devise, to imagine, to suspect, to meditate, to reason, and so on. Deliberation at the highest level means careful reasoning before forming an opinion or reaching a decision.

Deliberative traits appear in the speeches of characters especially as two principal sorts of thought. First is expedient thought, considering how to do something. Second, and more significant, is ethical thought, reflecting about whether or not to do something. Expedient thought is usually shorter in duration than ethical thought. Hamlet deliberates only briefly about how to kill Claudius, but he reflects several times at length about whether or not to kill him. Most of the soliloquies in the play are fascinating examples of ethical thought, such as when Hamlet finds Claudius praying and speculates about whether or not to kill him at that moment. In contemporary plays, deliberative traits, or thought-centered passages, most often occur in two kinds of scenes: (1) discussions between a character and a friend, and (2) arguments between two conflicting characters. For example in *A Streetcar Named Desire*, Blanche expresses certain kinds of thoughts when talking with her friendly sister, Stella, and she expresses far different thoughts in scenes of mild conflict with Mitch or violent conflict with Stanley. Ethical deliberation, weighing good and evil, is one of the basic components of serious drama, and it leads to the highest level of characterization—choice.

Decisive traits represent the sixth and highest level of characterization. They show a character deciding. In fact, these traits appear only in moments of decision. All major deliberations are crises, and every major decision is a climax. Deliberations can and should take up a period of time, but decisions occupy only a moment. Nevertheless, decisive traits are the highest level of differentiation, and they deserve that rank for three reasons. First, they are always composed of, or dependent upon, all the other five kinds of traits. In a sense, when decisions occur, they stand as form to the other character traits, and the other traits are material to the decisions. A character must first be an identifiable being with certain physical features. His basic needs and drives give him a certain attitude toward his environment and impel him to desire some things and avoid others. These goals are attainable or not. And the more difficulty he has in achieving his goals, the more he will deliberate about whether or not to try for them and about how to try for them. Thus, all the other stages of characterization contribute to a decision.

The second reason that decisive traits are the highest sort has to do with how one individual can best know another. Most people are recognizably male or female, but all persons possess unique physical traits, such as height and weight, that particularize them even further. But one can understand someone else more fully by knowing the basic attitudinal aspects of their personality, their disposition. The kinds of things a person wants and strives to get reveal his inner nature even more; in this respect, his active convictions also become apparent. Once an individual's motives are clear, his true nature is increasingly obvious as he considers whether or not and how to achieve his goals. But in action—and decision based on reason is the most revealing kind of action—the person reveals himself most fully.

The quickest and best way to know someone is to see that person make a significant decision. So it is in drama. Decisions, like deliberations, are of two main kinds, expedient and ethical. Expedient decisions have to do with choice of means, and ethical decisions concern specific ends. Deciding something expediently, such as whether to use a knife or a pistol for a murder, has little relation to good and evil, or rightness and wrongness of conduct. Ethical decision or moral choice, such as deciding whether to murder or not to murder, reveals the quantity of evil in a person and the quality of good in him. The key to the action in *Lysistrata* by Aristophanes is the ethical decision of the women to join the title character in attempting to end war; the comedy in the play arises from their decision about how to force men to stop fighting. The entire action of Brecht's *The Good Woman of Setzuan* proceeds from one decision to another, both expedient and ethical, by Shen Te, the play's protagonist. Most great thinkers, from Aristotle to Jean-Paul Sartre, have pointed out that moral choice actually comprises the individual. A man is the summary of his ethical decisions.

The third and final explanation of why decisive traits are the most important has to do with the relationship of character to plot. Decision, i.e., choosing or not choosing for a reason, is action. Not only is there mental activity in a decision, but also decision forces change. At the instant a character makes a choice, he changes from one state to another; his significant relationships alter; and usually he must follow a new line of action as a consequence of his decision. For example in *Arms and the Man*, when Bluntschli enters Raina's bedroom late at night, she is afraid and defensive. But when she realizes that he means no harm, she must then decide whether to expose him or to hide him. The moment she chooses to hide him, their relationship changes, and their joint action turns in a new direction. Thus, decision is action and leads to further action. Therefore, *at the highest level of differentiation, character becomes plot.* If plot is structured action, then character blends totally with plot in the action of decision. This is one of the reasons that no one can absolutely separate form (plot) from content (character) in drama. Such is the most essential connection of character and action, personage and plot.

In order to differentiate characters fully, a playwright should be aware of the six kinds of traits and know exactly how to employ them in a play. Although when thinking of individual characters the word *traits* is appropriate, the dramatist should recognize that the characterizations must be drawn *within* the play. He can best think of traits as specific items about the characters which can be concretely inserted in proper and specific places in the playscript. He cannot efficiently characterize the personages simply by having their traits in his mind. He should write the traits down before the drafting begins, and then while he writes the dialogue, he should carefully put the traits into the play. Even after completing a first draft, the playwright can usefully read the play at least once to count the traits of each character. A good director or a skilled critic will do this; the playwright should be equally sure of the characters. No one can fully know a play without identifying what traits each character possesses.

Any character's traits appear in two basic ways. For each character, traits are apparent in the speeches and activities of that character himself, and sometimes in the actions and declarations of other characters. The playwright can get various traits into the play in sequences of physical action, in whole beats of dialogue, and in single speeches. For example, in the first scene of *The Good Woman of Setzuan*, Brecht used all three methods to characterize Shen Te, the protagonist. That scene not only establishes the total basis for the action, but also it makes Shen Te's character clear. Unskilled playwrights often fail to devote enough beats, paragraphs of dialogue, to the development of characterizations. In general, everything a character says or does reveals traits, but each significant trait of each major character should have at least one beat of its own in the play.

At the same time, a playwright should consciously see to it that he uses traits only when truly needed and only as many as needed *for the specific action.* This is the problem of magnitude in characterization. For the sake of the action, it is unimportant to know what Oedipus likes for breakfast, but it is important that he is a male of a certain age, that he has a limp, that he is somewhat impulsive and given to fits of temper, that he wants the best for the kingdom, that he relentlessly pursues truth and justice, that he decides to do what is necessary for the good of all men even though he himself may suffer, and so on. Too many playwrights seem to think that sheer numbers of traits or multiplicity of details will make full characterizations. The fact is that traits unrelated to the action will make the character confused and confusing. Not every character needs every kind of trait. Any one of the six levels may be enough to characterize the personage, depending upon the function of that personage in the play. To illustrate from *The Good Woman of Setzuan*, the following characters reach mainly the listed level of characterization; the scene in which they first appear is also identified:

Biological: Old Whore, 3
Physical: Policeman, 2
Dispositional: Family of Eight, 1
Motivational: Mrs. Shin, 1; Mr. Shu Fu, 4
Deliberative: Three Gods, prologue
Decisive: Shen Te, prologue; Wong, prologue; Yang Sun, 4

A playwright should carefully consider and consciously choose the kind and number of traits for each character. He best achieves the textual virtue of providing insight into character by letting the traits of the characters occur in action, thus permitting character and plot to fuse properly. The beginning playwright should avoid taking the character's traits for granted; he must be sure they get into the play. Too many playwrights leave the majority of a character's traits to the inventiveness of the actor who will play the role. Additionally, a writer should remember that unless the traits he uses directly affect the action in some way they are irrelevant. The appropriate traits of each character will do more than merely differentiate him from others. They will give each character credibility, clarity, right focus, unity and probability in relation to the action, and proper magnitude in the whole.

The matter of using type-characters relates to the craft of differentiation by assignment of traits. From the outset, a playwright should understand that more critics than authors worry about types. Some critics seem to take pleasure in accusing writers of developing type-characters rather than real people. What they mean, perhaps, is that certain authors fail to link characters and action in a credible manner, or that certain characters have only common traits. A playwright can skillfully formulate individual, peculiar, or distinctive characters and still be accused of dealing in types. When an author gives a character only the quantity and kind of traits necessary for a certain action, many times some critic wants to know more than those essential traits; hence, he hurls the term *type-character* at the writer. The playwright, then, must recognize the differences between critics' use of terms and authors' use of them.

With reference to characterization, *type* implies the possession of traits that a number of individuals hold in common, qualities distinguishing them as an identifiable class. Type refers to designations of kind, sort, nature, or description. When one individual well represents a type, he possesses inherent and essential resemblances rather than obvious superficial similarities. Hamlet represents the Renaissance prince as a type; Oedipus early Greek tyrants; Willy Loman a generation of American workers. The adjective *typical,* when applied to a character, means having possession of the nature of a type, group, or class of human beings. It may also mean that in one character all the essential characteristics of a group are collected, epitomized, or symbolized. Characters as types are, in a simple sense, characters as symbols. If a playwright acquires the skill of composing type-characters of the sort described, then he should probably be praised rather than damned.

Stereotypes, however, usually are to be avoided in building characterizations. A stereotypical character is one that conforms to a fixed or generalized pattern; it is an oversimplified type-character. The traits assigned to a stereotypical character are too obvious and unselective. They reflect choices of a uncritical judgment. In *The Weavers,* Gerhart Hauptmann made each of his weavers simultaneously typical and individual, but all are types, whereas in *Stalag 17* by Donald Bevan and Edmund Trzcinski, the characters are more nearly stereotypes of Allied prisoners of war and Nazi guards. Many nineteenth-century melodramas feature stereotyped characters, e.g., *Under the Gaslight* by Augustin Daly.

Another term related to representative characterizations is *archetype.* An archetype is an original from which all other individuals of the same type are copied, a prototype. For first-rate characterization in drama, most personages should to some extent be archetypical. If a playwright has sufficient insight and originality, each of his characters will be all new constructions and not patterned after any other characters in any other plays. The idea for creating an archetypical character will only occur to the writer who is willing and able to discern the essence of certain types of people in his own world.

All effective characters in plays are partially types; they are to some degree universal, recognizable to many people, otherwise they will fail to communicate anything or to render action appropriately. But every character needs also some degree of distinctiveness, some unique traits, if the character is to avoid being a stereotype. At best, each character in every play is, therefore, both universal and particular. And the simplest and most effective way to make a character universal is to motivate his action, to link him causally with certain activity. Character universality exists, inherently, in the relationship of character to action. And unique traits, too, are best rendered in a character's actions rather than in his assertions. A writer can in this manner strive to formulate archetypical personages. Characterization is first a matter of devising credible agents to execute the action, and second, contrasting one agent from another as fully as necessary.

CRUCIAL QUALITIES

In addition to the kinds of differentiating traits, one or more of six qualities may act as an aid to characterization. Not every one of these is positively essential to every character, but most major characters to some degree might possess each. The six crucial qualities are: volition, stature, interrelation, attractiveness, credibility, and clarity.

Volition, the first, is especially useful as a quality in a protagonist or an antagonist when the writer wishes to compose a story or to initiate dynamic action. It is an important quality for precipitating conflicts in a play. Volition is willpower, or the capacity for making events happen. It has to do with resoluteness, the energetic determination to carry on an action to its conclusion. More importantly, volition is the power of consciously determining one's own action. It is the active mental factor which impels men to make decisions. Characters exhibit volition when they think and perform in some of the following ways. A character should have an objective, and most of his desires should relate to that goal. He should be consciously aware of both desires and objective. Other characters should recognize his objective too. Multiple objectives cause confusion in a character, unless two clear and conflicting objectives form an active dilemma at the heart of the action. A character should not, if his volition is to remain strong, shift major objectives during the course of a play. Further, the character should suffer in some way from lack of the objective. He should make a plan to achieve his objective, and then take risks both for the sake of the plan and the objective. He may well foresee certain penalties, sacrifices, or threats on the way to reaching the objective. Of great significance to volition is the fact that a character should make his own decisions. Most beginning playwrights let someone other than the intended volitional character make the decisions. Finally, a volitional character should influence the decisions of others. Such characters as Oedipus, Lysistrata, Hamlet, and Harpagon in classical drama plus Shen Te, Sergeant Musgrave, Bernarda Alba, and the Madwoman of Chaillot in modern drama well illustrate volitional characters. Man's greatness stems directly from the demonstrated quality of his volition.

Stature is the second crucial quality, one particularly suitable for a protagonist. It harmonizes with and partly depends upon volition in a character. Stature is not just greatness, but rather strength and intensity of character. It is a quality that pushes one character to prominence above others; it gives a character supremacy. Convictions are the fundamental materials for stature. In order to achieve notable stature, a character should have strong convictions. At best, these beliefs are intelligent, admirable, and universal. The character should hold his convictions as more important than himself. His specific objectives should be a concrete and dynamic facet of his convictions. He should suffer because of his convictions and yet never put them aside. And to make all this believable, the subject character should have at least one moment of weakness, doubt, fear, loss of control, or error in judgment. If a character is too perfect, he will possess less stature. Technically, a character should voice his convictions early in the play, and he should always mention them prominently in every crisis. Sophocles' Oedipus and John Osborne's Martin Luther are two characters with unusually good stature in their respective scripts. The stature of any character is directly proportionate to the number, kind, clarity, and strength of his convictions.

Interrelation, the third quality, makes a direct impact on each character's stature, identifies each as sympathetic or not, and furnishes the basic matter for situation in drama. Interrelation is the number and kind of involvements one character has with all others in a play. A playwright will naturally establish certain ties between characters without really thinking about the quality of interrelation. Sometime during the writing of a play, however, the playwright should consciously examine the relationships between his characters. Often he will find it necessary to emphasize and clarify them. The following possibilities can be chief aids to rendering interrelation as a major quality both of a specific character or an entire play. A major character should enact at least one sequence of warmth, affection, or love toward one or more other characters. And he should receive the same from at least one other personage. Others should in some way respect him. He should be consistent in his attitudes toward others, favoring or opposing appropriate ones, and changing only when forced to do so. Interrelation is an especially significant quality in the characterization of Sophocles' Antigone and Brecht's Mother Courage. Interrelation, as an important character quality, suggests dynamic relationships, either intimate or discordant.

Attractiveness, the fourth among these qualities, is commonly considered to be a commercial device applicable only to pieces of cinematic entertainment. The writers in the movie industry, however, learned about the quality from such playwrights as Sophocles, Goethe, Ibsen, Chekhov, and George Bernard Shaw. The following factors enhance the attractiveness of individual characters; their sympathetic qualities can thus be heightened. A character, especially a protagonist, can possess attractive traits in any of the six categories mentioned in the previous section. He may be biologically and physically attractive. A positive disposition, rather than a negative one, will help. His objectives should at least be acceptable, but might better be estimable. His deliberations and decisions about reaching objectives should be ethically admirable. For all these, illustrative action best establishes attractiveness in a character. Further, he will appear sympathetic if his friends are attractive, and he will be even more admirable if his opponents are unattractive. Next, a significant but often neglected dialogue technique should be employed. For continual attractiveness, he should frequently express opinions and attitudes. His likes should, of course, be attractive, and his dislikes should be unattractive. Characters possessing exemplary attractiveness are: Prometheus in *Prometheus Bound* by Aeschylus, Juliet in *Romeo and Juliet* by Shakespeare, Bluntschli in *Arms and the Man* by Shaw, and Shen Te in *The Good Woman of Setzuan* by Brecht. The quality of attractiveness, when handled adroitly, permits a playwright to control the potential for empathy surrounding his characters. He can thus make them worth caring about. Finally, attractiveness depends greatly upon the moral purpose of a character. If his moral purpose is admirable, he will appear attractive. If it is not, he will tend to be unattractive. And if it is unidentified, he will likely possess neither positive nor negative appeal. The playwright should always arrange details to control attractiveness as he wishes.

The fifth character quality of importance is *credibility*. It is sometimes called life-likeness or verisimilitude. Credibility in characterization relates first, as was demonstrated in Chapter 3, to probability in plot. Hence, it depends somewhat upon devices of preparation—e.g., exposition, plants, and pointers—and the establishment of causal sequence. If the reason for a character's action comes out before he performs the action and if one action is a causal result of an antecedent action, then the character will be believable. Credibility will also be heightened if a character is consistent throughout the play, even if inconsistency is a given character's control quality. The following items are further considerations in establishing verisimilitude. A character's actions should be consistent with his physical and social environment, and with his background. His actions should be proportionate to his motivations. His motives need to be clear as revealed by himself and others. Always, credibility is relative to the overall context of the play. What is credible in the character of Snow White is not what is credible in Hedda Gabler, and vice versa. Hence, appropriateness in character is also a factor of credibility. Among the Greeks, Sophocles was most skilled at establishing internal credibility in such characters as Oedipus, Electra, and Antigone. Among contemporary dramatists, Arthur Miller, Jean Anouilh, and John Osborne are particularly adroit with character credibility. The key to believability in characterizations, for any playwright formulating any kind of play, is that the characters should strive for the probable and necessary. As Aristotle pointed out, whatever a personage does or says should be a probable and necessary consequence of his total character and of his foregoing experiences in the play.

The last of these six crucial qualities is *clarity*. Whatever a playwright may intend his personages to characterize and however he constructs them, he should make them lucid in feeling, thought, and action. It does a character no good for the writer to think about his traits or qualities unless he also sets those items concretely in activities or words of the play itself. Scenario character studies are fine exercises for a dramatist, but he must incorporate various aspects of each character in the script. Too often playwrights fail to devote enough beats of dialogue to the solitary purpose of depicting some single trait or quality of one particular character. Such factors as those that follow provide the playwright with the means to invest the characterizations with clarity.

All major traits and qualities of a character should be demonstrated in action, not merely claimed or discussed. Furthermore, one of each character's traits should stand out above all others. Each character should have the opportunity to make his reactions clear during a crisis. The characters should strongly contrast with each other, at least insofar as the play's overall probability permits. Those are the major considerations, but some less important techniques may also help. A character's social relationships ought to be clear before or during his first scene. Each major character should have exhibited all his most significant traits by the end of the first third of the play. Both the character himself and others should talk about the character's feelings, motives, traits, capacities, and abilities. Minor characters are always important reflectors for major characters. Also, a few traits or qualities of each major character should be indicated but not developed in action. Examples of clarity in characterization are bountiful in Shakespeare's plays and in those of Strindberg, Shaw, Chekhov, and O'Casey. Complexity of character is not nearly so important as clarity.

The worst thing that can be said about any drama is that it is a bore. Such a comment ordinarily results because the play has too little dramatic action for its length. The next most unfortunate remark about a play would be that its characters are uninteresting or not worthy of attention. If that comment is justifiable, then the playwright has not so much failed in his choice of characters as in his treatment of them. A playwright should know his craft well enough to be able to make any character interesting, and he can best accomplish that by employing some of the factors mentioned in this discussion of six crucial qualities. These are the chief means for making characters admirable—whatever their station, profession, or attitude. The danger, however, in building characters by assigning them qualities is that the playwright may know what qualities a character ought to have and yet not get them into the play itself. The best defense against this error of omission is for the writer to design an appropriate number of dialogue beats, or units, for each quality of every major character. After a play is drafted, the playwright should analyze it to identify the places where the script exhibits individual character qualities of volition, stature, interrelation, attractiveness, credibility, and clarity.

THE PROTAGONIST

Whenever a writer formulates a play and puts it into dialogue, he deals not simply with a non-related group of individual characters but with a collectivity. Each personage plays out his particular destiny in direct connection with the collective forces of the action, the other characters, and the significant thoughts. The interrelated energies of all the characters tend to overwhelm each individual. As a result, a playwright naturally strives to make one, two, or some small number of characters dominant in the play. Indeed, most plays focus primarily upon one character. The problem of focus in drama, in fact, directly represents one of the most challenging problems of the twentieth century. The collectivization of man in contemporary mass society reduces the individual to a usable digit. To render a character more prominent than others and more important than society itself is to portray the value and dignity of the individual. Such portrayal comprises, in most cases, the nerve center of a drama.

Most sets of dramatic materials will result in a clearer play if one character is focal. Such a character is a *protagonist.* Some writers prefer another term, such as hero, central character, focal character, or even leading role. But hero suggests the protagonist of a melodrama; central character a relatively inactive individual; focal character a victim; and leading role something flashy for an actor. The term *protagonist* implies involvement in an extended struggle and passion of some sort. To the ancient Greek playwrights, protagonist probably meant first or chief actor; for most modern dramatists, it means the character receiving the most attention from the playwright, the other characters, and eventually the audience. The protagonist is the character with the most volition, the one who makes events happen and propels the action. The protagonist's problem, more than that of any other character, is centripetal to the play's entire organization. A protagonist is also a key element of the story as explained in Chapter 3. In that regard, he is the chief agent for the reestablishment of balance. Ordinarily, a protagonist is one individual, but group protagonists are possible and sometimes necessary. Furthermore, the protagonist makes the major discoveries and decisions in a play. He usually delivers the most speeches, is on stage longest, and engages in the most activity, or at least has the most things done to him.

In tragedy, the protagonist is usually more good than evil, or at least potentially so. Oedipus, Hamlet, and Blanche Dubois are well-known tragic protagonists. In comedy, the protagonist is most often either the leader of the normal people in the play, one who falls prey to anormal situations; or he is the central anormal character, one who creates the comic situations and acts as butt for the jokes. Bluntschli in *Arms and the Man* is an example of the former type of protagonist, and Lysistrata and Harpagon are examples of the latter type. In melodrama, the protagonist is usually a good hero who suffers but finally wins. Bo in *Bus Stop* by William Inge and Johnny in *A Hatful of Rain* by Michael Gazzo are two typical protagonists in American melodramas. Jimmy in John Osborne's *Look Back in Anger* acts as protagonist in a well-known British melodrama. Occasionally, a villain acts as protagonist in a melodrama that is filled with mostly "evil" characters. Two good examples are Regina in *The Little Foxes* by Lillian Hellman and Martha in *Who's Afraid of Virginia Woolf?* by Edward Albee. In didactic drama, the protagonist is either the desirable example or the despicable example. Hence, his personal ethos is a positive or a negative means to persuasion for the play. Everyman exemplifies the former; Sol Ginsberg in John Howard Lawson's *Success Story* the latter; and Brecht's Shen Te both in one. Exemplary plays containing group protagonists are Gorki's *The Lower Depths*, Hauptmann's *The Weavers*, Lope de Vega's *Fuente Ovejuna*, and *Waiting for Lefty* by Clifford Odets. These descriptions of typical protagonists of certain forms of drama are only epitomizations, and the characters cited are merely selected examples. The potential variety of protagonists is genuinely infinite, as the following will suggest: Medea, Cyrano, Peer Gynt, Liliom, Mother Courage, and Sergeant Musgrave. The characteristic which all protagonists share is formulative centrality in a plot.

The next most important figure in the majority of plays is the *antagonist*. Although a play can well exist without one, an antagonist lends clarity and power to a dramatic structure. The primary function of an antagonist is opposition to the protagonist. An antagonist usually best represents the obstacles. If his volition is approximately the same as or greater than that of the protagonist, the resultant crises and conflicts will be more dynamic and can more easily reach an optimum level for the specific material. An antagonist frequently is responsible for initiating the protagonist's, and the play's, crucial problem. An antagonist is often the leader of a group that opposes the protagonist. The antagonist, also, is likely to face both expedient and ethical decisions. He is usually second in number of speeches, amount of stage time, and degree of activity. Other commonly employed terms for antagonist are villain, opposer, and chief obstacle. In the various dramatic forms, many antagonists are as significant as their companion protagonists.

In tragedy, the antagonist is usually more evil than good. Among all antagonists, Shakespeare's Iago is probably the most well-known and surely the most purely evil. Other representative antagonists in tragedies are Creon in *Antigone*, Claudius in *Hamlet*, and Stanley Kowalski in *Streetcar*. In comedy, an antagonist is not necessarily good or evil, normal or anormal; he is ordinarily the character who entangles the protagonist in the comic situation or with whom the protagonist conflicts. Some well-known comic antagonists are Katharina in *Shrew*, Cleante in Molière's *The Miser*, and Eliza Doolittle in Shaw's *Pygmalion*. Most antagonists in melodrama are totally evil; they are properly called villains; and they deserve the punishment they usually receive. Among American melodramas, notable antagonists are Mat Burke in Eugene O'Neill's *Anna Christie*, Harold Goff in Irwin Shaw's *The Gentle People*, and Big Daddy in Tennessee Williams' *Cat on a Hot Tin Roof*. In didactic drama, the chief antagonist takes the opposing position to the protagonist; the alignment in most didactic plays resembles that in most melodramas. If the protagonist is evil, then the antagonist is good, and vice versa. Three examples of antagonists from American didactic plays of the 1930's are Inspector Feiler in *Marching Song* by John Howard

Lawson, Prescott in *Black Pit* by Albert Maltz, and Lem Morris in *Stevedore* by Paul Peters and George Sklar. Two outstanding antagonists from European didactic plays are Hwang Ti in *The Chinese Wall* by Max Frisch and Yang Sun in Brecht's *The Good Woman of Setzuan.* As with a protagonist, there are no universal rules about the nature of an antagonist, but to establish one is to heighten the stature of a protagonist and to enliven an entire action.

Three other special kinds of characters are the foil, the raisonneur, and the messenger. Each can perform useful functions in a play, but none is absolutely essential as a singular agent. Their functions, however, must be filled by some character in nearly every play. The choice about them is whether in a given play they are worth a whole character or whether their function can be combined with others in a multiplex character.

A *foil* is a minor character who stands as a contrasting companion to a major character. The specific functions for a foil are potentially diverse. He may possess strongly contrasting and partially complementary traits by comparison with his superior companion. If the protagonist in a melodrama is smart and always serious, his foil might be a bit stupid and lighthearted. The foil provides a major character a close associate with whom he can discuss problems and plans; hence, the foil is a means to deliberation in drama. The foil can help build a major character's stature by talking sympathetically about him to others. Also, the foil can perform jobs unsuitable for the major character. When the foil suffers or profits from the actions of the major character, the action gains strength through implication. Some well constructed foils are Horatio in *Hamlet*, Ben and Charley in *Death of a Salesman,* and Staupitz in *Luther* by John Osborne. Foils can usefully serve both positive or negative characters, protagonists or antagonists.

A *raisonneur* is a character who speaks for the author. In a broad sense, all the thoughts and words in any play come from its author. Most writers, however, try not to impose their thoughts on their characters, at least in key deliberative or attitudinal speeches. But most dramatists like to insert some of their favorite reflections. Hence, playwrights sometimes establish one character whose sole function is to speak for them. The contemporary practice is to spread author-reflections among several characters. When this occurs, of course, only the writer himself can be certain about which thoughts are truly his and which belong to the characters. Some clear examples of the raisonneur are The Contemporary in *The Chinese Wall* by Frisch, Tom in *The Glass Menagerie* by Tennessee Williams, and The Father in Pirandello's *Six Characters in Search of an Author.* Each of these characters, however, serves more than this one function.

The *messenger* has been an important functionary in drama from Aeschylus to Anouilh. Because certain incidents appropriately happen offstage and because these must sometimes be reported, the carrying of news and descriptions is a necessary activity. Again, either a single or a multiple purpose character can fulfill such a function. The messenger is an obvious device in most Greek tragedies, but those playwrights usually chose to handle such characters directly and simply. They were, of course, skilled dramatists, and they sometimes made their messengers more complex. In *Oedipus the King* for example, Sophocles used a simple messenger to tell about Jocasta's suicide and the reaction of Oedipus, but he made another messenger, the Corinthian Shepherd, more complex and more thoroughly involved in the basic action. Messages also abound in *Hamlet*. Polonius and Ophelia carry some, but Osric is a functionary messenger whom Shakespeare must have had great fun in characterizing. Modern plays containing an unusual number of messages are: Sean O'Casey's *The Plough and the Stars*, Brecht's *Mother Courage*, and John Arden's *Serjeant Musgrave's Dance*. The messenger is a carrier of news, but more significantly he precipitates discovery in other characters. In this special way, messages directly contribute to a dramatic action.

A discussion of various kinds of characters would be incomplete without some mention of the *narrator*. Actually, however, narrators appear in plays infrequently. The skilled playwright avoids them; he prefers to dramatize his material rather than have some agent tell it. Most of those who employ narrators are beginners, novelists, didacticists, or writers of pageants. Many beginners use a narrator because they feel the urge to speak directly to the audience or because they cannot devise a connected action and must make transitions some way. Novelists or short-story writers who try playwriting have the habit of employing narrators of various kinds in their fiction. Many fail to understand the directness of drama, and a fictional storyteller is a contradiction of the immediacy of drama. Authors of dianoetic plays often find that the narrator is a handy device for explanations, exhortations, and appeals to an audience. The narrator is thus justifiably useful in didactic drama. Pageant writers frequently resort to a narrator because the quantity of factual material and the episodic nature of the story demand it. There are other reasons, too, why a dramatist may wish to use a narrator, but most writers learn by experience that narrators tend to spoil plays. The following are examples of skillfully employed narrators in didactic dramas: Wong in *The Good Woman of Setzuan*, The Contemporary in *The Chinese Wall*, and the Loudspeaker in the Living Newspaper *One-Third of a Nation*. Tom in *The Glass Menagerie* and the Stage Manager in Thornton Wilder's *Our Town* are two examples from mimetic dramas. An old saying among professional playwrights is: "Don't use a narrator unless you have to, and if you have to, don't write the play."

To formulate characters for a play, a writer needs to know and control the various kinds of agents according to their appropriate functions. Each should be clear, receive the proper emphasis, and be developed to the necessary complexity. One of the chief difficulties of playwriting is the creation of characters that are at once precisely functional, credibly life-like, and imaginatively stimulating. Thus, the work of characterizing requires a thorough understanding of the craft plus a penetrating vision into the nature of human life itself. The rational part of a man is small and precariously situated by comparison with the subterranean forces of life within and around him. When characters in a play function causally in relation to the action, the reasoned connections represent the triumph of the author over natural disorder. And even in works depicting human alienation or grotesquerie, the characters simultaneously symbolize man's troubled nature, rational consciousness, and creative vitality. Dramatized characters are, at best, man's full recognition of the human self.

CHARACTERS IN ACTION

Dramatic action occurs within specifically delimited circumstances, and thus the action, in turn, strictly determines the requisite characters. Sometime during the writing process characters may suggest action; however, once the play becomes an organic whole they are absolutely subservient to its demands. The only way a dramatic action can happen is through the external behavior of characters, and external behavior is limited to the bodily and the vocal. Furthermore, action in drama is usually interpersonal; one character does or says something to another. All the items, then, of each character's external, interpersonal behavior are activities. And all active doings and sayings of all the characters materially comprise the total action of a play. Although a play's action can be capsulized in a single sentence, the action is actually the sum of the activities in the play. In order for a character to serve a dramatic action properly, all his activities should be useful, appropriate, credible, probable, and consistent.

Universality in drama is another important quality related to both plot and character. Some theorists, Allardyce Nicoll for example, claim that universality is a general atmosphere in the play, a mist enveloping the story, or a coloration of the characters. Despite such mystic views, universality is not an indefinable, philosophic quality of a play that permits thousands of people to recognize the typicality of the story or characters. Universality in drama exists in the causal relation between character and action. Life for any human individual is composed of singular incidents which may or may not be connected. Drama, however, is an action made up of singular incidents coherently related by some sort of causality, whether progressional or abstract. In linear plays, one activity or event follows another causally; all the incidents form a chain of internal antecedents and consequences. In abstract plays, the activities or events comprise a configuration, or a non-sequential pattern. But in both kinds of plays, characters are universal whenever they are carefully limited by a playwright to depict an aspect of human nature which is rigorously tied to an action. A character is universal because he is an item in a unified and organic whole; a human being is singular because he is, potentially at least, a free individual not necessarily related to any particular line of action. To the playwright, universality of plot and character is not a matter of striving for philosophic effect in an audience but rather the craft of binding characters to their consonant action. And, finally, for universality to appear in a drama, what happens with the characters must in some respect—either rationally or imaginatively—be true to human experience.

A play's action also decrees the number of characters to be used in a play. Such considerations of economy and necessity should arise in relation to any work of art, especially in drama. The situations and events require certain characters, and only certain characters would do or say certain things. By using only essential characters, a dramatist can make a stronger play than by employing a larger group of semi-necessary ones. While composing a scenario, the writer should determine the number of characters necessary for the action. To insert extra characters while writing dialogue is almost always unwise. The skilled contemporary dramatist combines several functions within each character. The fewer characters in a play the better, provided the action is fully served. Some of today's dramatists put together small-cast plays for the sake of marketability; the cost of producing a small-cast show is much lower than that of a large-cast production. But such business considerations have nothing to do with the inherent requirements of a particular dramatic action. A play is best served when only essential characters appear.

The characters of most plays should possess an awareness of time, but strangely many playwrights fail to permit their characters this imaginative freedom. In the mind of normal human beings, thoughts about the past, the present, and the future mingle until they are nearly inseparable. Likewise, characters gain credibility when the playwright permits them to have memories and hopes. The scripts of most beginning playwrights are nearly always devoid of memory passages. Two plays by two young American playwrights can serve as examples of time awareness in characters: *Slow Dance on the Killing Ground* by William Hanley and *Summertree* by Ron Cowen. The characters of both plays often reflect about the past and contemplate the future.

The choice of names is the final major consideration among all these principles of characterization in drama. The most important functions of any character's name are identification and epitomization. Names differentiate characters from one another simply but efficiently, and at best a name should capture the image of a character. Any sensitive playwright will know how in everyday life each person's name affects that person. Also, he will be aware of the social and cultural associations of various first and last names. All his knowledge of names should help him determine his choices for his play. In addition to social and metaphorical images, every name has an acoustic impact. The sum of the individual sounds composing a name helps to determine its emotive effect. For instance, a first name beginning and ending with a plosive—e.g. Bob, Bart, Ted—has more acoustic strength than one with two nasals, two fricatives, or two vowels—e.g. Myron, Seth, Ira. The characters' names should always, of course, serve the play. Sometimes realistically denotative names are desirable—Willy Loman, Stanley Kowalski, Abraham Lincoln. In some plays, however, invented names are more appropriate—Ragpicker, Gogo, Big Daddy. Even non-names may be properly concomitant to the characters—Mr. Zero, He, K. Comedies sometimes employ satiric or witty names—MacBird, Sir Jasper Fidget, Reverend James Mavor Morell. Any large phone book is a wonder-house of names. A playwright will enhance his characterizations by consciously deciding what kind of name each character should have and then by selecting a name that suits his purposes. In every case, a name should be appropriate, credible, and functional.

Truly dramatic characters do not exist alone. They are causal factors in a plot, and they are socially interactive. These principles are as true for abstract plays, such as Jean Genêt's *The Balcony*, as they are for causally sequential plays, such as *A Streetcar Named Desire* by Williams. To create characters, then, a playwright composes not merely a number of solitary individuals but also a social environment, a society in miniature, a mythic microcosm. The interactional nature of behavioral control has always fascinated dramatists. Every human has potential for good and for evil, and each society attempts to maximize the former and inhibit the latter. Man, therefore, is constantly involved in a socialization process. In a play as in everyday life, the mutuality between a society and individuals produces harmony and conflict, balance and crisis, suspense and climax. Societies have authority structures, roles of conduct, and forms for relations; plays correspondingly possess action structures, functional roles of activity, and agent-object relationships. All these conditions produce anxiety within individuals. Anxiety—whether defined as a high drive state or as a state of diffuse fear—operates as a common human motivation. It is a necessary antecedent to expedient or ethical choice. Thus, the human pattern of drive-anxiety-choice points to the problem-solving nature of conflict behavior. Conflict defines a problem, and the character strategies for its solution involve instrumental and noninstrumental responses plus temporary escapes or enduring solutions. By solving problems, people increase their ability to master themselves and their world. Hence, a play establishes an action containing problems and conflicts. It employs characters who carry out the functional activities of response, struggle, and decision. In such ways, character and action are interlocked. Thus, drama presents universal human problems, and depicts strategies, solutions, or results that reveal the most corrupt or the most admirable aspects of human nature—all packed into one art object.

For most of the twentieth century, certain critics and teachers have argued about the relative primacy of plot and character in drama. Even some writers have joined the controversy. This dispute is like an argument about the relative food value of cabbage leaves and coleslaw. First, most of the participants fail to distinguish between plot and story. Second, they forget that a playwright is an artist constructing an object, and not a psychologist recording case histories. Third, they misunderstand the form-matter relationship of plot and character. Character is the concomitant material of plot (organization) in drama. They can never really stand separately in any play. The structural relationships of a play as an artificial object apply even to biographical or autobiographical materials. Although some dramatists may be stimulated more by character and others more by form, it is a logical contradiction to claim that one is more important than another in a play. Both are essential.

Exemplary plays for a study of character have appeared throughout this chapter. Some, however, are worthy of close scrutiny for separate principles of characterization: the fascinating but non-human nature of characters, *Who's Afraid of Virginia Woolf?* by Edward Albee; traits of contrast and differentiation, *A Streetcar Named Desire* by Tennessee Williams; crucial qualities of volition and stature, *Becket* by Jean Anouilh; the protagonist, *Mother Courage* by Bertolt Brecht; economy and necessity, *Waiting for Godot* by Samuel Beckett; tragic characters, *Hamlet* by Shakespeare; comic characters, *The Miser* by Molière; melodramatic characters, *Look Back in Anger* by John Osborne; and didactic characters, *The Good Woman of Setzuan* by Bertolt Brecht.

A dramatist, then, is not only a maker of plots but also a builder of characters. He develops myths, and he reflects images of human nature. The playwright contributes to man's progressive process of self-liberation by exercising his creative power in constructing the world of a play. It is an ideal, a world full of action and friction, a world involving characters in contrast and conflict, and a world moving from drive through discord and ultimately to chaos or harmony. The world of a play demands functionary characters, but ones that represent man's search and struggle for meaning in life.

5: *THOUGHT*

Of course, in my plays there are people and they hold to some belief or philosophy—a lot of blockheads would make for a dull piece—but my plays are not for what people have to say: what is said is there because my plays deal with people, and thinking and believing and philosophizing are all, to some extent at least, a part of human behavior.

FRIEDRICH DUERRENMATT
"Problems of the Theatre"

Dramatic art is something more than one human method of making order from the chaos of life or of investigating the behavioral nature of man. Drama also makes meaning. A man constructs a play as a unified and as a meaningful object. But the question of *how* a play "means" is controversial. The ideas in any drama, or the meanings derived from any drama, proceed from thoughts. The playwright, therefore, should consider the various ways through which thought relates to drama. If a play's ideas begin in a playwright's mind and end in an audience's consciousness, the trajectory of thought must occur in the play itself. What, then, are the ways that thought appears in drama?

THE THREE LOCI OF THOUGHT

To understand thought in relation to drama, one must first be aware of its three locations. Significant differences exist between thought as one set of materials in a plot and thought as conceptualizations previous to or resulting from a play. The three loci of thought in relation to drama are the playwright, the play, and the audience. In fact, the state of being of any drama, in its wholeness or as sets of separate parts, has likewise a tripartite actualization—creator, object, and receptor. But the perception of thought in each of these loci is more confusing, perhaps, than the perception of other dramatic elements.

The first locus of thought is the mind of the playwright. Thought occurs in the playwright's mind in two guises: (1) thoughts about the construction of the play, and (2) ideas to be communicated by the play. Both types of thought exist before the completion of the play as an object. The architectonic thoughts are apparent in each unit or in the whole. During the first scene of *Macbeth*, Shakespeare's intent was probably to capture the attention of the audience; the idea in the play, however, arises in the three Witches' cogitation about when and where they will meet again. The thoughts the playwright has in mind about structure and technique for any unit of the play are only incidentally communicated. These formative thoughts usually interest only the play's producers and critics. The playwright's conceptual ideas about ethics, morality, or truth usually grow in his mind before the play is completed, and he ordinarily attempts to communicate them with the play. He may or may not succeed, however, in getting his ideas into the play. Also, he may decide to present them in the play either mimetically (subsumed in the action) or didactically (controlling the action). He might, for example simply imply the ultimate meanings within the play's action, or he might state the meanings directly in set speeches by major sympathetic characters. Sophocles made meaning by implication in *Oedipus the King*, but the anonymous author of *Everyman* used set speeches for moral instruction. George Bernard Shaw wrote a series of notable examples of thought in relation to drama as it exists in its first locus. The prefaces he affixed to his plays indicate that there was a great deal going on in his mind that he did not weave into his dramas. Thus, thought about a drama must occur first in a playwright's mind as considerations about composition and as potential materials of the action.

The second locus of thought in relation to drama is in the play itself. Plot is structured action, and as such, it contains thought because all the ideas and arguments in the play, taken together, are materials of the plot. For example, whenever a character thinks, he engages in activity, and a series of small activities make up a larger action. A playwright can hardly avoid putting ideas in a play; after all, an organic combination of characters and events always reveals some view of human behavior and of life value. To put it another way, one can discover the overall meaning of many plays by asking the following analytic questions: What are the two forces in conflict? Which force wins? Why? In Shaw's *Arms and the Man*, the two forces are represented by two sets of characters: (1) the realistic group—Bluntschli, Louka, and Nicola—who consider war to be a sham and love a straightforward relationship; and (2) the romantic group—Raina, her parents, and Sergis—who think of war as an heroic adventure and love as a pretentious game. The force epitomized in Bluntschli wins by puncturing the others' romantic bubble, by convincing them of life's realities, and by getting a mate as a prize. Shaw established the "why" of Bluntschli's victory—and that of Louka and Nicola—by demonstrating that war is cowardly, love is devious and woman-oriented, and that life is quite a mundane business.

Another fact about thought's locus within a play is that it is the content of the characterizations. In this regard, a functional definition of character, especially in tragedy, is: a series of choices that impel action. Since intellectual reflection usually precedes choice and since choice itself occurs in the mind of an individual, thought is thus material to character. The simplest definition of thought in drama, however, is anything that goes on inside a character—sentience, feeling, recognition, knowing, reflection, cogitation, and decision. Functionally, thought often appears in characters in three states. It may occur in a character as feelings or desires. In *A Midsummer Night's Dream* when Bottom awakens from his dream, he has certain feelings about it which he states as thoughts, and he has a desire to communicate his experience which he also states as an idea. Second, a more complex level of thought in a character is thought as deliberation—i.e., expedient or ethical reflection. Ibsen's Peer Gynt articulates a number of deliberative speeches, such as his consideration of repentance near the end of Act III, scene 3. Third, the most complex instance of thought in a character is decision—i.e., expedient or ethical choice. In *The Taming of the Shrew*, Petruchio decides to woo Katharina "with some spirit" when he sees her. And King Lear makes what turns out to be an ethical choice by deciding to let "truth" be Cordelia's sole inheritance. Thought may also appear in a play as a universal, not a philosophical apothegm but rather the causal relationship of a character and an action. Othello's actions, for example, connect with his inner conflict between love and jealousy.

Thought within a drama, therefore, is the material and the process necessary for characterization. It is, thus by extension, another material part of plot. It occurs in what the characters say—e.g., in Cassandra's prophecies in Aeschylus' *Agamemnon*. It also inheres in what the characters do—e.g., as in Strindberg's *The Ghost Sonata* when Old Hummel reacts fearfully upon seeing the Milkmaid. When thought arises in a speech or in an action, it should be true, i.e., a logical extension of, a particular character, and it should be probable in a specific sequence of action. This is the chief manner in which thought becomes conceptual in drama. Further discussion of the various ways that thought appears in drama comes in the third section of this chapter. A play's action is a consequence of thought in the characters. Hence, as a qualitative part of drama, thought is a necessary internal component of every dramatic construction.

The third locus of thought in relation to drama is in the audience. All the recognitions, realizations, and imaginings stimulated *by* the play are included in this category of thought. Audience, in this context, refers to more than only the people who witness a performance of a play. Audience also means anyone who reads it, for whatever purpose. Thus, the third locus of thought is in the minds of all the theatre artists who work at producing the play, performance spectators, reviewers, critics, scholars, teachers, students, and casual readers, in short all those human beings, other than the author, who contact the play.

Thought in the audience can be similar to thought in the playwright's mind and to thought in the play itself, but it also can be quite different. An individual may or may not be able to discern what thoughts the playwright intended to communicate, but each audience member can only be sure by reading an accompanying essay by the playwright or by talking with him in person. On the other hand, a person may be able to recognize a great variety of thoughts which reside in a play. He may discern the play's informational or ethical content. He can easily glean informative thoughts, for instance, about dope addiction from Michael Gazzo's *A Hatful of Rain* or about alcoholism from William Inge's *Come Back, Little Sheba.* Few readers or spectators would miss the simple ethical ideas about courtship and filial duty in *The Taming of the Shrew.* H. D. F. Kitto's *Greek Tragedy* demonstrates a critic's recognition of the ethical and visionary thought in the plays of the three classical Greek tragedians. Any knowledgeable director's production script—e.g., Stanislavsky's prompt book for *The Sea Gull*—affirms how a director attempts before he begins rehearsal to understand the ideas in a play. Additionally, an audience member might recognize thought in the sense of discerning the structural ideas and practices of an author. Such recognition, however, arises rarely in an ordinary group of spectators. Even the best critics seldom write about the structural principles of any given play. Elder Olson, Kenneth Burke, and Susan Sontag are three exceptions who have written adroitly about structural matters in drama.

Thought most often occurs in an audience as identification of philosophic ideas; spectators deduce meaning. The bits of conceptual thought any particular person walks away with after meeting the play may or may not be *in* the play. Often a play stimulates such thought in audience members without really containing it. Francis Fergusson made such identifications in his *The Idea of a Theater.* He commented, for example, about the action of Racine's *Bérénice* as a demonstration of "the soul-as-rational" in "three passionate monarchs."[1] But Fergusson, to some degree, confused the purpose and action of an author with the intent and activity of characters. A further instance of thoughts in the audience is when audience members understand a character's simple deliberations which lead to activity. Every school child, upon seeing *Hamlet* easily realizes that in the famous soliloquy of Act III, Scene 1, Hamlet contemplates suicide. Probably the most common manner in which thought occurs in the minds of an audience is as personal thought excited *by* a drama, any drama. These thoughts arise in the spectators' minds, not necessarily in the play; such thoughts are singular and unpredictable. A typical instance of this is that college students often get angry when a bright young literature instructor, or a foggy old English professor, tries to impose thoughts *about* the play on an entire class. What thoughts arise in an individual watching, or reading, a play depend upon that person's heredity, education, age, physical state, mental health, cultural traits, social milieu, personal beliefs, and many other fortuitous factors. Indeed, what a play means to any individual may result from such conditions more than from the play itself. To witness a dramatic production is to participate in an experience of imaginative and intellectual provocation. Drama arouses thoughts unique in each individual. Also, conditions exterior to the structural and philosophical nature of the play itself can and do affect the way any play is produced. Thus, *Hamlet* can be and has been produced as a Marxian document or as a Freudian exemplum. That a play can arouse varied thoughts is illustrated vividly with the riots during the opening performance of John Millington Synge's *The Playboy of the Western World* at the Abbey Theatre or with the conflicting views expressed publicly about the plays of such authors as O'Casey, Shaw, and Beckett. Members of any audience—theatre artists, critics, or spectators—frequently use their experience with a play as a springboard for an examination of their own lives and beliefs. This is one reason why dramas appeal to people. It is probably a favorable quality in plays that they stimulate thoughts in others. But the power of a play to arouse thoughts still should not be confused with what thoughts are *in* the play as an object.

This discussion indicates, then, the three principal kinds of thoughts about a play and their respective loci: the playwright, the play, and the audience. Any play may be considered as one whole and complete speech. As such, thought stands as the statement (conceived by the playwright) which that speech (the play) makes (to the audience) through the action of the whole play. Thought, in this simple rendering, is what a play within itself "says." But with the word "says," one must remember that what a play "says" is *in the play*, not in the playwright nor in the audience. A play-

[1] Francis Fergusson, *The Idea of a Theater* (Garden City, New York: Doubleday & Company, Inc., 1953), p. 62.

wright can use thought to build characters and to construct a plot. But he should be certain that the thoughts he selects actually get into the play, either through direct statements or by implication. Any individual in a play's audience should attempt to understand what thoughts that play truly contains.

MEANING

Thought is crucial to drama and is itself naturally dramatic because it requires the action of thinking. Thought is an activity, involving a physical process, within the brain of a living creature. In its broadest interpretation, thought can include the mental processes of learning, retention, recall, cogitation, reflection, conception, imagination, planning, belief, reason, or argument. Whenever an individual engages in any of these mental activities, he is involved in action, and action is central to the creation of drama. A playwright best utilizes thought in two ways: as specific detail in characterizations and as overall meaning. Each type contributes directly to a play's structural action.

Thinking is both direct and indirect. Indirect thought is rambling, casual, or appreciative. An individual thinking indirectly is briefly cognizant of sensations, worries, desires, possibilities, and the like. Direct thought is reflective thought. It depends upon knowledge and reason, and it leads to meaning. Dealing with the necessary, it is more than simple mental activity. It is persistent and careful. A person thinks reflectively when reaching a well-founded belief that will serve as a motive for action. Direct, reflective thought aims at solving a problem, discovering a meaning, reaching a conclusion. The highest sort of direct thought arises when a decision must be made; thus, thought depends upon character and plot in drama. A playwright can usefully think of plot as a series of difficulties involving characters who must resolve them by employing direct thought. Thought of this sort is far more important to the construction of most plays than the imposition by a playwright of an overall thesis. In fact, the following sequence describing a typical human direct-thought pattern could well furnish the basic structure for a play.

A person goes about the habitual activities of getting through a day while using mostly indirect and semi-reflective thought. A direct-thought pattern begins when a difficulty (a problem or an obstacle) arises. The person must first face the difficulty by clarifying the conditions. Next, he conceives possible solutions to the problem, and this conceptual process may be perplexing. By applying his imagination, experience, or intelligence, the person tests each possibility. Mental examination is thus the third step. Then, he selects the apparently best solution and perhaps tests it once again by seeking confirming evidence that it will help him surmount the difficulty. In other words, he must try to verify his theory or belief by experiment, advice, or research before determining final action. The last step in the reflective process is the culminating decision about handling the difficulty. After that, the person carries out his chosen course of action. These logical steps of the reflective process are not all always present in everyday thought patterns, and thus an individual often makes behavioral errors. Everyone at some time or another makes hasty judgments and superficial choices. Naturally the quality of each individual's capacity for reason may be quite different from that of others. Human beings vary greatly in their ability to think, and so it is with characters in drama. Man's nature is impulsive, impetuous, and passionate; thought is slow, questioning, and deliberate. Appetites, drives, desires, habits, and haste always tend to be enemies of thought. Nevertheless, man's capacity to reason is his most powerful weapon in his struggle to survive and in his ascent to a genuinely humane existence. A human being must think reflectively in order to understand, appraise, criticize, predict, verify, and control. The playwright himself thinks reflectively in order to compose a play, and his play generates direct thought in an audience. But most importantly the play itself *contains* reflective thought —in the characters and in the very action itself.

A playwright should be aware of some of the basic postulates of formal logic. He may wish to use them, test them, or contradict them. Undoubtedly, he will want to study them. Five of the most fundamental principles are these: (1) identity: A is A, a thing is what it is; (2) contradiction: A is not non-A, a thing cannot both possess and not possess a given attribute; (3) excluded middle: A is or is not B, a thing possesses or does not possess a certain character; (4) sufficient reason: one can determine why everything is as it is and is not otherwise, if one possesses sufficient knowledge to give sufficient reason; and (5) uniformity of reasons: that which is sufficient reason in one case must then be sufficient reason in all cases of the same kind. These basic assumptions of logic are indemonstrable; they cannot be proved or disproved. And paradoxically, to question their validity is to assume an unquestioning position. Nietzsche and other sceptics, however, have challenged these principles by asserting that they are peculiarities of human thought rather than concepts about objective fact. In any case, a playwright needs to understand the principles and procedures of logic insofar as he wishes to employ them to help establish the intrinsic probabilities of a play.

Meaning is not the same as thought, although thoughts can be meaningful. Simply defined, meaning in drama is the complex of signification residing in a play. A play may or may not contain the same meaning as that intended by the playwright or as that deduced from the play by any particular audience member. Meaning, however, implies that interpretations are desirable. Meaning has to do with ideas conveyed from one mind to other minds, and in the case of drama the means of this communication is the play itself. Furthermore, meaning has to do with correspondences.

In drama, any situation, character, or action which suggests something else may be interpreted as having meaning. Drama, like other arts, attempts to extend beyond mere observation and description, and hence it is seriously, even in comedy, concerned with the meaning of things—especially of actions, people, and ideas. Thought generates meaning; it creates signification. Without the existence of thinking beings, things might exist but would only be what they are; they could not mean something else. They could not imply, represent, stand for, or symbolize. Insofar as a play has meaning, it contains symbols; a drama symbolizes life; and it suggests symbols for life. Meaning in drama, then, proceeds from thought in drama. It consists of concepts and ideas in the action and in the characters, and it is communicated outwardly from the art object by symbolization.

One special kind of meaning that poetry—whether lyric, epic, fictional, or dramatic—produces is symbolic meaning. A symbol is something that stands for something else. In poetic works, symbols operate through comparisons, relationships, associations, resemblances, and implications. Symbols range from the simple to the complex—such as the word *cat* which stands for a certain small furry animal or as the Latin cross that implies a body of meaning in the Christian church. In literary constructions, words, thoughts, characters, and actions represent connotatively and suggest imaginatively. Thus, nearly any item in a play can be symbolic. Symbols, like thoughts, must exist *in* a play, and a playwright must be more concerned about their presence there than in his mind or in the spectators' minds.

Symbols are imaginative shorthand. They make possible the inclusion of much more than the writer actually has space for. A symbol is like a keyhole. If one peeks through a keyhole, one can see part of the room behind the door, but more is there to be seen by the person who has better vision or who changes the angle of sight through the keyhole. Also, symbols are arbitrary, artificial, and natural. These types do not necessarily refer to the quality of various symbols, but rather how they enter a work. Arbitrary symbols are those common to a society which the playwright inserts into a play. A national flag and a wedding ring are simple arbitrary symbols. All plays contain some of these. Artificial symbols are most often literary borrowings. References to other characters exemplify such symbols. T. S. Eliot, for instance, used many artificial symbols in his lyric and dramatic poetry. Many playwrights use none. Natural symbols are those arising uniquely in one literary work. They are recurring items that come to be symbolic as the work proceeds from beginning to conclusion. Two superior examples of natural symbols, one in a novel and one in a play, are water in *A Farewell to Arms* by Ernest Hemingway and sight in Sophocles' *Oedipus the King*. Most plays hold a number of natural symbols. Symbols, whether consciously or unintentionally used by a playwright will affect the complex of meanings in a play and will add texture to the whole.

Motifs are also texturally useful in drama. These operate similarly to symbols but are not the same. There are two primary differences between symbols and motifs. First, although motifs often assume the appearance of symbols in a dramatic work, they do not carry precise meaning. A symbol stands for something else, usually something larger in scope, but a motif is self-generative and stands only for itself. A motif is totally suggestive. Whereas at least part of the meaning of every symbol ought to be comprehensible to every intelligent audience member, no two spectators are likely to get exactly the same meaning from a motif. Also, motifs are nearly always natural, i.e., unique to one work. Second, motifs must be repeated in order to have any impact; a symbol, on the other hand, can be operative when used only once. By means of repetition alone motifs accumulate affective power. The term *motif* is borrowed from music, and it is analogous to *theme*. Just as a recurring theme in a piece of music becomes identifiable and draws emotion, so in a play, a recurring motif—theme would be a confusing term—stimulates emotion and thought. Also, a musical theme and a poetic motif provide imaginative coherence in an appropriate art object. Since motifs are significant imaginative items in all poetic works, the following examples are apropos. Dust in George Orwell's novel *1984* and eyes and voices in "The Hollow Men" by T. S. Eliot are effective motifs. Examples also abound in drama. Some simple but fascinating motifs are: time in Strindberg's *The Ghost Sonata*, loans in Molière's *The Miser*, and food in Harold Pinter's *The Birthday Party*. Motifs are even more important to the dramatist than are symbols; a play's motifs are a chief means for originality, infinity of connotation, and imaginative convergence.

Theme is perhaps the most confusing, dangerous, and useless word that can be used in reference to playwriting as an act and plays as objects. Theme means so many different things to so many different people that it has come to mean practically nothing. If, however, one were to avoid words because of misuse and confusion, there would be scarcely any left to use. At best the word *theme* can mean the subject or topic of a drama, or perhaps a recurring melodic sequence, but it does not clearly represent either the complex of thoughts supporting a play's characters and action or the specific thoughts contained in most speeches. When most teachers and critics refer to theme in drama, they probably mean *thought*. The latter term serves much more satisfactorily in representing reason, logic, knowledge, reflection, and meaning. A playwright may indeed be interested in recurring sound systems, and if he is to use the word *theme* at all, it best fits such repetitive auditory patterns.

Thesis can sometimes be an appropriate term for use in relation to thought in didactic drama, but for that species of drama alone. Whenever a singular thought stands as the organizing control of a play, then the play is necessarily didactic, and the central thought can often be reduced to a thesis, or a "message." But only the weakest and most obvious didactic plays amount to mere platitudes. The great didactic works by writers such as Euripides, Shaw, and Brecht are more than merely thesis plays. They are dianoetic, or thought-controlled. A thesis can be deadly for the creativity of a playwright. It is much more likely to be useful to those critics who boil a didactic play down to one bony sentence. At best, thesis refers to a given playwright's intellectual position regarding a certain human problem extant in life and included in one play.

Another significant word a playwright should know is *subtext.* Not only should a writer understand the term, but also he should take care to write every speech and action of his play as a consequence of a specific subtext he has previously conceived. Subtext refers to the emotion or thought underlying all the words and deeds of each character in a play. Words and deeds are, after all, most usually symbolic of something a character wishes to communicate. Subtext is especially important to the final life of a play as a performed drama. The director, actors, and designers must understand the subtext, or they will misinterpret the play in its details and in its overall powers. Subtext need not be restrictive to the words and deeds, but it should be clear. Jean Genêt and Harold Pinter are two modern playwrights who join Shakespeare in the highly imaginative yet dynamic use of subtext. Those three playwrights, for example, make the subtext clear within a character but permit it to be quite connotative on the plot level. For a character to say "I love you," for instance, means almost nothing unless the words are said within a particular context, for a certain reason, and as supported by a specific thought. "I love you" can mean "I worship you," "I lust for you," "I feel protective of you," or even "I hate you." The subtext is the thought foundation for all the sayings and doings in a play. A dramatist rightfully can conceive any bit of subtext with expansive freedom, but he should always be consciously aware of what he is doing with it.

Universality in drama, as explained in the previous chapter, correctly refers to the causal relationships between characters and actions. But what about universal ideas in a drama? Certainly a vast number of plays contain thoughts that are widely applicable, recognizable, or meaningful. They are thoughts a play generates that seem to be true of at least some human beings in all cultures and ages. Universals of this sort appear in drama implicitly, explicitly, or both. They can exist implicitly in the credible and meaningful relationships of character to action. The rationale for the actions of a play's characters is an inherent thought complex which others can understand. For example, when in Sartre's *No Exit* the three characters have an opportunity to leave their room, they refuse to do so; and thus the play communicates certain thoughts about the universal human condition. Explicit universals, on the other hand, are more obvious. They most frequently occur in the speeches of specific characters. In Brecht's *The Good Woman of Setzuan*, for example, Shen Te directs a number of speeches to the audience, all of which contain one or more universally applicable ideas. And, of course, a play may contain both sorts of universals, implicit and explicit ones, as is the case with both Sartre's and Brecht's plays.

No discussion of meaning in drama would be complete without some reference to truth. The most important application of truth in drama is truth as *verisimilitude.* In this sense, a drama as a whole and in its parts gives the appearance of life. A life-like action or character accurately represents human life or human experience. Verisimilitude should not be confused with realism, which is merely one manner, or style, of rendering reality. Each play will approach life-likeness in its own way and will achieve it with a varying degree of success. Verisimilitude in *The Ghost Sonata* by Strindberg is far different from that in *Hedda Gabler* by Ibsen, or in *Six Characters in Search of an Author* by Pirandello. Verisimilitude stems mostly from probability on the plot level and from causation on the character level; it is made by the playwright but must ultimately exist in the play.

Second, truth in relation to drama can refer to *veracity* in the writer. Judgments about an author's truthfulness, accuracy, or correctness have to do with this kind of truth. But veracity, too, is relative. It has to do with both insight and honesty. For example, the sort of truth sought by Ernest Hemingway is truth of action, while that attempted by Jean-Paul Sartre is truth of concept. And both may be equally valuable.

A third kind of truth important in a drama, or any other art object, is *aesthetic truth.* It depends upon the consistency and verity of a play in itself. To achieve full aesthetic truth, all the parts of a play should agree and harmonize with each other. Functioning organically, they furnish the chosen powers and the appropriate beauty in the object as a structured whole. Aesthetic truth also suggests the quality of an art object as being exactly what it purports to be. *Hamlet* and *Oedipus the King* exemplify a high order of aesthetic truth.

Fourth, *factual truth* is sometimes crucial in a play. This is a special quality in biographical, and perhaps autobiographical plays, such as Robert Sherwood's *Abe Lincoln in Illinois, Luther* by John Osborne, or one of the outdoor pageant-dramas so prevalent in the United States. Factual truth may also be desirable insofar as a play presents information about a particular place or subject. *The Investigation* by Peter Weiss, for example, presents factual truth about certain people and events related to the Nazi atrocities performed on Jews at Auschwitz. T. S. Eliot used some facts about Thomas à Becket in *Murder in the Cathedral*; James Thurber and Elliott Nugent used factual information about Bartolomeo Vanzetti in *The Male Animal,* although the play features fictional characters. Factual truth, however, is not always a major necessity in plays about well-known historical figures. Only a few verifiable biographic facts exist in *Caligula* by Albert Camus or in Shakespeare's *Julius Caesar.* If a playwright decides to utilize factual truth, to whatever degree, he should perform his research carefully and establish a non-distorting context for the selected facts.

The fifth kind of truth significant to the playwright is *conceptual truth.* This is philosophic truth and is represented by general ideas and formal concepts—ethical, moral, economic, political, and the like. All message hunting critics and liberal arts literature teachers notwithstanding, didactic plays are far more likely to contain such truths than mimetic plays. Most playwrights seldom intend their plays as mere receptacles of essences, ideals, codes, judgments, propositions, platforms, or beliefs. And although most plays contain a few such conceptual thoughts, the thoughts are usually present as materials of character and action rather than as platitudinal messages. But didactic plays, such as *The Good Woman of Setzuan* or *The Chinese Wall,* definitely propound ideas. Truth in drama can occur in many guises, and a skilled playwright handles each consciously and separately.

Drama is undeniably an art form through which men make meaning out of human existence. Nevertheless, meaning results from the thought inherent in any given drama. Each play is meaningful in an unique way. Thus, formulas and doctrines are of questionable value to playwrights and are unjust scourges in the hands of critics. Most significantly, every drama means—itself.

HOW THOUGHT APPEARS IN A DRAMA

The most crucial considerations for a playwright about thought in relation to drama are thought as material and thought as form. As Chapters 3 and 4 demonstrated, thought furnishes the most important materials for characterizations and thus contributes also to action or plot. Since thought is anything that goes on within a personage, characters can be identified, in this special sense, as complexes of thought. But thought functions as form too, even in a mimetic play. Thought is the form of the diction; it provides the subtext which the words symbolize and communicate. As both material and form, then, what are the various specific ways thought appears within a drama?

There are five main guises of thought in drama:

Statement
Amplification and diminution
Emotional arousal or expression
Argument
Meaning of the whole

The first of these is the simplest, the most common, and yet the most essential. Thought as *statement* implies the use of meaningful language (even nonsensical sounds when used for a purpose are meaningful). Characters make statements throughout a play, but thought at this simple level means statements of an indifferent kind, statements located in a play for their own sake. When such statements occur in a speech, the name for them is "the possible." Some speeches are necessary in a play, but others are merely possible. When thought appears as a general statement of an indifferent kind (the possible), the speech containing it should not be a poor or uninteresting speech. It could, for example, present information, memory, awareness, hope, or ideas. An instance of the possible exists in Theseus' speech about lovers which opens Act V of Shakespeare's *A Midsummer Night's Dream*; it is fascinating in its use of both thought and diction. Several characters in other Shakespearean plays, however, might with nearly equal credibility deliver the same speech. Jacques, Feste, Touchstone, Mercutio, or even Petruchio—personages from five different Shakespearean plays—might say:

> The lunatic, the lover, and the poet,
> Are of imagination all compact.
> One sees more devils than vast hell can hold:
> That is the madman. The lover, all as frantic,
> Sees Helen's beauty in a brow of Egypt.
> The poet's eye, in a fine frenzy rolling,
> Doth glance from heaven to earth, from earth to heaven;
> And as imagination bodies forth
> The forms of things unknown, the poet's pen
> Turns them to shapes, and gives to airy nothing
> A local habitation and a name.

To the Shakespearean scholar such a transplantation would be unthinkable, but few non-academic audiences would notice anything unusual about this speech as an utterance from any of the five characters mentioned. The point here is that sometimes an author gives a fresh and intelligent speech to a character because he wants that speech in the play, whether or not it grows essentially out of that specific character or action.

Thought also occurs often as *amplification or diminution.* This means using a speech to make something better or worse, more or less important, and the like. A notable farcical sequence in Shakespeare's *The Comedy of Errors* well illustrates thought as amplification in comedy. In Act III, scene 2, Dromio of Syracuse explains to his master, Antipholus, that a strange "kitchen wench" claims him as a lover. Dromio's amplifications of what the woman is like are traditionally called "the globe speech"; actors love this series of speeches because the comic effects overwhelm audiences with humor. The following portion indicates the use of amplification:

> DROMIO: Marry, sir, she's the kitchen wench and all grease; and I know not what use to put her to but to make a lamp of her and

run from her by her own light . . . but her name and three quarters . . . will not measure her from hip to hip.

ANTIPHOLUS: Then she bears some breadth?

DROMIO: No longer from head to foot than from hip to hip: she is spherical, like a globe. I could find out countries in her.

ANTIPHOLUS: In what part of her body stands Ireland?

DROMIO: Marry, sir, in her buttocks: I found it out by the bogs.

The final, touching speech in Chekhov's *The Cherry Orchard* contains diminution. The aged and infirm servant, Firs, has been left behind by the family, which has permanently departed. After finding himself locked in the empty house, Firs sits on the sofa and says:

> They've forgotten me. Never mind! I'll sit here. Leonid Andreyitch is sure to put on his cloth coat instead of his fur. (*He sighs anxiously.*) He hadn't me to see. Young wood, green wood. (*He mumbles something incomprehensible.*) Life has gone by as if I'd never lived. (*Lying down.*) I'll lie down. There's no strength left in you; there's nothing. Ah, you . . . job-lot![2]

This particular diminution inheres not just in the thoughts of Firs; the entire play at this time diminishes in sound and activity. After Firs stops talking and lies motionless, probably forever, silence is broken only by axe strokes cutting down the cherry orchard. In this case, then, the diminution enhances the entire meaning of the play. Amplification and diminution are means for expression of thought in drama. All humans exaggerate and rationalize, and so it is with characters.

A third basic guise of thought in drama is the *arousal or expression of emotion*. This includes the thoughts of a character as he attempts to arouse feelings in himself or others, and as he reacts to stimuli and then tries to express his own passion. The funeral orations of Brutus and Antony in Shakespeare's *Julius Caesar*, Act III, scene 2, furnish notable examples of the employment of thought in speech and action to arouse feeling. Both characters not only express their thoughts, but also they attempt to persuade others to emotion and actions. The expression of passion has always been one of the most fascinating things in drama. In Act III of Edmond Rostand's *Cyrano de Bergerac*, Roxane speaks the idea of most women about passionate expression from their lovers, and she articulates the basic need for emotional expression in drama. This dialogue occurs when Christian tries to woo her without the help of Cyrano:

ROXANE: Let's sit down. Speak! I'm listening.

(CHRISTIAN sits beside her on the bench. A silence.)

[2] Anton Chekhov, *The Cherry Orchard*, in *The Nature of Drama*, ed. Hubert Heffner (Boston: Houghton Mifflin Co., 1959), p. 570.

CHRISTIAN: I love you.
ROXANE: (Closing her eyes.) Yes, speak to me of love.
CHRISTIAN: I love you.
ROXANE: That's the theme. Improvise! Tell me more!
CHRISTIAN: I. . . .
ROXANE: Rhapsodize!
CHRISTIAN: I love you so much.
ROXANE: No doubt, and then? . . .
CHRISTIAN: And then . . . I would be so happy if you loved me!— Tell me, Roxane, that you love me.
ROXANE: (With a pout.) You offer me broth when I hope for bisque! Tell me how much you love me.
CHRISTIAN: Well . . . very much.
ROXANE: Oh! . . . Explain to me your *feelings*.

How much more moving for Roxane, and the audience, when Cyrano soon thereafter pretends *he* is Christian! Cyrano is no more honest nor passionate than Christian, but what a difference in his expression of thought:

CYRANO: . . . I love you. I suffocate with love.
I love you to madness. I can do no more; it's too much.
Your name rings in my heart like a tiny bell.
And every time I hear "Roxane," I tremble;
Each time the tiny bell rings and rings!
I treasure everything about you, every movement, every glance.[3]

Emotional arousal and expression, then, are always fundamental means to thought in any play.

The fourth way thought most often appears in drama is as *argument*. This term, however, implies much more than mere verbal conflict. It also connotes deliberation, proof and refutation, and cognition (awareness and judgment). Conflict—whether arising episodically as in Thornton Wilder's *Our Town* or extending throughout as in Albee's *Who's Afraid of Virginia Woolf?*—is one element in drama that always provides interest and suspense. Most authors render conflict as physical or verbal opposition, and verbal conflicts—within one character or between two or more—occupy a great deal of performance time in most plays. All verbal conflict rests on a foundation of thought.

An entire line of dramatic theory, beginning in the 1830's with Georg Wilhelm Friedrich Hegel, points to the concept of the centrality of conflict in drama. Hegel wrote of tragic conflict as another instance of the con-

[3] Author's translation with Richard Reney.

tradition of thesis by antithesis to be resolved only in higher synthesis. The typical argument in tragedy, for example, should be between rival ethical claims. Embellishing Hegel's basic conception, Ferdinand Brunetière pronounced in 1894 the influential idea that drama at best depicts a conscious will striving toward a goal. Although many American playwrights took up this idea, John Howard Lawson's *Theory and Technique of Playwriting* brought the conflict theory to a peak of popularity in the United States during the 1930's and 1940's. "The law of conflict," although limited as a total expression of the principles of drama, is useful to any playwright who wishes to use thought as material for deliberation and argument, or proof and refutation. After all, deliberation requires thought and is a form of internal argument. And most arguments involve proving or disproving an idea. The *agon* in *Oedipus the King* between Oedipus and Teiresias typifies thought's appearance as deliberation and argument. Oedipus both shows and vocalizes his deliberation about how to receive the accusation which Teiresias makes about Oedipus' own guilt. Their argument rises to intense conflict when Teiresias accuses Oedipus of the murder of the former King and when Oedipus incorrectly discovers that Teiresias and Creon are conspiring against him. Thought also appears in that particular scene as discovery and decision. From Prometheus' tenacious argument with Hermes in *Prometheus Bound* to the vitriolic verbal clashes in Pinter's *The Homecoming*, thought as argument has taken a central place in the crisis scenes of most dramas.

Last, thought usually appears in plays as the overall conception; the whole action of a play can be considered as an ideational speech. Thus, thought is *the meaning of the whole*. Every play in some way reflects its author's vision, or philosophic overview, which informs the whole. H. D. F. Kitto, for example, demonstrated in his *Greek Tragedy* that the entire action of *Oedipus the King* reflects Sophocles' total vision of life. He maintained that Sophocles believed the universe is not irrational but rather is based on a *logos*, a law that shows itself in the rhythm, the pattern, or the ultimate balance in human existence. Every detail in *Oedipus* indicates that as a speech it says "piety and purity are not the whole of the mysterious pattern of life, as the fate of Oedipus shows, but they are an important part of it."[4] Kitto's discussion illustrates how a critic can deduce the overall thought of a play. Not all members of an audience, however, will be so astute. Thus, if a playwright cares at all about the idea advanced by his play, he can usefully take the time to check the play as a total argument.

Thought in drama begins as sentience in a character. All the individuals in a play must in some way be aware and perceptive. They exist in a state of consciousness, and they receive impressions. Thus, any act of aware-

[4] H. D. F. Kitto, *Greek Tragedy* (New York: Doubleday & Company, 1955), pp. 148–49.

ness, any conscious recognition, any response to stimuli on the part of a character is thought. Even a motion of the hand indicating recognition of one person by another results from a simple act of thought. Various acts of sentiency show varying levels of awareness in a character. And the kinds of awareness a character exhibits serve to indicate what sort of personage he is.

Desire is a result of sentience and a stimulant of deliberation. Sentiency is awareness of sensation, an elemental kind of suffering (unpleasant or pleasant). Any sensation causing awareness, if sufficiently intense, may cause pleasure or pain and result in feelings, emotions, and passions. These are themselves elements of thought in drama. The resultant culmination of a feeling or an emotion is epitomized in desire. Desire may be a barely conscious physical need—e.g., thirst, hunger, sex—or it may be cognitively thought out—e.g., Hamlet's plan to catch the King. In this way, desire in drama depends upon awareness, sensation, and feelings, and it leads to—sometimes forces—deliberation. Of course, all this depends upon the nature of the sentient character. In some characters—often in comic or villainous ones—desire results merely in habitual activity, rather than in deliberative action. Whether desire leads to habitual or to deliberative action, it is a kind of thought.

Deliberation may be, and often is, reflection about ways and means of satisfying desire. This is expedient deliberation. Or, a character's deliberation may be about the ethical nature of the desire itself and of the moral nature of its possible satisfactions. This is ethical deliberation. Both types, and especially the latter, are high levels of thought in drama. Deliberation usually begins with maximizing and minimizing. It often proceeds in passages of fully formulated thought to emotional arousal and eventually to an argument. In a play, each is a progressively higher formal level of thought—thought more completely formulated as an element of a plot. Each of these thought levels also represents a progressively higher level of characterization.

Decision is the highest level of thought in drama; it is also the best characterizing element; and it is a significant part of plot. With decision, thought becomes character, and both become plot. It is precisely in a moment of decision that the three become one and make action. Thought as deliberation leads to and ends in decision, expedient or ethical. Decision is a moment of focus in which thought and character become action—i.e., initiate change. Recognition of desire as a result of sentiency is a kind of discovery (a change from ignorance to knowledge). Recognition *is* thought. Dramatic decisions are, in turn, based on discovery; therefore, they are based on thought. They may also be based upon deliberation, which is thought as argument. At this point, all that is true of formal rhetoric, the art of persuasion, may profitably apply to drama; the functional rhetorical principles appear in this chapter's next section on thought in the didactic play. On the level of thought as argument, rhetoric and poetics are complementary disciplines.

The ways that thought can appear in drama relate to the five qualitative types of thought discussed in this section. The potential application of each type is infinite when a playwright sets out to create a given character performing a given action. How thought most frequently arises and what functions it serves in mimetic drama should now be clear.

THOUGHT IN A DIDACTIC PLAY

Didactic plays are a special kind of drama, and their uniqueness results mainly from their persuasive use of thought. Although they too involve an organization of a human action (a plot), the principles of action in them are combined with and modified by the principles of persuasion. Chapter 3 explained the theoretic form of didactic plays. A didactic drama is functional as persuasion, and this function, rather than powers, governs the construction of the whole, the play. Didactic drama is allied with rhetoric perhaps more than with poetics. But since the function operates through a constructed action, elements of poetics are utilized in the construction. Hence, a didactic play may tend towards a tragedy, a comedy, or a melodrama (some more, some less). Any study of didactic drama, therefore, must recognize that thought often appears in didactic plays in similar ways as it does in mimetic drama. Thought serves a didactic drama in two functions. First, thought works as a qualitative part, acting as form to diction and as material to character and plot. What has been said in the previous section, then, applies to didactic as well as to mimetic drama. Second, thought operates as the control of a didactic drama; plot and character, in the mimetic sense, are material to it. Thus, the principles of thought in didactic drama closely resemble the principles of thought in rhetoric, active persuasion by a human being.

This discussion, then, shows how some rhetorical principles work in some well-known didactic plays, and it collects ideas about how thought functions as the controlling element in didactic drama. Further, it identifies the ways thought works in rhetoric which are also applicable to its employment in didactic plays.

Since Aristotle's *Rhetoric* best explains all the most important ideas about persuasion, the rhetorical principles described here come from that work. In Book I of the *Rhetoric*, the three modes of persuasion, commonly called proofs, are the first significant and applicable items: (1) the personal character of the *speaker*, i.e. *ethos*; (2) his power of stirring *audience* emotions, i.e. *pathos*; and (3) the *speech* itself as evincing truths or apparent truths by means of persuasive arguments, i.e. *logos*.

With *ethos*, the personal character of the chief personages affects the persuasive power of didactic plays. If an audience believes in the speaker, it will tend to believe in what he says. In dianoetic plays, the protagonist is often attractive, admirable, and sympathetic. This is true of the title characters of such didactic plays as *Mother Courage* by Bertolt Brecht, *Golden Boy* by Clifford Odets, and *Antigone* by Jean Anouilh. Everyman, in the play of the same name, evidently appeals today—it is frequently produced—as much as he did in the late fifteenth century. Even the name Everyman assists the ethical credibility of that character, at least in the way the author desired. Everyman possesses a properly admirable nature, for most people, as an allegorical archetype. His relationships with others, such as Goods and Good Deeds, correspond to similar abstract relationships in the life of everyone in most any audience. And a heavy portion of widely accepted Christian doctrine inheres within Everyman; for example, repentance is his central action. A contrasting method for using *ethos* as a means of persuasion is to make the central character's opponents unsympathetic and evil. Thus, persuasive plays frequently approximate the form of melodrama. The Spartan characters in *Andromache* by Euripides provide apt examples. Menelaus, who is cruel and trecherous in his connivance against Andromache, and Orestes, who murdered Neoptolemus, are opponents as oppressors. They act as persuasive elements *for* Andromache, and they arouse indignation against the power philosophy in which they themselves believe. Furthermore, *ethos* can serve in a reverse manner; the central character can possess negative *ethos*. What he believes in and represents can be made distasteful in a didactic play through his evil nature and activities. This is exactly how Regina works as a negatively persuasive element in Lillian Hellman's *The Little Foxes*. Regina depicts some of the basic evils of exploitive and rapacious capitalism. In such ways, then, *ethos* as the nature of characters is one form of thought in didactic drama.

Pathos, the power of stirring audience emotions, is the second rhetorical proof. For persuasive plays, it is much more audience-oriented than are the emotive powers in a mimetic play. Bertolt Brecht often handled the device of persuasive pathos with satisfactory didactic effect in *The Good Woman of Setzuan*. Shu Fu, an "evil" capitalist, breaks the hand of Wang, the "good" water seller. As Wang writhes in pain, Shen Te, the sympathetic prostitute and heroine, tries to help him. When she asks the bystanders to testify against Shu Fu, they refuse, and Shen Te is distraught. By that time in the scene, Brecht carefully has aroused pity and outrage; he then has Shen Te say, half to the bystanders and half to the audience:

> Unhappy men!
> Your brother is assaulted and you shut your eyes!
> He is hit and cries aloud and you are silent?
> The beast prowls, chooses his victim, and you say:
> He's spared us because we do not show displeasure.
> What sort of a city is this? What sort of people are you?

When injustice is done there should be revolt in the city.
And if there is not revolt, it were better that the city should perish in fire before night falls![5]

Thus, Brecht aroused emotion for the sake of more effectively communicating thought, and he used thought to transform the emotions aroused by the play into emotions and thoughts in the spectators about their actual lives outside the theatre. He attempted to establish a persuasive pattern of emotion arousing thought inside the theatre for the sake of pushing emotion into thought, decision, and action outside the theatre. This is the chief way thought can affect a didactic play in the matter of *pathos*.

The third rhetorical proof, *logos*, as the power of proving a truth or an apparent truth, can also work effectively in a didactic play. It is an especially useful principle for arranging the proof and refutation in rhetorical arguments between characters. George Bernard Shaw mastered this method of using thought. In *Misalliance*, he demonstrated with the argument between Tarleton and his daughter, Hypatia, the schism between parents and children. It begins when Hypatia asks Tarleton to buy young Joey Percival for her husband. At the same time, she forces old Lord Summerhays to reveal that he has proposed to her.

TARLETON: All this has been going on under my nose, I suppose. You run after young men; and old men run after you. And I'm the last person in the world to hear of it.
HYPATIA: How could I tell you?
LORD SUMMERHAYS: Parents and children, Tarleton.

The scene goes on, with Shaw continually igniting ideas about the imbroglio between parents and children, until Tarleton and Hypatia are literally shouting. Finally the defeated father says:

TARLETON: . . . I cant say the right thing. I cant do the right thing. I dont know what is the right thing. I'm beaten; and she knows it. . . . I'll read King Lear.
HYPATIA: Dont. I'm very sorry, dear.
TARLETON: Youre not. Youre laughing at me. Serve me right! Parents and children! No man should know his own child. No child should know his own father. Let the family be rooted out of civilization! Let the human race be brought up in institutions.[6]

[5] Bertolt Brecht, *The Good Woman of Setzuan*, in *Parables for the Theater: Two Plays*, trans. Eric and Maja Bentley (Minneapolis: University of Minnesota Press, 1948), p. 44.

[6] George Bernard Shaw, *Misalliance* (New York: Brentano's, 1914), pp. 95–96 and 100.

Thus, Shaw caps the argument with a climax and explosively makes his point.

In Chapter 3 of the *Rhetoric*, Aristotle divided rhetoric into three kinds: political, legal, and ceremonial. He determined these by identifying the various speaking situations, purposes, and types of listeners. Although these three types of speeches do not correspond exactly to all the types of didactic plays, their unique principles to some degree affect the structure of didactic plays.

Political, or deliberative, oratory attempts to persuade an audience to do or not to do something; it is exhortation. *The Trojan Women* by Euripides, as a play, corresponds to this type. The play's central idea is that some men, those who engage in war, are responsible for miseries of other human beings. Euripides successfully communicated a tragic idea about the results of war through his argument against a policy of conflict and against war itself. Political drama, like political oratory, exhorts an audience about their future activity; it deliberatively argues that one course of action is more expedient or ethical than another. Anti-war dramas—such as *Bury the Dead* by Irwin Shaw, *Viet Rock* by Megan Terry, and *The Chinese Wall* by Max Frisch—furnish one sort of example. Also, plays about strikes and revolutions are usually rhetorically deliberative in nature—e.g., *Marching Song* by John Howard Lawson, *Waiting for Lefty* by Clifford Odets, and *The Cradle Will Rock* by Marc Blitzstein. The third sort of play persuasive in the deliberative manner is the documentary—such as the Living Newspapers of the American Depression theatre and such pieces as *The Deputy* by Rolf Hochhuth and *The Investigation* by Peter Weiss.

Legal, or forensic, oratory attempts to accuse or defend for the sake of justice. Didactic plays, too, sometimes deal with accusation and defense, or justice and injustice. *Fuente Ovejuna* by Lope de Vega and *The Weavers* by Gerhardt Hauptmann, plays with a "collective hero" or "group protagonist," illustrate dramas of indictment. In *Fuente Ovejuna*, a village full of staunch, colorful, and courageous peasants resists the injustices of feudal overlords even to the point of killing one. When many peasants are consequently tortured, to a man they place the responsibility by giving only the name of their village, Fuente Ovejuna. Finally, the King intervenes and protects them. Thus, the play argues concurrently for central authority and damns arbitrary rule by feudal noblemen. *Everyman* is also a play of defense; Everyman endeavors to gather help to prove the worth of his life and thereby achieve salvation. Of course, courtroom plays provide the most obvious examples of legal principles entering didactic drama. Many authors have written plays directly and persuasively articulating justice and injustice. One of the earliest is the third play of Aeschylus' *Orestia, Eumenides.* This play, less mimetic and more rhetorical than its companions in the trilogy, shows Orestes finally confronting the Furies in an open trial before the Aeropagus. When the court vote is a tie, Athena enters as *deus ex machina* to resolve the action by acquitting Orestes and by persuading the Chorus of Furies about the justice of the decision and the necessity of courts. Aristophanes also used the court as a subject and the search for justice as a pattern in his play *Wasps*. Two well constructed didactic court-room plays of the 1930's are *They Shall Not Die* by John Wexley and *Judgment Day* by Elmer Rice. And certainly Shaw and Brecht used forensic principles in *Saint Joan* and in *Galileo*. Forensic drama, in conclusion, attacks or defends somebody or something on the evidence of past actions or conditions in order to establish the justice or injustice of some action or circumstance.

Ceremonial, or epideictic, oratory praises or censures a man or an institution by proving him worthy of honor or deserving of blame. In a like manner, plays can also be epideictic. In addition to the medieval mystery and miracle plays that venerated biblical personages or saints, many later passion plays have praised Jesus. Numerous biographical plays lean toward the didactic as ceremonial pieces; even some of Shakespeare's history plays incline toward the didactic proportionately as they censure or praise historical figures. Epideictic plays about political subjects often praise one figure while damning another. *Bread* by Vladimir Kirshon exemplifies the Soviet version of this type. Written at a time when the Kremlin was collectivizing agriculture, the play praises Mikhailov as the incarnation of the Party line, and it censures Rayevsky as the embodiment of political heresy in the ranks, and Kvassov, as an archetypical profiteering landlord. A number of playwrights wrote outstanding epideictic dramas for the American stage during the 1930's. *Paradise Lost* and *Awake and Sing!* by Clifford Odets represent what might be called didactic plays of awakening. They moderately praise good, but unenlightened, characters and severely censure the system responsible for the chaos in society. Another set of Depression plays censures one individual, in each case a decadent capitalist: *Success Story* by John Howard Lawson, *Gold Eagle Guy* by Melvin Levy, and *Panic* by Archibald MacLeish. And before and during World War II, anti-Nazi plays were common; *Till the Day I Die* by Clifford Odets, *Margin for Error* by Clare Booth, and *Watch on the Rhine* by Lillian Hellman are leading examples. Although written in dialect, *Black Pit* by Albert Maltz is the most skillfully structured of all these leftist epideictic plays of the thirties. In summary, ceremonial oratory, with its concern for the present, praises or censures in order to prove a man, an institution, or a system worthy of honor or dishonor. Thus, from the discussion of deliberative, forensic, and epideictic principles and examples, it should be apparent that didactic plays can and do benefit from rhetorical principles of these three types of oratory.

One further set of rhetorical principles is germane to this discussion of how thought operates in didactic plays. In Book III of the *Rhetoric*, Aristotle discussed persuasive disposition, i.e. arrangement or overall organization. In

Chapters 13 through 19, he explained the best sort of disposition of the material of a speech by identifying the following parts: (1) exordium, or introduction; (2) exposition, or statement; (3) proof, or argument; (4) peroration, or conclusion. This rhetorical arrangement is useful in some didactic plays, and many playwrights have employed it. *The Chinese Wall* by Max Frisch exhibits such a pattern. The play opens with a clear-cut introduction, entitled "Prologue"; it is spoken by The Contemporary directly to the audience. He explains the given circumstances of the action to follow and what to expect in the way of form, characters, and ideas. Next, in scene 1, comes the statement, again enunciated by The Contemporary:

> We can no longer stand the adventure of absolute monarchy... nowhere ever again on this earth; the risk is too great. Whoever sits on a throne today holds the human race in his hand.... A slight whim on the part of the man on the throne... and the jig is up! Everything! A cloud of yellow or brown ashes boiling up toward the heavens in the shape of a mushroom, a dirty cauliflower —and the rest is silence—radioactive silence.[7]

The action of the play, forming the lengthy argument, from that point rises in intensity until a revolution occurs in scene 21 and until, in scene 22, Brutus stabs the two business leaders who symbolize the profiteers behind every tyrant. After this dual climax comes the peroration in two scenes, 23 and 24. Romeo and Juliet reveal that personal love is the only hope for mankind and the only solution to the problem of human survival. The Contemporary and Mee Lan, the Chinese princess whom he has loved, end the play with a declaration confirming the truth of what has been presented and pleading for love as understanding.

Hence, the leading theoretical means of persuasion—such as the three modes of rhetoric, the three kinds of speeches, and the method of dianoetic disposition—can be useful as working principles in didactic plays. With each of these principles, thought is both the material to be organized and the form of the organization itself. The great didactic playwrights—such as Euripides, Shaw, and Brecht—have adroitly utilized the best of the rhetorical principles. These principles provide a playwright with the means for originality and experimentation. There undoubtedly has arisen in our time a critical attitude which would condemn without examination all plays of didacticism or propaganda. The rich tradition of drama as instrument, the many eloquent defenses of drama as teacher and the structural potentialities of thought-controlled drama argue convincingly that the anti-didactic

[7] From *The Chinese Wall* by Max Frisch, English translation © 1961 by James L. Rosenberg (New York: Hill and Wang, 1961), p. 28.

critics are wrong. Undoubtedly there are many badly written didactic plays, but without the tradition of persuasive thought-oriented plays, our drama would lose one of its ancient but still productive sources of energy. The distinction of a dramatist can be measured by gauging not only his overall vision but also his immediate reflective perceptions.

Thought as sensation, reaction, idea, deliberation, argument, and overall meaning is present in all plays, even those which seem to have no philosophic intent. By placing specific characters in a milieu and by permitting them to participate in given events, a playwright always indicates his vision and reveals what he considers significant about human behavior. Thought is a necessity for him in the work of creation and for the audience in perceiving and understanding his play, but most importantly it is a necessary ingredient of the art object he makes.

Many modern plays furnish instances of the functional and pertinent use of thought in drama. The following well exemplify the most important principles: meaning in symbols and motifs, *Christopher Columbus* by Michel de Ghelderode; truth as verisimilitude, *The Brig* by Kenneth Brown; thought as possible statement, *A Man for All Seasons* by Robert Bolt; thought as amplification and diminution, *Serjeant Musgrave's Dance* by John Arden; thought as emotional arousal or expression, *The Queen and the Rebels* by Ugo Betti; thought as argumentative verbal conflict, *Becket* by Jean Anouilh; thought as meaning of the whole, *The Physicists* by Friedrich Duerrenmatt; thought as motivation and deliberation, *The Just Assassins* by Albert Camus; didactic structure, *The Caucasian Chalk Circle* by Bertolt Brecht; rhetorical principles in didactic drama, *Andorra* by Max Frisch.

A playwright, like other artists, demonstrates man's incessant struggle to discover true ideas, good actions, and beautiful objects. Science represents the endeavor to uncover natural laws in order to benefit mankind. Religion represents the attempt to find and to live by a moral system which will permit the continuing existence of humanity. Art represents the venture to penetrate the bleak and disordered mass of everyday experience to render a vision of balance and harmony that gives value to existence. To reach these three goals, the struggle for which man can assume only by choice, each person endlessly battles his chief opponents—time, mystery, and death. Why should art stand apart from the true or the good? Does the artist, in practice, avoid philosophy and morality? And if he cannot avoid them, how should he handle them? An artist creates best with imagination, sensitivity, and intelligence. He uses thought, and he makes thought. Thought—as a process, a power, or an item—always occurs in relation to every drama. Drama, then, is not just a beautiful object designed to stimulate an aesthetic reaction; it is also an exploration of the nature of man and the morality of life.

6: *DICTION*

. . . if you make it up instead of describe it you can make it round and whole and solid and give it life. You create it, for good or bad. It is made; not described. It is just as true as the extent of your ability to make it and the knowledge you put into it.

ERNEST HEMINGWAY
"Monologue to the Maestro: A High Seas Letter"

Diction refers to all the words a playwright uses to make a play. Just as a wooden frame, a piece of canvas, and quantities of paint are the concrete materials a painter uses to make a painting, words are the concrete materials of a writer. One quickly learns, whether by reading about playwriting or by attempting to write a play, that for a writer words are only a means to an end, the creation of the play as art object. More specifically, diction in drama is the material of thought. Thoughts in characters within plots must exist before words can be put on paper. Dialogue in drama is a means of expressing thoughts which characters employ as they participate in an action. Because diction is subsumed in plot, character, and thought in drama, playwrights normally compose scenarios before they write dialogue. Nevertheless, words are essential for the best kind of drama. Thus, with reference to the activity of a dramatist, playwriting is a making with words.

The simplest definition of diction is patterned words. A playwright selects, combines, and arranges groups of words in speeches that within a play perform certain functions. Although the playwright puts the words together, what each character says depends upon what that character feels and thinks. Dialogue, then is expression in words. In drama, this is diction's primary function.

THE PROBLEM OF EXPRESSION

The playwright's first concern is what writers often label "the problem of expression." This problem, when analyzed, means that an author consciously establishes principles of selection and arrangement of the words. Then, he proceeds to write consistently within the limitations imposed by the chosen principles. In order to arrive at such stylistic principles, a writer ought to consider the context into which each word will fit. In drama, four major determinants make up the overall context. First is the thought or feeling to be expressed; specifically, this is the subtext. Second, the nature of the character who is speaking, especially his motivations (whether explicit or implicit), controls the expression. Third, the given circumstances of the situation in which the speaking character is involved affects the expression. And fourth, the effect to be achieved by the expression in the play—i.e., the intent—shapes the expression. These determinants, as context, control the selection and arrangement of the words.

Another phase of the problem of expression confronts a contemporary playwright just as it did Sophocles or Strindberg. Once a writer recognizes all the limitations of the context, he identifies the appropriate mode of expression. He decides to what degree the diction of his play is to be lyric or prosaic. He selects the style that best fits the play—all the specific principles of diction that control word choice, grammatical structure, rhythmic arrangement, quantity of pauses, repetition, and richness of imagery. In short, a playwright chooses the proper blend of poetry and prose; he finds a balance between heightened expression and verbal verisimilitude. With the modernist passion for showing things as they are, contemporary playwrights usually set themselves the task of representing life through the authentic speech of men. Most employ the language of prose.

The use of prose, however, complicates the problem of expression. Everyday diction makes possible intense verisimilitude and the effect of authenticity. It is the enemy of rant and bombast. But it forces an author to rely upon action and probability, psychology and suspense more than upon dazzling imagery and verbal pyrotechnics. The speech of men's daily lives —with its elisions and hesitations, its repetitions and iterations, its moans and cries—limits the verbal expressiveness of characters. But the same sort of speech also contains the pathos and the sting, the sob and clutch of the absolutely human being. With common expression, a playwright must accomplish two difficult tasks: (1) to reach an elevation of spirit through expression and (2) to make exciting verbal effects. Because drama is a concentrated verbal form requiring economy, selectivity, and intensity, it inherently moves toward lyric expression. The greatest characters express their dramatic insights and react to their conflicts in poetry; thus, most of the greatest dramas feature great poetic diction. How, then, can a contemporary playwright simultaneously achieve both verisimilitude and elevation in diction? Although realistic dialogue was the most frequently used style from the late nineteenth century to the middle of the twentieth, such dialogue now appears to be an unsatisfactory answer. As written by a few masters, it provided impact and authenticity, but as executed by many it was flat and contrived. The best playwrights of the twentieth century, especially those writing since 1950, have learned that heightened prose, or free verse, is more functional and dynamic. Playwrights are learning—especially since television and cinema have taken over common speech and flooded audiences with its banalities—that to devise intelligent and imaginative verbal styles is not only their right but also their artistic responsibility. The stark non-imagist poetry of Brecht, the convoluted and fascinating dialogue of Tom Stoppard, the sparse and resilient diction of Harold Pinter, and the lush and symbolic expressions of Jean Genêt are only a few pertinent examples. A playwright, then, should approach the problem of expression with intelligence and imagination, and above all he should carefully avoid imitating other writers. This is an age of verbal innovation.

Some authorities claim that since 1950 language in drama has been devaluated. Perhaps it would be more correct to say that it has been revaluated. One of the most significant features of verbal style in drama in recent years is that many playwrights now use language as symbolic contradiction nearly as much as they use it as straightforward statement. Especially in the works of the so-called absurdist playwrights—Beckett, Ionesco, Adamov, Genêt, and others—what happens to characters transcends and contradicts what is said by them. In the straightforward realist play, language is a means of communication and plot revelation; in many of the new abstract plays, it often prevents communication and denies plot revelation, in order to make human revelations. Nor is the diction of such dramas necessarily less poetic. Despite the devaluation of diction as common language, in the plays of some the diction is thus all the more important as melodic and expressive poetry—e.g., the works of Michel de Ghelderode, Georges Schehade, and Maria Irene Fornes. In short, the new abstractionism in drama encourages a playwright to write, to set down direct diction rather than to utilize language as a pedestrian means for telling a titillating story.

The style of the diction in any drama is the manner of the characters' expression. Dramatic dialogue is a specialization of ordinary speech. As each character speaks, he may try to communicate some information, but he mainly attempts to express feelings, attitudes, and interpretations. Thus, the basic impulses and materials of dramatic dialogue are identical with those of "pure" lyric poetry. And the motivations and methods of both poetic diction and dramatic dialogue reach deeply into human experience. Style in drama, however much intensified or distilled, reflects universal human habits of thought and feeling. To express all the nuances of human experience and to relate all the words in a play organically, a playwright must to some degree create more than mere informational or scientific dialogue. The nature of the art compels him to employ a style that is clearer, more interesting, and more causally probable than common speech. A playwright, therefore, necessarily writes every speech as poetry.

Recognizing the poetic nature of dramatic diction, however, is not enough. A playwright not only needs to understand the best organizational theories of plot, character, and thought, but also he needs to study the structural principles of diction. Hence, this chapter treats some of the most important ones—selection of words, arrangement of phrases and clauses, construction of beats, and use of punctuation. It also identifies desirable qualities, the selection of titles, and considerations for revision. With each of these, the discussion deals with basic principles, specific practices, and common errors. Writing good dialogue requires knowledge, skill, and infinite patience. As Hemingway and other great writers have pointed out, writing is hard work.

WORDS, WORDS, WORDS

A word is a combination of one or more speech sounds symbolizing an item of thought and communicating a meaning. Words are the components of auditory language, just as physical signs are the components of visual language. Chapter 7 treats speech sounds as the materials of words, and Chapter 8 deals with the visual language of drama as the accompaniment to words. Here the concern is with the selection and arrangement of words as materials of emotion and thought.

The key criteria for the selection of words in drama are clarity, interest, and appropriateness. A clear word successfully communicates its meaning. Every word symbolizes an object or an idea. Each should stand, qualified by the context, for something that can be understood between individuals. Since an audience cannot reread a character's speech, most words should be simple and common. A playwright should consider the contemporary associations of each word. Clear words are meaningful words, but verbal meaning has several aspects. The plain-sense part of word meaning is the referential fact. Intention, another aspect of verbalization, is the effect the communicator wishes to produce. Attitude has to do with meaning insofar as the communicator has feelings to express about himself, his subject, and his audience. Context, of course, always determines the specific meaning of a word. The context of each word in a play involves the parent sentence, the subtextual thought, the character, and the circumstances of the plot. Meaning can be further qualified as the denotation and connotation of words. Denotation refers to a word's literal meaning or meanings, especially its concrete referents. A word's connotations are all its suggestive or associative meanings; that the connotative meanings of any word are infinite is apparent in the work of lyric poets and in the fact that all words constantly acquire new meanings. The final complicating aspect of clarity of word meaning in drama is that with each word a playwright must communicate with his audience, and the character who speaks each word must communicate with other characters in the play. The meaning of a word may not be at all the same in each of these two spheres. Clarity is essential in both. Most writers realize that precision in word choice is an intellectual attribute requiring constant attention within themselves. A playwright should consciously weigh every word for denotation, connotation, meaning, and clarity.

While a playwright obtains clarity by using commonly understood words in an aurally comprehensible sequence, he should still select as many words as possible that are, by themselves or in combinations, interesting. The ordinary words of daily life when used unimaginatively are stale and boring. Words are more interesting when used for both literal and figurative purposes. Words without a definite literal meaning are apt to be fuzzy and easily misinterpreted; those without imaginative or emotive associations are likely to be dull. The greatest writers of American realistic prose select words that function dually, and they also balance gray words with colorful ones. As Aristotle pointed out and as the next section of this chapter will detail, a playwright can best achieve interesting diction by using metaphors.

Appropriateness, the third major criteria for word selection, refers to the social nature of language. A word may be clear and interesting yet unusable because it is inappropriate for the speaker, the hearer, the milieu, or the occasion. Each word admitted into a playscript should be both objectively and subjectively appropriate. Words function objectively in drama as structural materials. The four basic word functions, or parts of speech, are: (1) nouns and noun substitutes to name objects, people, events, and situations; (2) verbs to indicate conditions and actions of a subject; (3) modifiers (adjectives, articles, and adverbs) to qualify the items named and the actions asserted; and (4) relaters (prepositions and conjunctions) to connect single words, phrases or clauses. A playwright will do well to remember that English operates on the objective level as a noun-and-verb language. Subjective appropriateness has to do with how well a word fits the human context, as contrasted to the structural context. Considerations about the character and his emotions are the controls. The appropriateness of any word is its effectiveness within a specific structural and emotive environment.

Choice of diction also depends upon the size and range of the playwright's vocabulary. The writer's working vocabulary should be accurate, and although it need not be phenomenally extensive, it should be constantly developing. A genuine interest in words—their sounds, meanings, functions, and associations—is the first requisite for vocabulary development. This half of the twentieth century is more "the verbal age" than the "age" of anything else, and a playwright, like other writers, should be a verbal expert. The study of diction and the activity of building a strong vocabulary are not simply matters of memorizing words. They involve a study of word meanings, functions, uses, abuses, changes, and effects. Since the total English vocabulary probably exceeds a million words, a writer's potential work with diction is unending. Because more than half the words in any unabridged English dictionary are not of English origin, the writer can profitably study word origins and roots. Most contemporary writers who use English believe that Anglo-Saxon words are the most dynamic. But a knowledge of Greek and Latin derivations is also essential. A writer should distinguish between his recognition and his active vocabularies. Since the former is normally three times the size of the latter, the easiest way for anyone to develop a large compositional vocabulary is to pull words from his reading vocabulary into active usage. This is best accomplished when an author writes slowly and reflectively with a dictionary close by. Reading widely, while using a dictionary as a constant study companion, is also helpful. A writer should never hesitate to use a book of synonyms. Not only will such an aid furnish an occasional apt word, but also while handling such a volume a writer learns new words and stores others for active use.

A special vocabulary problem is unique to playwriting. A dramatist is subject to dual vocabulary work. First, he is limited by his own active vocabulary, and second, he must delimit the vocabularies of his various characters to the words and idioms appropriate to each of them. Too often playwrights err in writing dialogue by using their own vocabularies with no thought to the store of words of each character. One of the best ways to control the vocabulary of any given character is to read only the speeches of the subject character in exclusive sequence and to revise the diction for consistency and probability.

Although this section is meant only to introduce the study of individual words, one final matter must come to any playwright's attention—*spelling*. For some strange reason, probably having to do with a combination of laziness and ignorance, many beginning playwrights apparently consider correct spelling to be unimportant. Perhaps they think that actors only have to say their words not read them, or that a director will correct their spelling before the play is handed to a company, or that an editor will perform this nose-wiping service before publication. Misspellings in a manuscript, other than an inevitable typo or two, indicate that the author is uneducated, sloppy, or stupid. What manufacturer would turn out an automobile with one square wheel, with a missing steering column, or with one segment left out of the exhaust system? The writer's pride of creation or his sense of professionalism should encourage him to make as perfect a piece of composition as he can. A young writer should learn to spell before attempting to write anything and certainly before mailing a half-crafted piece to someone else. Every writer should strive to perfect his command of language, and he can best do so by studying it methodically and technically.

PHRASES, CLAUSES, AND SENTENCES

For a writer to understand diction fully, he should not only study the problems of word choice but also the attributes of effective syntax. Grammar can be thought of as the systematized study of the forms and functions of words. Syntax is a branch of grammar that deals particularly with the relations of words in phrases, clauses, and sentences. Single words standing alone can do little work; any playwright who has written a dialogue exercise composed of a series of one word speeches can attest to this fact. When words stand in coherent relationships to one another, they form basic units of discourse and thus make possible expression and communication.

Although clarity, interest, and appropriateness are the major objectives for every aspect of diction, the following are the major objectives of word groups: unity, coherence, emphasis, and color. A unified sentence expresses a complete thought; its parts cohere to produce that unity. In order for a sentence to be clear, it features *one* element, for the sake of which all the words are arranged in a selected order. The emotive color of the focal element and of the sentence as a whole generates interest as an effect. Since English is a noun-verb language, these crucial sentence qualities are best achieved by the careful selection and disposition of a subject (an item named) and a predicate (something said about the item). The core of every sentence is the finite verb because it provides the kind of unity that signals a complete thought. Finally, a sentence is also a sound unit. A complete sentence always provides a certain pitch change at its end and a terminal pause thereafter. Special items within a sentence may receive vocal stress, and the emotive color of the human voice, when the sentence is read aloud, affects the whole. Thus, the music of live sentences also provides a means to unity, coherence, emphasis, and color.

Do all these considerations about syntax apply to dialogue? Since speeches in a play represent spoken conversation, should a playwright worry about grammar? Is there a difference between spoken and written diction? The answer to all three questions is an emphatic *yes*. Knowledge about rules of grammar and principles of syntax are essential to the playwright as a craftsman. He must know them intimately in order to control them. Because there is a vast difference between written and spoken diction, a dramatist's problem as a writer is specially complicated. His prose, or verse, is in fact written diction, but it becomes spoken diction in performance. At best, it should be excellent both on the page and in the theatre; at worst, it must provide some illusion of human conversation. Oral speech is direct expression; dialogue should have equal directness and even greater intensity. In daily conversation, people speak in rapidly flowing words and tumbling sentences. Speech is constantly qualified by the voice and supplemented by the face and body. But dialogue is not an exact transcription of daily speech. The lazy locutions of plain talk are not necessarily desirable. Familiar metaphors may be acceptable but are seldom preferable in a play. Dialogue should own the virtues of simplicity and clarity, but this does not mean it should be simple-minded, repetitious, and common to the point of dullness. A playwright should actively construct the diction of his play; he should not be controlled by his own speech habits. Effective dialogue, like effective writing of any sort, is a distillation.

The principles of word structures which follow are all applicable to playwriting. They should be considered, however, in a special context—that of time and immediacy. The diction of a play, like everyday speech, is meant to be heard rather than seen. What works in a novel on a printed page will not necessarily work in a theatre. With auditory diction there is so much less time for reflection on the part of the audience. In this special sense, dialogue should be more simple than formal prose and more quickly comprehensible than most lyric poetry. Also, dialogue exists more for the sake of thought, character, and plot than for its own sake. Further, it should be written to stimulate and to be expressed by intonation, stress, pause, inflection, blocking, business, gesture, facial expression, and the other devices of the theatre. All this is packed into the dialogue and served by the dialogue. Neither dialectical nor rhetorical diction, neither fictional nor scholarly prose, neither epic nor lyric verse is the same as dialogue. Dialogue is not precisely exposition, argumentation, description, or narration, and yet it may occasionally fulfill such functions. Dialogue is the chief and special *means* of drama.

To construct functional sentences, a playwright must know basic grammar, and that knowledge is taken for granted here. With dialogue specifically in mind, however, it is worthwhile to review some fundamental syntactical patterns. In a loose sentence, the essentials (subject and verb) come before the modifying elements. For example: "Pete walked cautiously, moving across and to the edge of the pier which was weathered and cracked with age." A periodic sentence contains the opposite arrangement; the essentials come at the end. "Moving cautiously across the pier which was weathered and cracked with age, Pete walked to the edge." The modifiers appear first, and the unit is not grammatically complete until the final word. An example is this: "After running thirty yards, sitting on the uneven slope, and jamming the rifle to my shoulder, I fired." In a balanced sentence one segment matches another in syntactical arrangement. These examples illustrate: "Life is short; love is shorter." "Men love to play, but women play at love."

Not all sentences in dialogue, however, need to be complete or extended grammatical wholes. The three most common types of sentence fragmentation in dialogue are the exclamation, the elliptical sentence, and the broken sentence. An exclamation is one word or a small group of words expressing abrupt emotion. An elliptical sentence is an abbreviation of a complete sentence with only a part expressed but the rest clearly implied. A broken sentence in dialogue is left incomplete because of an interruption. The following lines contain examples of each:

> TOM: Hey. That looks neat—your hair over one eye like. . . .
> CHERYL: Yeah, they call me. . . .
> TOM: One-eyed jack.
> CHERYL: Damn!
> TOM: What?
> CHERYL: Oh, nothing.
> TOM: Swallowed your gum, I suppose?
> CHERYL: Kiss me.
> TOM: But all these people!

Dialogue should naturally contain many fragmentations, but they should be carefully controlled and well balanced with complete grammatical units. One without the contrasting relief of the other can be monotonous.

Many other fundamental principles of syntax have important application in dialogue. Seven are appropriate for discussion here: (1) climax, (2) suspense, (3) end position, (4) structural emphasis, (5) repetition, (6) contrast, and (7) interest. They may suggest vague qualities to the beginner, but the seasoned author knows exactly what they are and how to utilize them in individual sentences. All these principles should be considered as relative to what is often called "normal" sentence order: subject + verb + indirect object (if any) + direct object or other verb complements (if any). In normal order, adjectives precede their substantives; adjectival units follow their substantives; and adverbs and adverbial phrases are movable. This example has only the essentials: "You give me the rifle." And this shows the same essentials in normal order plus modifiers: "John, my stupid friend, you give me the pretty little rifle with its lovely telescopic sight, right now."

Climax in sentence structure results from an order of increasing importance. It is most easily accomplished through the climactic arrangement of words, phrases, or clauses. Usually, the principle of climax appears when three or more such elements comprise a progressive series. An unusual example of sentence climax occurs in Shaw's *Arms and the Man.* Near the play's end, Bluntschli characterizes his impression of Raina in an exclamatory sentence with two climaxes, a minor one first and then a major one: "She, rich, young, beautiful, with her imagination full of fairy princes and noble natures and cavalry charges and goodness knows what!"

Suspense in grammatical units is another type of climactic order. Like suspense on the plot level, the principle in sentences is best effected by a hint, a wait, and the fulfillment. The beginning makes a hint about something, but the middle postpones it until the end. Suspense in a sentence automatically creates unity, coherence, emphasis, and interest. "Without thinking about who might be watching, Turner slowly put out his hand and tenderly stroked Helen's bare shoulder." Suspense of a less obvious sort also appears in a series that moves toward a climax, in periodic structures and in extended grammatical patterns. For example, the following series of negatives creates suspense about the possible positive at the end: "I love you not because of your beauty, not because of your youth, not because of your wealth, nor even because of your skill in bed, though all that helps, but because you love me so intensely."

The end position principle is also crucial for playwriting. It combines the principles of climax and suspense, and it is a chief tool for the dramatist who wishes to write dialogue of intense impact. Stated simply, the principle is that the most important word, idea, image, or expression should come at the end of a sentence. It applies not only to sentences but also to beats, segments, scenes, and acts. Sentences with strong endings are not so likely to appear in a first draft as in a revision. Since a writer usually thinks in nouns and verbs, he will probably conceive the subject and verb first and then think of modifiers. Subordinate qualifiers, if necessary, are best placed in a sentence center. Participial phrases are especially common offenders. For example: "The shark worked toward me, swimming in half circles and becoming gradually more frenzied." How much stronger and clearer is this revision: "The shark, swimming in increasingly frenzied half circles, worked toward me." Every skilled writer of plays in English uses the end position principle consciously and constantly. Shakespeare, for instance, employed it in each of these great sentences from Act V, scene 5, of *Macbeth*:

> Tomorrow, and tomorrow, and tomorrow,
> Creeps in this petty pace from day to day,
> To the last syllable of recorded time;
> And all our yesterdays have lighted fools
> The way to dusty death. Out, out, brief candle!
> Life's but a walking shadow; a poor player,
> That struts and frets his hour upon the stage,
> And then is heard no more: it is a tale
> Told by an idiot, full of sound and fury,
> Signifying nothing.

Although a playwright will want to make variety with occasional weak sentence endings, he should habitually compose units of climactic strength.

The principle of structural emphasis, sometimes called phrasing, ought also to be ingrained in any writer. It is this: The more significant the detail, attitude, or idea, the more important a structure it should have. Additionally, the principle admits considerations about proportion: the more an item's importance, the more extension and completeness it should possess. Items of least importance should appear in phrases, more important ones in clauses, and the most important ones in sentences. Brevity is not in itself a virtue, nor is extension; both are proportionately relative to significance. By using structures of appropriate length and completeness, a writer should ration attention to the various ideas expressed.

Repetition is a principle allied to emphasis and proportion. It too is a necessity in dialogue. Serving multiple functions—unity, clarity, and emotive effect—repetition is essential to patterned prose. A writer should develop the skill of using repetition, but he must remain alert for monotony and wordiness. These stem primarily from too much repetition or repetition of the wrong items. This principle can apply to individual words, structures, or ideas. The following are some specific functions of repetition; the first two are accompanied by illustrations from Shaw's *Arms and the Man.* Repetition often creates intensification: "Now, now, now, it mustnt be angry." Frequently it makes comedy: "Nothing keeps me awake except danger: remember that: danger, danger, danger, dan—Where's danger?" Repetition may also provide information, unity, and clarity. In most plays, for example, the dialogue repeats the names of characters several times, early, as informational reminders. Unity often comes from such a simple device as repetition of a pronoun. Repetition of nearly any item naturally produces clarity, but it may be boring if the repeated item has little importance. The principle of repetition sometimes produces awkwardness or boredom in beginning or unpolished scripts, and so a playwright should handle it with caution.

Contrast, or variety, is always a significant principle of any art, and it applies to sentence construction in dramas. Few writers establish contrasts without some conscious effort. The most elementary sort of contrast is that of length. Other types involve changes in phrase, clause, and sentence structure. Variety of word choice is the most common. In dialogue, fragmentary sentences and interruptions in the midst of long and complete sentences are good means for contrast. Most practicing playwrights mentally examine each potential sentence unit for contrast with what has gone before. A part of a writer's very nature should be to see similarities and differences in all things. A playwright too should strive for variety in diction.

Interest is not so much a single principle as an ever present goal. Many well-known devices provide the means: figures of speech, periodic sentences, parallel structure, quotations, wit, irony, and examples. The conventional is likely to be the most serious threat to the creation of interest. A stream of ordinary ideas expressed in ordinary syntax with ordinary words makes boredom. Interesting sentences are most likely to be suspenseful and periodic. Fairly short sentences tend to be more understandable than longer ones, but contrast ought to be the control of juxtaposed sentences of varying lengths. As a general guide to average lengths, technical prose often averages thirty words per sentence, popular prose about twenty, and dialogue about ten. Parallel structures, so long as they are credible and not too frequent, boost interest. Other useful devices are syntactical transposition, vowel and consonant patterning, and occasional epigrams. Setting two opposed items in one sentence works well. Finally, analogy is as useful in dialogue as in expository prose. Analogies usually draw parallels, convert abstractions into concrete items, and simplify the difficult by making the unknown comprehensible. An analogy is a comparison of two or more things; it indicates how they are alike in a number of respects.

In addition to these seven principles, three qualities are especially germane to dialogue sentences. These are economy, liveliness, and rhythm. Economy in grammatical units is the simplest of the three. It can be achieved by the inclusion of every necessary word, phrase, or clause and by the exclusion of every unnecessary item. If possible, a clause should replace every wasteful sentence, a phrase every wasteful clause, and a word every wasteful phrase. Best of all, economy proceeds from omissions. What Ernest Hemingway often said about writing fiction applies to dialogue as well: Good writing means creating an iceberg of words; only a few words are visible; but many more are there under the surface. So it is with dialogue economy in a play. A writer should avoid superfluous words and delete every one that does not carry a burden of meaning. In plays, actors' physical actions can substitute for many words. Although dialogue has to be continually emotive, it should be absolutely economic.

Liveliness in any kind of writing has mainly to do with the imaginative use of diction. Dialogue as well should be imaginative, and it should stimulate imagination. This comes first from the use of concrete sensory words. The best nouns are the most specific—e.g., scissors, Laramie, John F. Kennedy. The best verbs suggest state or action—e.g., know, feel, hobble, sob, shiver. And the best modifiers provide sensory impressions—e.g., sticky, pumpkin orange, sizzling. Next, liveliness reflects life experience. This means that it is attitudinal and anecdotal. Both of these qualities can appear in single words or in larger units. Many speeches in any good play contain verbal indications of characters' attitudes and small stories about what they have experienced. For example: "I hate getting so nervous about telling him." "Remember when we walked up to Jan's porch and that jerk shined a flashlight in my face?"

Liveliness also comes from figurative language. Metaphor, the most common and yet the best figure, is an implied comparison. It shows how two things are alike in one striking respect. Metaphors most often appear in nouns, verbs, adjectives, and as personifications.

Noun: The sound of my heart was a *wing beat.*
Verb: The sun *painted* the trees October orange.
Adjective: Her *feather* kiss touched my cheek.
Personification: The *winking* moonlight *took* her by the hand.

Although metaphors are important for a playwright, they can be dangerous when ill-used. Mixed metaphors and trite ones are to be avoided. At best, metaphoric diction is imaginative, new, and clear. It should always appeal to the senses. Figurative language should seldom, if ever, be ornate or not causally related to the thought and character that support it.

Rhythm, a third quality, is not a vague "something" that a writer merely develops a feeling for, but rather it is a quality to be consciously brought out in each syntactical unit, and in larger units as well. Rhythm can occur on any of the quantitative levels of the poetic structure—i.e., sounds, syllables, words, phrases, clauses, sentences, beats, segments, scenes, acts, and whole plays. In a sense, rhythm is subjective because it depends on the ear more than upon a set of rules. Nevertheless, certain techniques are available to the writer who is willing to learn what they are, to practice how to apply them, and to invest his sentences with them. Rhythm in anything can be defined simply as stress pattern, as organized repetition of emphasis. The next chapter treats the controlled repetition of sound, but the discussion of rhythm here applies especially to verbal rhythm. To understand how rhythm operates in sentences, a writer should pay attention to words of importance and be conscious of all stressed syllables, especially those in the key words. The most meaningful words in each sentence are usually the verb and the nouns; they receive the stress as an actor makes meaning. So, the first way to control rhythm is to establish a pattern of spaced, meaningful words. Rhythm should serve meaning, and it should not call attention to itself. Next, the *idea* of a sentence—whether state, activity, or concept—should affect its rhythmic arrangement. This too is a subjective matter, but it should be an ever-conscious one. The rhythm in a sentence should make possible a proper and unimpeded reading. A writer should normally avoid the stuttering rhythms of diffuse everyday speech and yet convey its fluency. All dramatic dialogue appropriately utilizes some metrical rhythms, and a playwright should control these too. Many plays would best be rendered in the regular rhythms of metrical verse, but a playwright must technically master them first.

For prose sentences of dialogue, the sentence essentials—of subject, verb, and objects—usually fall in normal order; therefore, sentence rhythm often results from the selection and placement of modifiers. Each sort of modifier —whether word, phrase, or clause—provides a special rhythmic effect. Single word modifiers tend to make a sentence staccato or emphatic. The modifiers in the following example do so: "Pam was a small, curt, angry girl." Phrase modifiers create a more complex and smoother rhythm, as the following sentence shows: "Through the terminal shuffled men with briefcases and with overcoats but without faces." Clause modifiers make rhythm of greatest extension and weight. "The girl Jimmy expected to meet walked to him, set down her suitcase, and kissed him so passionately that everyone turned to stare." Some writers classify sentences as loose, periodic, and balanced. In a loose or informal sentence, first come the sentence essentials and then the modifiers: "He made his decision after pondering the financial advantages, considering the loss of friends, and discerning the benefits to his reputation." In a periodic or suspenseful sentence, the modifiers precede the essentials: "Running, skipping, and sometimes trudging through mud puddles, Tina hurried home." In a balanced or formal sentence, grammatical units of the same order are juxtaposed: "One does not make love to a body; one always makes love to a personality." Each of these sentence types, regardless of length, depends on a different rhythmic structure. Since so many fragmentary sentences occur in dramatic dialogue, the rhythm of plays often becomes quick and dynamic. A series of sentence fragments, however, eventually makes a rhythmic whole. The following sequence from *Rosencrantz and Guildenstern Are Dead* by Tom Stoppard illustrates this principle:

> ROS: He's the Player.
> GUIL: His play offended the King—
> ROS: —offended the King—
> GUIL: —who orders his arrest—
> ROS: —orders his arrest—
> GUIL: —so he escapes to England—
> ROS: On the boat to which he meets—
> GUIL: Guildenstern and Rosencrantz taking Hamlet—
> ROS: —who also offended the King—
> GUIL: —and killed Polonius—
> ROS: —offended the King in a variety of ways—
> GUIL: —to England. (*Pause.*) That seems to be it.[1]

Other considerations also help a writer control rhythm. First, emotion is likely to be more clearly rhythmic the more intense it becomes. Second, the basic difference between prose rhythm and verse rhythm is that in prose there is far less regular repetition of pattern. Third, well-ordered rhythm means clearer sentence structure. Fourth, a writer can best control rhythm by consciously arranging pauses as well as accents; this has to do with both punctuation and stage directions. In conclusion, rhythm is most useful in making emphasis, meaning, tone, contrast, and emotional expression.

[1] Tom Stoppard, *Rosencrantz and Guildenstern Are Dead* (New York: 1967), p. 117. Copyright, 1967, by Tom Stoppard. Published by Grove Press, Inc.

While selecting and arranging words, a writer needs also to remember a few negative principles about substandard and wordy diction. Vulgarisms, including obscenities and illiteracies, are not usually effective as communication. They work best as devices for characterization, attention, or shock. A contemporary playwright need not avoid them so long as he uses them for what they are and realizes their probable effect on an audience.

Slang is usually substandard language. It is often expression coined by very ordinary imaginations or fragmented from more precise language. It consists of words or units with wrenched, twisted, or altered meanings. Most slang expressions come into use because of someone's desire to be bizarre, and at first such expressions are comprehensible only to a limited group. When a slang item comes into general use, its greatest virtue, novelty, automatically disappears. Thus, most slang is useless to a writer because it so rapidly grows stale or becomes unintelligible. Although jargon is ordinarily more academic or technical, about the same things can be said of it. A playwright should avoid both slang and jargon except as he may wish to characterize with them. Even then they are dangerous; they are likely to characterize the play, rather than a character, as unimaginative, pedantic, or stupid. To give the flavor of slang to any character's speech, a writer had best create original expressions.

In addition to slang and jargon, a dramatist should avoid trite diction. Ready-made phrases usually come to mind more quickly than precise original ones, but the latter are more valuable to a verbal work. Although idiomatic language is desirable, it should be fresh. Ideally, a writer shuns any expression he has ever heard before, especially those he has heard often.

Redundancy is also a major problem for all writers. Experienced professionals as well as inexperienced beginners constantly struggle to avoid wordiness. The following errors lead to excesses in sentences; the accompanying examples are typical. (1) A sentence may express the same thing twice. "Thinking rapidly and hastily, she hit upon what you might call a bright idea or a concept." (2) Redundant modifiers, especially adjectives or adverbs, ruin many sentences. "He is a very great man, mainly because his mind is so active and dynamic." (3) Many sentences contain superfluous words. "Probably because he thought that John was the sort of man who could get mad pretty fast, he hit him." (4) Compound prepositions are unnecessary and clumsy. "With regard to the letter, forget it." (5) Double negatives are anti-conversational. "I am not undecided." (6) The excessive predication of units beginning with "that," "which," and "who" frequently spoil dialogue sentences. "Jane told me that she was sorry." (7) A sentence with too many abstract nouns ending in -tion, -ness, -ment, -ance, and -ity sounds ponderous. "His one suggestion of relevance was that the dream was a rejection of reality." (8) If every noun and verb has its own modifier, the sentence will make a singsong effect. "The tall men with black hats quickly walked across the broad street." (9) Two or more sentences with the same arrangement, if placed next to each other, are likely to sound ludicrous. "I came outside. I walked down the steps. I turned to wave goodbye. But she wasn't looking at me." These are but a few of the dangers threatening a playwright as he composes sentences.

The final matter of importance in this discussion of unit construction is stylistic level. People from various social, economic, ethnic, or occupational groups tend to speak differently. Social environment and education affect each person's word choices and syntactical patterns. Further, each individual alters his diction to fit varying situations and companions. One's vocabulary, grammar, and pronunciation are not the same for a five-year-old child as for a lover, a university president, or an Eskimo. Each person has a characteristic level of diction, but he varies that level with each conversation. Characters in plays should possess an identifiable level of diction, and it too should vary in different segments of dialogue.

Diction levels are of two sorts, social and stylistic. Some grammarians distinguish between standard English and substandard English; they identify each as a social dialect. Standard English refers to the stylistic level of educated people. Because of their occupations and personal interests, these people associate mainly with each other. Thus, standard English, primarily a written language, is the dialect of certain social groups and economic classes. Substandard English normally refers to the stylistic level of uneducated people. Unlike those who attend college or who achieve an equivalent education, the uneducated segment of society primarily employs a spoken language. They acquire most of their verbal skills from listening and speaking; educated people acquire most of theirs from reading and writing. Standard English is used by such people as teachers, lawyers, journalists, and many office workers; substandard English is used by such people as young children, mechanics, and most persons engaged in manual labor. A creative writer, be he novelist or playwright, should be fluent in both general dialects and should be able to use all the subtle variations of each as he characterizes various individuals. Although he should be certain about what is verbally right and wrong, he should endeavor to write at different diction levels as they are appropriate to character and situation. Most important, a dramatist should always write what sounds like human speech. The most dangerous likelihood is that he may employ only his own verbal level for all his characters as they appear in every situation.

Generally, plays feature standard, if often fragmented, English. As with prose, dialogue of this sort may be written formally or informally. The difference between these two stylistic varieties has less to do with correctness than with appropriateness. A writer should vary the stylistic level of his characters' diction in vocabulary and syntax. He should identify each word or expression as popular, learned, idiomatic, colloquial, or slang. Popular words are common in the speech of everyone—e.g., *house, is, the, very.* Learned words appear most often in written discourse—e.g., *rendezvous* rather than *date, endeavor* rather than *try, arduous* rather than *hard, imprudently* rather than *carelessly,* etc. Idioms, the natural and semi-logical expressions of a language, are necessary in dialogue, so long as they do not quantitatively overwhelm other word units and do not substitute for key figures. The following are examples of common, useful idioms: *call up, fall sick, get away with, get behind, get off, get on, get up, hurry up, in time, keep up, put up with, set up, tear up, watch out.* Colloquialisms are not necessarily incorrect nor undesirable. The following are some common ones: *awfully* for *very, expect* for *suppose, out loud* for *aloud, over with* for *completed,* and *who* for *whom.* Slang is highly colloquial and frequently idiomatic, but as previously explained, a writer ought to avoid it. Most dialogue operates best at an informal level. Although a playwright occasionally may wish to use formal passages for their strength and eloquence, most of the time he should strive for the simplicity, liveliness, and wide comprehensibility of informal diction.

WANDERING AMONG THE PUNCTUATION MARKS

Too many writers think of verbal composition simply as the work of putting individual words on paper. The previous section indicates that the structure of word groups is important, and this section deals with the devices that demark such word groups. For every writer, especially a playwright, punctuation marks are significant materials. If a playwright commands them, he can control verbal structure, and he can precisely set the timing, tempo, and rhythm of his play's dialogue. Moreover, he can by this means affect the vocal delivery of the actors. In one sense, a play's punctuation marks are symbols to communicate the playwright's wishes to the actor. If a dramatist does not know the functions of punctuation, he can never write with assurance or clarity. Although this section takes for granted a minimal knowledge of punctuation, it treats the twelve most useful items of punctuation. It presents some basic information about each symbol, and it shows the particular applications of each in drama. Understanding punctuation is part of a writer's craft, and it demands periodic review.

Punctuation is a system of marks in written language used to clarify meaning, to indicate organizational units, and to identify pauses. For a playwright, punctuation is one means of recording oral discourse. In a play, punctuation is more than a convenience for a reader; it is a method to set proper vocal phrasing, emphasis, and rhythm. Each mark symbolizes a set of specific kinds of pauses or inflections. Punctuation is organic to play-writing, not mechanical or arbitrary. The rules of punctuation are conventions based upon established usages, but they are nearly always related to grammatical principles. Even though different writers may vary somewhat in punctuation usage, the fundamental principles remain the same. And all writers should keep in mind three general ideas about punctuation. (1) Correct punctuation aids rather than impedes good writing. (2) The primary purpose of punctuation is communication. (3) Ideally, a writer should keep punctuation to a minimum, but he should always use whatever is necessary. It is better to have too much than too little.

The most important punctuation marks for the playwright are:

Comma	,
Semicolon	;
Quotation marks	“ ” or ‘ ’
Parentheses	( )
Colon	:
Apostrophe	’
Hyphen	-
Dash	—
Period	.
Question mark	?
Exclamation point	!
Ellipsis	. . .

All twelve can be grouped as interior, introductory, special, or terminal.

The interior punctuation marks are the comma, the semicolon, quotation marks, and parentheses. The *comma* encloses, separates, or clarifies. Because it has so many uses, the comma is the most common and the most troublesome of all the marks. Commas set off parenthetical constructions, appositives, nonrestrictive modifiers, nouns of address, inverted elements, and direct quotations. They separate main clauses connected by a coordinating conjunction, all elements in a series, and contrasting elements. They are sometimes used merely to make a passage clear and thus avoid a confused reading. The following sentences illustrate the enumerated functions and show the accepted practice for playwriting:

> I remembered, fortunately, that Johnny, the mug, was her brother.
>
> Frank, the men in this compound, some of whom are damned smart, will all try to kill you.
>
> Except for Mary, the girls I've dated in this town have been pushovers, not virgins.
>
> I ran to the door, but the stranger was standing outside with Jim, Dan, and Pete.
>
> He turned to me and said, "Before I finish firing, you better be gone."

A playwright should not use commas excessively, and he should especially avoid using them as substitutes for periods and as marks of interruption between closely related elements. A dramatist should take care to put a comma before every element in a series of three or more. He should also use commas appropriately to separate elements in dates, addresses, place names, and long numbers. Finally, he should know that actors generally give more length, emphasis, and fall in pitch to comma pauses than do other people.

The *semicolon* carries more force than the comma. It signifies a longer pause, but it does not carry terminal emphasis. Entirely a mark of coordination, it belongs only between elements of equal rank. The semicolon has three major functions. It stands between related coordinate, or independent, clauses not otherwise joined by a conjunction. It separates two clauses when the second begins with a sentence connector or a conjunctive adverb—e.g., also, furthermore, however, indeed, so, then, thus, yet. And it sets off elements in a series when they contain internal punctuation. The following sentences demonstrate its correct use:

> I noticed her toes; they looked like ten pink shrimp.
> She patted Jimmy's cheek, because he was nice to her; so in return he patted her fanny.
> The three dates on which the defendant visited home were June 10, 1952; October 2, 1956; and June 14, 1961.

Most playwrights fail to make full use of the semicolon, especially in the first function.

Quotation marks are occasionally necessary in dialogue. Most often they appear when a character makes a direct quote or refers to an essay, song, or poem title. The other functions of quotation marks, double or single, are seldom required in a play.

> The hostess stood up at midnight and said, "All right, everybody take off your clothes."

Regarding punctuation sequence, this general rule will help: Quotation marks always follow periods and commas, and they usually follow other marks.

Parentheses have an important function in playscripts, but are seldom found in the dialogue itself. Parentheses arbitrarily enclose all stage directions. The following passage shows the various manuscript locations of directions and the necessary enclosures:

(JIM and MARY sit on a bench.)

MARY: (Laughing.) I simply couldn't remember. Could you?

(Pause. She does not look at JIM.)

Jim, what's wrong?

The *colon* is the one introductory punctuation mark. In plays, it is primarily a symbol for introducing a formal series. These, of course, are infrequent in dialogue.

> This is his deposit record for last week: Monday, ten

thousand dollars; Wednesday, three hundred; Thursday, fifty; and Friday, ten thousand again.

The three special punctuation marks are the apostrophe, the hyphen, and the dash. Although these are most likely to occur within a sentence, their functions are of a different sort than the other interior marks. The *apostrophe* forms possessives, as in these words: girl's, doctors', and Jones'. The *hyphen*, more a mark for spelling than for punctuation, shows that two words or two segments of one word belong together, for example: well-known, twenty-two, and pre-Socratic. The hyphen also indicates syllabic division of a word broken at the end of a line. And sometimes playwrights use hyphens to signal that a word should be spelled aloud: "Drop dead, d-e-a-d!"

The *dash* is a transitional symbol, and it is especially useful to a playwright because he often employs fragmentary and interruptive units. The dash, however, is much abused by writers who use it indiscriminately for periods, semicolons, or colons. When an author employs it too frequently, he probably fails to understand the correct application of these other marks. Too many dashes create a choppy, incoherent effect. The five major functions of the dash are: (1) to indicate a break or shift in thought; (2) to set off a pronounced interruption, usually making parenthetic, appositional, or explanatory matter stand out clearly or emphatically; (3) to secure suspense; (4) to stress a word or phrase at the end of a sentence; and (5) to summarize or complete an involved construction. The sentences below demonstrate each function:

> Now this is what—you'd better listen to me.
> Two of my best friends—Dudley and John—got me out of there in a hurry.
> We waited—one, two, three minutes—before he came into the light.
> What I want is—you.
> Step by agonizing step, without shoes and with bleeding feet, often stumbling and sometimes falling—she walked nearly two miles.

A dash should not be used in place of ellipsis to indicate a sentence broken off or interrupted by an external stimulus or another character. Furthermore, it should not be used to designate a pause. In a typed manuscript, a dash consists of two hyphens with no space before or after. When handled with ingenuity, the dash is useful in dialogue, but a playwright should not utilize it indiscriminately.

The four terminal marks of punctuation are the period, the question mark, the exclamation point, and ellipsis. The *period*, the strongest of all the punctuating symbols, is not often misused. Playwrights employ periods in about the same manner as other writers, but they probably apply them more often to fragments. The major functions of periods are: (1) to end a declarative sentence, (2) to end a mildly imperative sentence, (3) to punctuate abbreviations, and (4) to terminate fragmentary sentences when the meaning of the fragment implies a grammatical whole. These sentences illustrate:

> I shall always remember Ghost Hill.
> Mark, don't forget to help Steve out if he gets in a fight.
> Dr. Begley removed Sean's tonsils.
> Very funny. I suppose you think I like to get out of bed on a cold morning like this. (Shivering.) Freezing.

The *question mark* is also generally the same for playwrights as for others. It ends a direct question, whether or not the sentence is grammatically complete. But indirect questions do not require it.

> Will you remember to bring me a big bag of bubble gum? Please?
> Phil asked how many rabbits I killed this season.

Playwrights sometimes use a question mark at the end of what might ordinarily be a declarative sentence; they do so to indicate an interrogative lift to the ending. This device should be used sparingly and with care.

> I'm supposed to think of an answer?

The *exclamation point* is another useful mark for a playwright, but it is often misused. It should never substitute for other punctuation; it has its own appropriate functions. First, it ends imperative and exclamatory sentences. Second, it follows isolated words, phrases, or clauses that express strong feeling. And third, it may demark an interjection within a sentence.

> Don't forget to call every single day!
> That's sickening!
> No! For me? I can't believe it!
> Amazing! the whole sky looks alive.

Ellipsis is a modern playwright's special device. Because he is likely to compose interrupted, fragmented, and suspended sentences, he should be acquainted with all the work ellipsis can perform. Unfortunately, many editors and most publishing houses have their own style for rendering ellipsis. The writer will do well to learn its correct usage and ignore the various forms it takes in published plays. Ellipsis, sometimes called suspension points, consists of three spaced periods. It is incorrectly used if there are more or less than that number or if each of the three does not have a space on either side of it. It can, however, be combined with some other punctuation marks—comma, semicolon, colon, period, question mark, and exclamation point. There are three major functions for ellipsis in plays. Internally, it indicates a full pause within a sentence. Externally, it stands at the end of an interrupted sentence, and it begins that sentence when it resumes. Also externally, when combined with a terminal symbol, it marks the conclusion of a broken or unfinished sentence. The following examples demonstrate each function:

> I'm trying to remember . . . all the things we did together.
> Finally, Carol returned to . . . (Another character speaks.)
> . . . show me what Max had given her.
> I'll never believe that she. . . .

Because ellipsis is so often misunderstood and misused, the sequence below further demonstrates its work in dialogue:

> LINDA: If you could only . . . leave him alone.
> MARY: You're acting ridiculously.
> LINDA: No, I'm not. I just want . . .
> MARY: Would you please stop blubbering?
> LINDA: . . . you to stop seeing him.
> MARY: I will not!
> LINDA: Please, Mary, I'm begging. . . .
> MARY: Shut up!

Unless a playwright uses ellipsis carefully and sparingly, it can harm a play or indicate the author does not know his craft. But it provides a means for communicating some effects not otherwise possible. It works best for suspension or full interruption.

Professionals thoroughly know the foregoing information about punctuation. But if one reads many manuscripts, it is obvious that most beginning and intermediate playwrights do not. Novices usually rationalize that they need not worry about such "trivia," and many practiced writers apparently need a review. A playwright should as carefully control the symbols of pause and enclosure as he does the selection and grouping of words.

MECHANICS

In addition to the somewhat technical matters of punctuation, a few considerations about the mechanics of diction are also appropriate. Chapter 11 discusses all practices concerning manuscript format. Here the concentration is upon acceptable stylistic practices having to do with capitalization, abbreviations, titles, italics, numerals, and dialect spelling. As in the discussion of punctuation, the only considerations described are those applicable to dialogue.

The standard practices of *capitalization* are the same for plays as for other verbal compositions. The following capitalization principles are not all the ones a writer should know, but they are evidently those that playwrights should review. A capital letter begins the first word of every sentence or of every fragment that stands for a sentence. Capitals begin the names of people, races, tribes, and languages. Only when used as titles should names of offices be capitalized. The names of seasons are capitalized only when personified or when they carry special connotations. Capitals begin the names of days, months, and holidays. They also begin names of specific institutions, governmental segments, and political parties or units. North, south, east, and west are capitalized when they designate exact geographical areas, not directions. Initial capitals are appropriate for adjectives formed from proper nouns. Playwrights should give special attention to capitalizing nouns that refer to specific persons, especially relatives. The four sentences below show the various correct ways to handle such words as *mother*:

> Well, Mother, you can hug me too, can't you?
> But Mother told me to, Daddy.
> Everyone should love their mother.
> My mother said I could.

The last sentence demonstrates that such nouns are not capitalized when preceded by a possessive. It is unnecessary to capitalize whole words or sentences to indicate emphasis or shouting; a stage direction would be clearer.

Abbreviations, too, can appear in dialogue in the standard manner. But a playwright should remember that an actor will be speaking rather than reading; thus, most of what an actor must say should be spelled out. A few examples illustrate the best practice: *February 10*, not *Feb. 10; University of Minnesota*, not *Univ. of Minn.; Sunday*, not *Sun.*; but *Dr. Oswald* is preferred to *Doctor Oswald*; and *TVA* to *Tennessee Valley Authority*. Slang abbreviations, such as *n.g.* for *no good*, should be avoided as meticulously as slang itself and for the same reasons.

Titles that occur in dialogue should be treated in the normal manner. Capital letters begin each major word. Quotation marks enclose the titles of essays and poems. Underlining designates titles of books, movies, periodicals, plays, songs, and works of art. An example is this:

> Sharon always cries when she hears "Trees" or rereads any part of *Gone with the Wind.*

Words to be printed in *italics* should be underlined in a manuscript. In addition to marking certain sorts of titles, a playwright should italicize: foreign words; names of ships, planes, and other craft; letters when referred to as letters and words referred to as words; and to emphasize a word or a unit. This last function of italics is an important one for a dramatist, but he should use it with discretion. A playwright who underlines too many words is like an actor who shouts too much; both project senselessly and communicate only headaches.

Considerations of clarity, economy, and ease of vocal presentation should always influence the presentation of *numerals* in a play. A few principles best aid the writer of dialogue. A playwright should write out numerals that require only one or two words—e.g., twenty-two, one hundred—but he should use figures for other numerals—1,150; 36.5. Figures are also appropriate for addresses, dates, room numbers, telephone numbers, time designations when followed by A.M. or P.M., and groups of numbers in the same sentence or speech. A number at the beginning of a sentence is always written in words.

Another mechanical matter of extreme importance is how to represent dialect in dialogue. Should a playwright attempt to render pronunciation by means of spelling? No! Several considerations should guide a writer's practice in this matter. First, he should write a paragraph in an introductory stage direction to explain what dialect applies to given characters in the play. This will be all the information the actors, the director, or the reader will need. Competent actors can reproduce dialects far better than most authors; dialects are part of an actor's craft. Second, a playwright should render the idioms appropriate to the dialect without using unconventional or elliptical spelling. Idiom is far more important to oral verisimilitude than is spelling. Of the two following sentences, the second would be preferred:

> Yee'd nivver 'spect thet thar's more'n one way to leek a caaf.
>
> You'd never expect that there's more than one way to lick a calf.

The first sentence is nearly unintelligible; the second, while requiring the actor to handle the dialect knowledgeably, is at least understandable. It is all right for a playwright to use established words, such as *ain't*, or to drop an occasional letter, as in *runnin'*. Third, contractions are always acceptable and often desirable. In fact, actors will perform the work of making contractions if the author does not. Fourth, the main thing to remember in rendering dialect is that the play must be readable. Producers, directors, actors, and contest judges will either refuse to read or else read with great distaste any play containing variant spelling. This is a fact that every playwright must learn and accept. George Bernard Shaw, who was as interested in dialect as any playwright, presented good advice on this matter in *Pygmalion*. For Eliza Doolittle's entry at the beginning of the play, Shaw wrote her first two speeches in dialect. Then, he explained in a stage direction that his desperate attempt to reveal her dialect without a phonetic alphabet was obviously unintelligible. From that point on, he spelled Eliza's words conventionally, but her speech is not likewise "purified" until later.

The mechanics of writing *stage directions* is another matter often slighted by some playwrights. A few important ideas about diction provide correct guides. The major, overall function of stage directions is to provide essential information to the production people about how they should present the play visually and vocally. Stage directions include *all* words other than the dialogue. It is crucial for the dramatist to compose directions for the theatre artists rather than for a reading public. Poor directions usually result from an author's ignorance of what they should be or his confusion about to whom they should be directed. The guiding principles for stage directions are: (1) Only necessary stage directions should be admitted. (2) They describe physical actions that take the place of dialogue. (3) They indicate the proper reading of a word or a speech when the dialogue does not suffice. (4) They suggest pauses and other considerations of timing. (5) They explain specific matters about visual elements of production, such as scenery, properties, and lighting. A playwright should strive, however, to avoid stage directions and work the necessary bits of information into the dialogue. For example, if a character refers to a sofa, a fireplace, or another character's age, then those things need not be mentioned in a stage direction. A writer should use full sentences and normal spelling and punctuation for directions; of course an occasional descriptive or qualifying fragment —such as *pause, hesitantly, laughing*—is permissible. Often a stage direction is a more adroit way to describe what a character does than is some crude attempt to innovate words. For example, it is better to write: "He laughs," than to write: "Ha, ha, ha!" In stage directions, contractions are inappropriate. A writer should avoid *to be* verbs, and he should not use trite or literary diction. Each sentence of stage directions deserves as much concentration and artistry as any sentence of dialogue. A dramatist, however, should be cautious about using stage directions for novelistic purposes. Good dialogue never needs editorial comments, psychological explanations, or literary embellishments. The final thing to be remembered is that stage directions should not tell the director, actors, or designers how to do *their* jobs. Chapter 8 treats this matter at length, and Chapter 11 explains the rendering of directions in manuscript form. Adroit stage directions are vital, and a playwright should exercise artistry in their composition.

BEATS—PARAGRAPHS OF DIALOGUE

A playwright puts words together to make sentences, puts sentences together to make speeches, and puts speeches together to make beats. A beat of dialogue is similar to a paragraph of prose or a verse of poetry. A beat, as a thought unit, treats one particular topic. Although beats are not mechanically designated by indentation or spacing, a playwright should know where each beat in his play begins and where it ends. As with the writing of paragraphs, the composition of beats is to some extent a matter of subjective rather than objective judgment. Nevertheless, a dramatist can structure beats with some logic. Causality or configurative pattern control beats as much as they control an entire play.

The functions of beats are multiple, and they are best explained by the identification of various types and their purposes. The three basic types are plot beats, character beats, and thought beats. To each of these, however, belong a number of more specific kinds.

There are six major kinds of plot beats. First is the story beat; it is devoted to advancing the story and thus usually has to do with one of the story elements, such as the disturbance or complication. Second, some beats are for preparation. These establish the beginning of the suspense sequence, set pointers or plants, or present significant foreshadowing. Third, although somewhat similar to a preparation beat, expository beats reveal information about past circumstances. Fourth, some beats present conflict; these are crisis beats. Fifth, mood beats are often necessary for the establishment of emotional circumstances in the play. Sixth, some beats contain reversals. These are climactic to a sequence beginning with suffering, passing to discovery, and ending with a reversal beat.

Character beats are units of dialogue functioning to reveal one or more traits of a single personage. It is difficult to write a beat that focuses on two characters at once, but it can be done. The most important characters naturally have more beats devoted to them. There are four primary kinds of character beats. First, dispositional beats show some basic personality bent of a character. Second, some dialogue units provide reasons for actions, or they provide the opportunity for characters to voice desires. These motivational beats normally occur before a resultant action takes place. Third, deliberative beats are the most frequent of all the character beats. In these units, a character thinks aloud. Such thoughts may be reflective or emotional, but some expression of emotion is nearly always present. This sort of beat relates closely to the thought beats discussed in the next paragraph. Fourth, decisive beats are those in which a character makes a significant decision.

Thought beats are units devoted to the expression of characters' thoughts. First, because thought is anything that goes on within a character, emotive beats are the most frequent of this type. Whenever a character talks about how he feels, the unit reveals some degree of suffering. Second, reflective beats contain cognition, deliberation, or discovery. Third, informative beats present most of the subject material of a play. Fourth, exaggerational beats mostly contain speeches that maximize or minimize something. Although these may actually be emotive, reflective, or informative, they should be considered as a separate type. Fifth, probably the most important type of thought unit is the argumentative beat. Such beats naturally contain conflict, and they involve, however informally, proof and refutation.

The three foregoing paragraphs describe the specific kinds of beats a playwright can use. It should be obvious, however, that the various types are interlocked. Any given beat, although it has one major objective, will undoubtedly serve other purposes as well. A playwright needs to develop the skill of composing focused beats, but he must also learn to control the secondary functions of the beats he constructs.

Several principles apply to the structure of most beats, and by thinking of these while composing dialogue and while revising, a playwright will better control his material. As with prose paragraphs, beats have an almost infinite number of potential arrangements. Thus, only a few principles apply to all of them.

Every beat has a beginning caused by a *stimulus*. Some initiating factor causes the characters to do or say something. This initial stimulus most often is the entrance or exit of a character, a question, a change of scene, an item of information, or a physical action. Although some initiating stimuli are surprising, the most effective ones are somehow causally or imaginatively related to something that has gone before.

Each beat should contain a *rise*. After a stimulus, some character or combination of characters naturally responds with rising intensity and increased activity. The response can be vocal or physical, or both. The rising segment of the beat truly contains the action of a play insofar as its detailed changes are concerned. A beat rise nearly always implies change of some sort. The most interesting rising segments tend to be crises.

The third structural element of each beat is *climax*. Every beat should have one, and it should be identifiable in one sentence or in a single physical action. Beat climaxes usually are moments when something is settled, performed, implied, or decided. Always they can be characterized as a peak of interest. The control in a prose paragraph is usually a topic sentence, but the control in a beat is most often a climactic sentence or a piece of activity.

The final principle in beat structure is *ending*. This is not so crucial nor so frequent as the others. Most often one beat simply interrupts another. But some beats, such as those at the close of a scene may contain endings. Composing an ending is primarily a matter of personal taste. A playwright, however, should know when a beat ending is necessary, and he should decide what he wishes the ending to accomplish.

The *transitions* between dialogue beats are also important for a writer to control. A beat transition is the causal, imaginative, or emotional connection between the close of one beat and the stimulus of the next. The most frequent transitions in most well written plays are causal in that they are credibly related to something that has gone before. Providing such credibility for beat transitions is one of the difficult skills of playwriting. Too many beats related only by coincidence make a play hard to follow or appear contrived. Some authors, however, emphasize discontinuity by using surprising, shocking, or free association transitions. Although contrasting transitions are desirable, the beat relationships in any one play should be generally consistent with the logic or pattern of that play as a whole. For example, the transitions in Edward Albee's *Tiny Alice* are of one sort, those in Megan Terry's *Comings and Goings* of a completely different sort, and yet in both cases these authors exercised consistency and control. Most of the transitions in Albee's play have causal connections between the antecedent and consequential units; the dialogue moves from topic to topic in a motivated manner. In Terry's play, the transitions are "transformational"; i.e., the actors transform themselves consistently from character to character between units. In her work, the connections are not causal but imaginatively configurative. Whatever the progressive logic of any play may be, that logic is most apparent in its beat transitions. A playwright cannot give them too much attention.

To achieve clarity in beats, a playwright should invest them with completeness, unity, and coherence. Completeness requires a comprehensive idea about the purpose of an individual beat and a thorough execution of that idea. An idea acts as the control of every beat. Striving for completeness in a beat may sometimes lead to overwriting, but more often it will permit full development. Lengths of beats will vary greatly. Some beats, such as those in long single speeches, will tend to be short, and some, such as those in major crisis scenes, will tend to be long. There is no predictable beat length for certain types or functions. Every beat should assume its own proper length. A playwright should develop a sensitivity to proper development in beats. In beginning scripts, beats tend to be far too short and underdeveloped.

Unity in beats has to do with purpose, and *in every beat there are two purposes*. This dual nature of dialogue beats is not only difficult for a writer to learn to handle, but also it is one of the most universally misunderstood qualities for directors, actors, critics, and students. The two purposes can be summed up like this: (1) The *author* tries with every beat to accomplish something, and (2) in each beat at least one *character* is trying to carry out an activity. The first purpose is *author intention*, and the second is *character action*. In any given beat, a verb best identifies each of these two purposes. For example, Act I, scene 1, of Shakespeare's *Macbeth* is a single beat. The author's intention is to capture the attention of the audience with the machinations of the Three Witches. The characters' action is to agree upon a time, place, and object for their next meeting. In composing every beat, a playwright should mentally identify his own purpose *and* that of the involved characters. Only by doing so can he be certain that the beat will have satisfactory unity both as a dramatic whole and as a structured action. A final consideration about unity in beats is that a writer should ruthlessly adhere to the one idea that acts as the control of the beat.

The original conception of a beat should not be greatly altered in its execution.

Coherence in dialogue units depends first upon their having some identifiable order. If a playwright conceives the organization of each beat before he writes it, or if he revises each one for orderly progression after he drafts it, then he can weave the elements of each beat into an organic unit. The actors' voices will flow naturally from sentence to sentence, and the meanings will come clear. Coherence in beats results from causal relationships between sentences. One sentence should stimulate another, and one speech the next, until the unit ending is reached.

The opening scene of *Macbeth*, a well wrought beat, provides a good example:

> FIRST WITCH: When shall we three meet again
> In thunder, lightning, or in rain?
> SECOND WITCH: When the hurlyburly's done,
> When the battle's lost or won.
> THIRD WITCH: That will be ere the set of sun.
> FIRST WITCH: Where the place?
> SECOND WITCH: Upon the health.
> THIRD WITCH: There to meet with Macbeth.
> FIRST WITCH: I come, Graymalkin!
> SECOND WITCH: Paddock calls:—anon!
> ALL: Fair is foul, and foul is fair:
> Hover through the fog and filthy air.
> [*Exeunt.*]

This beat can be analyzed as a plot beat generally and specifically as a preparation beat. As such, it establishes a suspense sequence pointing to the future. Secondarily, it sets the mood for the whole play; it also introduces the Witches. The author intention is to capture interest, and the characters' action is to decide. The beat opens with the stimulus of a question. The rise contains answers and qualifications. The climax is the naming of Macbeth, and the ending is the disappearance of all the involved characters. The transition which follows is a place leap and a scene break. Although this beat is only an example, it truly represents the basic structure within most beats in most plays.

Beats can assume many shapes and sizes. If a play is to have variety, it must have variety in its dialogue units. Every time the active characters in a scene enter or exit, change emotionally, or take up a different subject, a new beat comes into being. A play's overall style is largely dictated by the organization of the beats. Furthermore, even the transitions between beats affect the style. For example, in a number of Megan Terry's plays "transformations" occur between many of the beats. The actors are one set of characters talking about one subject, and suddenly they become different characters talking about something else or repeating their previous words in a new situation.

Another matter in the composition of beats is the balance in each between economy and multiple function. A beat should perform its work economically, and to do so, each should focus on one subject, serve one intention, and contain one action. No word, sentence, or speech should be present that could possibly be omitted. It is natural, however, to assume that any given beat, though well focused, will perform secondary functions. Good playwrights have always known that part of the craft of dramatic writing is compression. This requires that every item in a play perform not only its primary function but also secondary ones as well. That this is true does not negate the principle of economy. It is the secret of dramatic economy. A beat can be economic and serve secondary functions if it first does its singular job with dispatch and imagination. This paradox points to the principle of implication in dialogue.

The final principle regarding beats is dramatic rhythm. Although rhythm in a play occurs on many levels—e.g., in a series of words or in a series of scenes—the most significant rhythmic units are beats. The structural, and emotive nature of a play's beats plus their typical length and the frequency of variation among them—all these elements greatly affect the play's rhythm. The climactic rise and fall, as each unit is performed, is the heartbeat of a play. Audiences may not be greatly conscious of this rhythm, but if any emotion arises in them, it connects with this basic rhythm. A playwright can control this rhythm by making certain that each beat contains a climax, by controlling the length of the beats, and by varying the beats to the desired extent.

Beats are undoubtedly the most significant blocks in the diction structure of plays. How strange, then, that so few beginners know of their existence and that so many current professionals fail to master the craft of handling them! It is equally dismaying that relatively few actors and directors pay close attention to the beats of a play. The good ones have always used them to discover performance intentions. Stanislavsky, for example, knew about them and emphasized their importance in *An Actor Prepares*, and Bertolt Brecht when speaking as a director often stressed their importance. Contemporary production people could well heed their examples. And contemporary playwrights should study the construction of beats early in their careers and consciously work with them from then on in every play they construct.

SEGMENTS, SCENES, AND ACTS

Although single words are the basic means for the construction of any play, beats are the essential structural units. There are, however, three larger units of importance: segments, scenes, and acts. While these are in some ways more closely allied with considerations about plot, this discussion connects them with diction because each consists of one or more beats. A playwright should be no less aware of their necessity, their functions, and their organizational principles than he is of sentences and beats.

A *segment* is the next largest unit above beats and is made up of a group of them. Segments are necessary because well composed beats naturally adhere to each other. Whereas each beat possesses one subject, intention, and action, the structure of a segment permits several of each. The approach to the composition of segments ought mainly to involve considerations of coherence. One bracketing activity, as characters do or say a series of things, ties a series of beats together and thus transforms them into a segment. An example of segmentation occurs within the opening of Shaw's *Arms and the Man.* Three segments, each containing several beats, make up the first scene, the one between Catherine and Raina. The first beat is Catherine's discovery of Raina in the doorway, the second beat is their discussion of that day's battle and the involvement of Sergis, and the third beat is their talk about ideals. Segments, such as this one, are important quantitative elements in every play. And each should contain one bracketing action, intention, and climax.

Scenes are more obvious quantitative units, but they are of slightly less significance than beats. Nevertheless, they serve the progressive arrangement of a play. Although a playwright may wish to designate some dialogue sequence of arbitrary length as a scene, the most functional scenic divisions, during the writing, are *French scenes.* A French scene begins with the entrance of one or more significant characters and ends with the exit of one or more. Such entrances and exits naturally end the final beat and the final segment of a scene and provide a transition to the next ones. If two people sit talking in a hotel room and a third person enters, the conversation naturally changes. The Ghost's first entrance in *Hamlet*, for example, causes a break in Horatio's conversation with the soldiers, initiates a new French scene, and begins a new track of action. Just as segments bracket a series of beats, scenes enclose a group of segments. A scene is not only an organizational unit emphasizing coherence, but also it is a small enough portion of dialogue that a playwright can comprehend and deal with it as one compositional piece. This is not so true with acts. Scene divisions may sometimes occur without an interruptive entrance or exit; this is especially true of long plays containing only two or three characters. In such cases, the playwright should note the major divisions of dialogue and compose them as scenic units. Scenes, like beats and segments, should flow naturally into one another. And the best means for this requires one scene to contain a motivational item of preparation for the interruptive entrance that will eventually end it. Scenes, too, should contain their own sort of central action, intention, and climax—all of a relatively broad nature.

Acts, the largest compositional units in a play, are natural results of beats, segments, and scenes. As extended quantitative units of diction, they depend on the smaller units for their verbal structure. In a sense, act composition is, however, more closely allied to considerations about plot, and in this regard they stand as form to the lesser units. But as sequences of dialogue, they are of secondary significance. Many playwriting books stress act organization, but most of these books attempt to discuss acts as separated from plot or diction. To say the least, such treatments are misleading. On the plot level, most acts are simply quantitative portions of action or of story. True, a writer should exercise some care in a two or three act play to give each act a proper interest level. Each should point to the next, and no single act should totally overwhelm the others. Naturally, most acts deserve an initiating element, a crisis, and a climax. But the overall *structure* itself, as the chief qualitative part in a drama, determines the frequency, location, and size of each plot element. To create an act is first to create a plot. And as the writer composes dialogue to execute the plot, he easily discovers the major rhythmic blocks of the play. These, then, become acts or extended scenes. Because of all these considerations, the potential variety of act arrangements is infinite. A playwright should, of course, give attention to act designation, but it is not the overriding matter many have claimed.

Three other principles also apply to these quantitative units. First, it is possible that a beat can be long enough to become a segment, a segment long enough for a scene, and a scene long enough for an act. These terms and the sequences of dialogue they represent are somewhat arbitrary and may be interchangeable. Second, these units should act as rhythmic controls in a play. They create rhythm insofar as they are individually climactic. At best, each contains a high point of interest. Third, a playwright should not write for the sake of creating beats or scenes, but he should be able to employ them for the sake of better setting down the qualitative elements such as plot, character, and thought. Of all quantitative units, the beat is the most important.

TITLES

Since titles are verbal constructions, they are rightly considered in this discussion of diction. Titles are important for many reasons. A title can affect the whole of a play if it operates in the writer's mind as an epitomization of the whole. Even a working title, which the author carries mentally, affects his creative consciousness. A title, as an item of communication, should symbolize the play by catching its emotive quality. It functions to identify the form, material, style, and purpose of the work. And it can express the overall meaning. It attracts attention and excites curiosity in the minds of producers and audiences. It is the mnemonic symbol representing the whole, a symbol that people must be willing and able to carry in their minds.

Any author establishes criteria for selecting a viable title, but certain qualities are universally applicable. A title should represent the whole work by providing an imaginative image, rather than merely a verbal one. A title should be informative and not misleading. It communicates a play's mood, style, and subject matter. Thus, a title should itself fit the style of the whole. A title should also be unique, either as a fresh image or a new use of an old one. Surprising titles arouse unusual interest. Titles projecting sensory perceptions are especially vivid. Furthermore, a writer should analyze a title for its elements of sound. The quality, variety, and composition of individual sounds have much to do with a title's aural impact. These features are appropriate for most titles.

Many kinds of titles are available to a writer. The following twelve and their accompanying examples are the most frequently employed: (1) a leading character's name, *Mother Courage*; (2) a mood, *Look Back in Anger*; (3) an image, *Long Day's Journey into Night;* (4) a trait of a leading character, *Rhinoceros*; (5) a quotation, *The Little Foxes*; (6) a situation or event, *Murder in the Cathedral*; (7) a place, *Tobacco Road*; (8) a description, *The Man in the Glass Booth*; (9) an object, *The Chairs*; (10) a meaning, *The Just Assassins*; (11) an item of humor or irony, *Oh Dad, Poor Dad, Mamma's Hung You in the Closet and I'm Feelin' So Sad*; (12) a literary allusion, *Who's Afraid of Virginia Woolf?* A writer ought habitually to note titles of other works, not in order to keep up with the latest modes but to develop a feeling for aptness in titling practice. Also, he ought to read lyric poetry—slowly—to develop a lyric poet's awareness of the imagistic use of words.

Most writers devise a title for a play as soon as they begin working on the material; this one is usually a temporary, working title. Usually, many titles occur to the author during the writing process, and he may try out several before he makes his final selection. The final title is normally the result of careful thought after the play is finished. Writers no doubt compose or find more titles than they discover in bursts of inspiration. If a writer will creatively apply the principles discussed here, he can capture the essence of his drama in a title.

Diction is the material cause of any scripted play. It is the means for verbalizing thoughts. All the words of a play, taken together, make up the thoughts, and all the thoughts comprise the characters, which in turn are the materials of the plot. When used meaningfully and clearly, diction can lift drama to its most effective levels. Although action is possible without words and although plays can proceed on a low verbal level, the most thoroughly developed plays depend on effective diction.

As with any kind of writing, good diction in drama requires not only adroit composition but also skilled revision. Revision demands that a writer be a critical reader. At least four readings, accompanied by appropriate rewriting, lead to the efficient revision of any draft of dialogue. The first reading should aim at correcting and polishing phrases, clauses, and sentences. The second reading checks clarity of thought. The third reading works on fullness and consistency in the characterizations. And the fourth reading focuses on larger problems of structure and story, action and plot. This last reading may even indicate that blocks of the play need to be eliminated and new ones added, or it may reveal that a new draft is needed. In any case, a play can seldom be effectively rewritten without a number of readings by the author.

Every playwright must learn that handling diction requires practice. A meticulous study of examples can, however, enforce an awareness of techniques that no amount of daily practice could provide. The plays listed below are excellent ones to study for the accompanying principles; they are all contemporary plays originally written in English and thus reflect current practice: selectivity in prose dialogue, *Who's Afraid of Virginia Woolf?* by Edward Albee; selectivity in verse dialogue, *The Old Glory* by Robert Lowell; vocabulary, *The Lady's Not for Burning* by Christopher Fry; formal diction level, *Luther* by John Osborne; colloquial diction level, *Dutchman* by LeRoi Jones; formal composition of sentences, *The Devils* by John Whiting; informal composition of sentences, *Blues for Mister Charlie* by James Baldwin; punctuation and mechanics, *After the Fall* by Arthur Miller; beats, *The Homecoming* by Harold Pinter; transitions, *Viet Rock* by Megan Terry; segments, *Lamp at Midnight* by Barrie Stavis; scenes, *A Streetcar Named Desire* by Tennessee Williams; acts, *Rosencrantz and Guildenstern Are Dead* by Tom Stoppard. Reading the works of these particular playwrights should reveal some of the possibilities of the English language in contemporary theatre.

The diction of a play includes dialogue and stage directions. Both should be clear, interesting, and appropriate. A play presents select and well arranged words to communicate information to an audience. The stage directions are for the artistic producers of the play, and thus indirectly communicate with the audience. The dialogue of a play should present items of plot and story, reveal the nature of characters, communicate thoughts, set moods, and form basic rhythms. Dialogue is always heightened speech, and the playwright is responsible for the degree and balance of its stylization. The qualities most prized in contemporary dialogue are directness and verisimilitude, rhythm and allusiveness.

7: *SOUNDS*

In a good play every speech should be as fully flavoured as a nut or apple, and such speeches cannot be written by anyone who works among people who have shut their lips on poetry.

JOHN M. SYNGE
Preface to The Playboy of the Western World

Dialogue represents spoken language. A dramatist writes words to be heard rather than seen. This salient difference between drama and most other kinds of verbal composition dictates that a playwright should concern himself with the phonics of human speech and the acoustics of human hearing. The melodics of language are as important to a playwright as to a lyric poet. The principles of sound in diction and sound in a theatre, then, are important concerns in dramaturgy. Every dramatist, writing for actors' voices and listeners' ears, is a composer of a special kind of music, the melody of human speech.

THE MUSIC IN WORDS

Among the six qualitative parts of drama, melody is the material of diction, and diction stands as form to melody. Individual sounds, thus, are even more basic materials in play construction than are individual words. A word, in this sense, is a formulated group of sounds, and groups of words create melodic patterns. Melody, as Aristotle pointed out, is at once the most pleasurable part of drama and the basic material of the literary part of the constructed play. Aristotle also expected music of the instrumental sort, and many contemporary theorists, Antonin Artaud for example, have suggested sounds of many kinds as part of the aural "music" of the theatre. Many contemporary playwrights have called for incidental music, sound effects, and special abstract sound patterns from the actors. John Cage has even made clear that silence in a theatre is filled with sounds. Melody, as the fifth qualitative part of drama, encompasses all the auditory material of a play—verbal, mechanical, incidental, and accidental.

A playwright should select words and arrange phrases with attention to their component sounds. As he writes, he should hear sounds as much as he visualizes letters. Letters, after all, are only symbols for sounds—often rather misleading ones. A play's musical effects are not mere embellishments; they help make the meaning. As an actor vocally produces tones and as he qualifies these with volume, stress, and timing, the human music contributes directly to the meaning of every sentence, thought, character, and segment of plot.

Diction is a pattern of sound which a playwright must learn to control. Since a span of time is required for the performance of a play, drama is a temporal art. A sequence of sounds—one after another, overlapping or simultaneous—is what the dramatist actually writes. Even when marking pauses in dialogue, he is working with silence to structure a pattern of sound. He should be expert, then, at using the signs and symbols of sound, and he should understand some of the basic principles of acoustics and voice production.

ACOUSTICS: TONE AND NOISE

Because one cannot see or touch sounds, too many writers consider them intangible. A few basic principles of acoustics, the science of sound, should be helpful. As with the presentation of other principles in this book, the discussion here does not aim to establish rules but rather intends to provide information, encourage awareness, and reinforce what a writer may already vaguely understand.

Sound is an acoustic event. It is any vibration, in a material medium, capable of producing an auditory sensation. Sound begins with an explosion. A stimulus causes a disturbance that spreads spherically outward and that changes the space and pressure relations among the particles which compose the substance containing the sound. The well-worn analogy of a rock dropped in a quiet pool still best illustrates how sound affects air.

Ripples spreading across the water from the rock's point of impact are something like sound waves spreading through air.

There are two main types of sound stimuli, periodic and aperiodic. Periodic stimuli cause more regular and rhythmic sounds than the aperiodic. The former produce *tones*, and the latter produce *noises*. Periodic sounds are tonal in that they have identifiable dimensions and composition; they normally proceed from vibrators. Tones come from such vibrators as strings, reeds, tuning forks, and human vocal cords. Aperiodic sounds are noisy in that they lack regularity and repetitive definition; they normally come from non-vibrators. Falling rocks, squealing tires, and certain human articulators produce noises. A playwright should especially distinguish between the human sounds that are tones and those that are noises. Sound involves an event of impact occurring in a sound field; the disturbance expands spherically as a wave; it is a space-time phenomenon; and it can be received by an auditory mechanism.

A tone is a regular sound caused by a relatively simple number of vibrations per second. Pitch, intensity, and quality are elements of the structure of a tone. *Pitch* refers to the frequency of vibrations—the greater the number of vibrations per second (vps) the higher the pitch. A normal human singing range extends from about 82 vps to 1,300, and a normal person can hear tones from 16 to 20,000. *Intensity* of tone is its loudness or softness, its volume. In acoustic terms, intensity is the amplitude (width) of a tone's vibrations. A vocal tone, for example, can remain at one pitch while being projected at varying intensities. *Quality* permits one voice to be distinguished from another; some people call it timbre, or color. Quality also appears as the difference between tones of the same pitch and intensity as produced by different instruments. It is the element of tone that helps distinguish between the sound of a trombone and that of a cello. Different qualities exist because elastic media, such as human vocal folds, vibrate as a whole and in parts. When a person's vocal folds vibrate, they make a primary tone. Each segment of each fold, however, vibrates of itself, producing a faint overtone that differs slightly in pitch from the primary tone. Thus, the vocal tone of any individual has a quality different from that of anyone else. Lastly, resonance also affects tone quality. The size, number, and shape of resonating chambers affect the nature of the primary tone. *Duration,* which some people include as a fourth characteristic of sound, refers to the time span of a sound rather than to its composition. It will be discussed in relation to rhythm and timing.

To complete this introduction to sound, a brief discussion of the third acoustic factor is appropriate. Sound begins with a stimulus, continues through tone or noise spanning space and time, and finally concludes in an acoustical detector. In relation to vocalized drama, this third factor is hearing.

Hearing is a process whereby the energy of sound is converted into physical sensation. Sound waves enter the human ear through the auditory canal and cause the eardrum (tympanic membrane) to vibrate. The vibrations pass along the ossicles—the hammer (malleus), the anvil (incus), and the stirrup (stapes)—of the middle ear to the oval window. The ossicle chains amplify sounds during the transfer. The resultant vibrations of the membrane of the oval window cause pressure waves in the fluid of the inner ear, and these affect the hair cells of the cochlea. Thus nerve endings are affected, and the acoustic nerve carries the sound impulses to auditory centers of the brain. Only when the sound impulses register in the brain is the sound "heard." The human auditory system is one of the most sensitive and discriminating of senses. It is able to distinguish between the whispering voice of a friend and the faint buzz of a fly across a room. The competent playwright will undoubtedly improve his craftmanship by understanding the capabilities and limitations of the hearing mechanism.

Because most writers are unusally sensitive to sounds—especially disruptive ones—some facts about sound intensity are appropriate. The intensity levels of sounds are measured on a decibel scale. One decibel (dB) is approximately the smallest change in sound energy recognizable by a human ear. A rise of one decibel between sounds amounts to a 26 per cent increase. The following list shows the comparative decibel levels of various sounds:

Quiet breathing	10
Whisper	20
Appliance noise in a home	40
Low street noise	45
Average office	50
One automobile—average	60
Conversation, three feet	65
Children in a home	75
Noisy factory	85
Heavy traffic	90
Riveter or jackhammer	96
Loud power mower	107
Subway train passing	110
Hammer on steel, two feet	114
Machine gun	130
Jet plane at takeoff	150

For most people, sound becomes painful at 120 dB, and for some at an even lower intensity. Steady exposure to sounds with decibel levels over 80 can produce hearing loss; temporary deafness often occurs due to sounds between 100 db and 120; and sounds above 150 dB can cause immediate and permanent deafness. A mixture of sounds, of course, increases the total impact of sound at a given time and place. Because of the ear's sensitivity, intense sounds produce more than merely the sensation of hearing. Extremely loud noises can cause capillaries to contract and thus alter the blood supply throughout the system. Such noises can alter the blood pressure. And they can upset a person's mental well-being. Because noise is disturbing, it can interfere with concentration, communication, and rest. Most noises are beyond the control of the individual, thus they can produce psychological frustration. Noise pollution has become a serious threat in modern cities. As the complexity and multiplicity of sounds increase in the contemporary world, playwrights might well demonstrate more awareness of sound in their dramas. In *The Theatre and Its Double*, Artaud stressed the significance of sound as noise as a province of the theatre; he suggested its use to shock audiences into a new awareness; and he called for sound hieroglyphs, sound patterns as physical and emotive symbols.[1]

PHONETICS

Man's ability to reason and then communicate in language gives him an advantage over other animals in coping with life. Human beings use speech as a form of communication to control their environment, to get along socially, and even to adjust themselves personally. Spoken language precedes written language in the sense that people speak sooner and more easily than they write. A dramatist is unique among writers because he constantly strives to capture the oral qualities of language and because his words will rightly be heard instead of read. His work is thus paradoxical because he himself is more writer then speaker. The difficulties of writing oral language skillfully cause many playwrights to fail. The foregoing chapters of this book have dealt with the activities of composition, with written language, but this chapter is more concerned with speech. Speech, for this discussion, can be defined as an oral mode of language; it is human communication by means of auditory signals.

Phonetics is a systematic study of the sounds of human speech as represented by precise symbols. It must not be confused with *phonics*, the teaching of spelling by means of the correspondences with speech sounds. A knowledge of both can be useful to a writer, but an understanding of phonetics should come first.

An individual vocal sound is called a phoneme. It is the smallest sound segment in any word. For example, the word *bet* contains three phonemes, as does the word *fought*. From these two words, it is apparent that the phonemic structure of a word does not always correspond with that word's spelling. Hence, the International Phonetic Association has devised an

[1] Antonin Artaud, *The Theatre and Its Double*, trans. Mary Caroline Richards (New York: Grove Press, Inc., 1958), pp. 89–90.

alphabet of symbols, each of which represents one sound and only that sound. It is called the International Phonetic Alphabet (IPA). George Bernard Shaw was so interested in the universal establishment of such an alphabet that he willed much of his personal fortune for its development and promotion.

A playwright, as a student of oral language, can employ a knowledge of phonetics in several ways. It enables him, like other writers, to control the melody and harmonics of each sentence he composes. With such knowledge, he can more astutely distinguish and record aural idiom, dialect variations, misarticulations, and mispronunciations. He will better understand how phonemes affect one another when placed side by side. And his phonetic awareness will help him acquire a more diverse active vocabulary. To understand the sound pattern of a word is more fully to understand the meaning of that word and its potential impact on hearers.

In English there are forty-five basic phonemes, and the twenty-six letters of the English alphabet cannot represent them accurately. By studying a list of phonemes, a writer will not become a master phonetician, but he should make several meaningful discoveries about the sounds of oral language. For example, there are at least twenty important vowel sounds; the five everyone learns in school are only for spelling. And there are twenty-five common consonants in oral English, while written English has only twenty-one. The following list indicates a phonetic symbol for a phoneme on the left, an English word in the center with that phoneme in italics, and the whole word in phonetic symbols on the right:

CONSONANTS

[p]	*p*et	[pɛt]
[b]	*b*ite	[baɪt]
[t]	*t*oe	[to]
[d]	*d*og	[dɔg]
[k]	*k*ill	[kɪl]
[g]	*g*row	[gro]
[m]	*m*ake	[mek]
[n]	*n*urse	[nɝs]
[ŋ]	runni*ng*	[rʌnɪŋ]
[f]	*f*ollow	[fɑlo]
[v]	*v*ery	[vɛrɪ]
[θ]	*th*in	[θɪn]
[ð]	*th*ere	[ðɛr]
[s]	*s*ail	[sel]
[z]	*z*ip	[zɪp]
[ʃ]	*sh*ow	[ʃo]
[ʒ]	vi*s*ion	[vɪʒən]

[tʃ]	*ch*op	[tʃɑp]
[dʒ]	*J*une	[dʒun]
[hw]	*wh*en	[hwɛn]
[w]	*w*ish	[wɪʃ]
[j]	*y*ou	[ju]
[r]	*r*isk	[rɪsk]
[l]	*l*ate	[let]
[h]	*h*ey	[he]

VOWELS

[i]	*ea*t	[it]
[ɪ]	s*i*t	[sɪt]
[e]	*a*te	[et]
[ɛ]	m*e*t	[mɛt]
[æ]	c*a*t	[kæt]
[a]	h*a*lf (Eastern)	[haf]
[ɑ]	f*a*ther	[fɑðɚ]
[ɒ]	w*a*tch	[wɒtʃ]
[ɔ]	*ou*ght	[ɔt]
[o]	bo*a*t	[bot]
[ʊ]	f*oo*t	[fʊt]
[u]	b*oo*t	[but]
[ʌ]	b*u*t	[bʌt]
[ə]	*a*bove	[əbʌv]
[ɝ]	h*ear*d	[hɜ d]
[ɜ]	b*ir*d (Eastern)	[bɜd]
[ɚ]	broth*er*	[brʌðɚ]

DIPHTHONGS

[aɪ]	*eye*	[aɪ]
[aʊ]	no*w*	[naʊ]
[ɔɪ]	bo*y*	[bɔɪ]

Obviously, a playwright cannot learn the phonetic alphabet by reading through it once. It is offered here for three reasons. For those who have never seriously considered it before, it is an introduction; for those who know it a little, it is a reminder that they need it for their work; and also it is necessary that the whole alphabet be introduced for the sake of the ensuing paragraphs. In any case, the appearance of the IPA in this book should suggest to every playwright that a knowledge of it can be an aid to his craft.

English is a vowel language. This means that people who speak English mainly use the vowel phonemes in words to express varying meanings. Vowels and diphthongs are elongated sounds that a speaker can alter in length and color. Vowels always involve vocal tone, resonance, and articulation. Physically, all vowels begin with vibrations of the vocal folds and require that the velum be raised and the mouth opened; the resultant tone always comes from the mouth. They are chiefly melodic and seldom noisy. In establishing musical patterns with words, a writer should first consider the arrangement of vowel phonemes, even though in spelling consonant sounds are more numerous and more obvious to the eye.

Vowels differ in placement, duration, color, and purity. Vowel placement refers to the articulation or final formation of each vowel sound by the position of the lower jaw, lips, and tongue. The front vowels are: [i], [ɪ], [e], [ɛ], [æ], and [a]. The central vowels are: [ʌ], [ə], [ɝ], [ɚ], and [ɜ]. And the back vowels are: [ɑ], [ɒ], [ɔ], [o], [ʊ], and [u]. The diphthongs—[aɪ], [aʊ], and [ɔɪ]—are two vowels glided together to form a sound approximating a single phoneme. Vowel duration is the length of time a vowel is held. Vowels are longer in stressed syllables, at the ends of words, and when they receive emotive accentuation. The color of vowels, mainly effected by changes in pharyngeal resonance, is their emotional overtone. Vowel purity refers to whether or not they maintain the same characteristics throughout their production. If a vowel phoneme is altered as it is being sounded, it is impure. A writer should be a student of the music of vowels.

The consonants of vocal English are phonemes that separate the expressive vowel sounds. Consonants act as interruptive, transitional, or divisional units in words and phrases. They modulate the flow of human sound, and sometimes they color the vowels located beside them. A vowel preceded or followed by a consonant is easier to hear than a vowel sounded alone. All consonants involve an alteration of the airstream by the articulators—lower jaw, lips, tongue, teeth, alveolar ridge, hard palate, and soft palate. With these, a speaker blocks and releases, constricts, or redirects the airstream, and thus he creates noise. Of the twenty-five common consonants, ten are voiceless noises, and fifteen are combinations of oral noises and voice tones.

Consonants can be identified and classified according to placement, type of sound, and voice involvement. There are six identifiable types of consonant sounds. *Plosives* require a blocking of the airstream and a release. They are potentially the loudest consonants. There are six plosives, three voiced—[b], [d], and [g]—plus three unvoiced—[p], [t], [k].

The three *nasals* are the only sounds in English that require the airstream to be diverted through the nose and the tone to be resonated in the nasal cavity. All three are voiced continuants: [m] as in *m*ouse, [n] as in *n*ose, and [ŋ] as in si*ng*.

Nine *fricatives* come from friction noises made by the airstream passing articulators. All are continuants, but only four require voice. The voiceless fricatives are: [f] as in *f*ood. [θ] as in *th*eta, [s] as in *s*eek, [ʃ] as in *sh*rimp, and [h] as in *h*ello. The voiced fricatives are: [v] as in *v*est, [ð] as in *th*ese, [z] as in *z*ero, and [ʒ] as in plea*s*ure.

Two of the consonants are *affricatives*, or combinations. In each, two other consonants stand together as a single phoneme. One is voiced, and one is unvoiced: [tʃ] as in *ch*ime and [dʒ] as in *J*im. Both begin as a plosive and end as a fricative.

Glides are consonant phonemes involving movement of articulators. Each of the four glides begins as a vowel-like sound and ends as a noise. The one voiceless glide is [hw] as in *wh*ip. The three voiced glides are: [w] as in *w*ish, [j] as in *y*oung, and [r] as in *r*isk.

The one *semivowel,* or lateral, is [l] as in *l*isten. It requires voice and is a continuant sound. The tongue tip rises against the postdental ridge, and the airstream is thus diverted laterally over the sides of the tongue blade. At the end of the [l] sound a glide occurs as the articulators recover to form the next sound.

Why should a writer be concerned with individual sounds? Why should a playwright be aware of single phonemes? Why should a young writer—older ones too—study phonetics or take a course in voice and diction if he has the chance? The answers to such questions are numerous, but two considerations outweigh all others. First, the study of phonetics is bound to increase a writer's understanding of human speech. It is essential for a dramatist to understand oral language as well as he understands written language. Second, the study of written English—in schools and universities—is directly in the tradition of the study of written Latin. Formal instruction in English first began at a time when Latin was the scholarly language of written communication. And the manner in which our schools teach English grammar is still related to that used in early Latin instruction. But in English speaking countries, Latin was not and is not the oral language of everyday speech. The inclusion of the study of oral language in education is relatively recent, and too few writers have received systematic instruction in oral speech. Also, oral English tends to be more dynamic in nature than written English; the former normally contains more Anglo-Saxon words and the latter more Latinate words. Thus, if a playwright wishes to be a thoroughly competent craftsman, he cannot ignore the sounds, rhythms, and melodies of his language.

RHYTHM

Verbal rhythm is not so much a technical matter as it is emotional expression. As a person's feelings grow more intense, his speech tends to become more rhythmic. In daily life, passionate expression has noticeable rhythm. The angry man, the mourning woman, and even the jolly drunk—their speech becomes more regular as their emotion rises to a climax. Rhythm in words reflects life. There is a regular pattern of tension and relaxation in physical labor, in the matching steps of two people walking side by side, in the rhythmic sensations of physical contact between lovers. The pulse is the vital rhythm of life; the heartbeat is man's rhythmic pattern.

Rhythm in speech is the ordered recurrence of emphasis in sounds and the placement of silences. Rhythm in drama means patterned time. Drama is an auditory and visual time art, and rhythm is one of its characteristics, at once structural and expressive.

Rhythm in diction, the subject here, requires sequential *stress* in words and *accent* in phrases. English is a language that features rhythmic stress, a continual variation between emphasized and unemphasized syllables. The patterns are continuing, if not always regular. Meter is the systematic rhythm of verse, and cadence is the controlled rhythm of prose. Because drama is materially an organization of words, the best kind of drama—i.e., the most fully organized—is verse drama. And the best kind of prose drama contains strong cadences. Non-rhythmic prose dialogue is inchoate and usually sounds gauche and always generates boredom. A playwright should listen to the rhythms in live speech and should handle stress more like a lyric poet than an essayist. Formal prose is inappropriate for a play because when spoken on a stage it is usually dull and unbelievable; it does not sound like the speech of humans. Several recent studies have shown that in human speech the time interval between stressed syllables of English tends to be uniform, and when too many syllables occur between syllables possessing natural stress, speaking rate is increased and confusion results. A-rhythmic speech is halting and awkward. Although a playwright need not arrange every series of words in a regular meter, he should note the stressed syllables in every sentence and test them aloud.

Rhythm in dialogue marks patterned and progressive movement of words. In a sense, this is the action of diction. Verbal rhythm involves succession and alternation of short and long, stressed and unstressed syllables, plus pauses of varying lengths. Rhythm in words should definitely relate to the character, reflect emotion, and make meaning. Nevertheless, a playwright can better control diction rhythm by knowing some techniques of its accomplishment.

Any small combination of stressed and unstressed sounds creates a metrical unit. Various kinds of such units when arranged regularly make *meter*, but when arranged irregularly they make *cadence*. Both are rhythmic. Metrical feet can be made up by single polysyllabic words or by two or more monosyllabic words. In English, there are four common types of metrical units.

The first is iambic. The following illustrations show an *iamb* graphically and in words:

. / above′, the man′

Other examples of iambic words are: believe, delight, forget, retreat, undress.

Trochaic meter is the second type. A *trochee* looks like this graphically and syllabically:

/ . moth′er, stop′ it

The following words are also trochaic: cabbage, heaven, puppy, Steven, thunder. The trochee and iamb always consist of two syllables.

The third type of metrical unit is anapestic. An *anapest* contains three syllables:

. . / resurrect′, I don't care′

Further examples of words that amount to anapests are: introduce, supervene, reassign, reproduce, unresolved.

Another type with trisyllabic structure is the dactylic unit. In a *dactyl*, the syllables follow this pattern:

/ . . pul′verize, all′ of it

Other dactylic words are: hexagon, nausea, punishment, ricochet, suffocate.

In addition to the four most common metrical units, three others can sometimes be useful for variety. The *spondee* consists of a two-syllable unit in which both are equally stressed:

/ / heart′ break′; out′, out′

The *pyrrhic* is the opposite; it contains two unstressed syllables:

. . si*lence of* the night, *in the* above

With an accent in the middle of three syllables, the *amphibrach* is another metrical variation:

. / . togeth′er, to see′ with

Because many speeches in a play are likely to be short, metrical units are more important in drama than beginning writers suppose. Seldom do playwrights try to establish an arbitrary and unvariable metric pattern and maintain it for the length of a play. Meaning should always supersede metrics. In all verse plays and in good prose plays, however, a playwright carefully structures the rhythmic effects. Modern playwrights sometimes choose to write drama in blank verse. Simply defined, blank verse is any metrical, unrhymed verse. The "heroic" blank verse of Marlowe and Shakespeare consists of any number of unrhymed lines written in iambic meter, usually iambic pentameter. When blank verse is used for a play, the writer should give special attention to variations in line endings and caesuras. He should have a variety of stressed and unstressed endings, and he should avoid ending too many metrical lines with the sense-close of a passage. A caesura (sense pause in the middle of a line) should occur in most lines of five metrical feet or more. Dramatic verse needs a full and varied use of caesura. Free verse is more common than strict blank verse in modern drama. Free verse is a kind of blank verse in which the lines are composed of varying kinds of metrical units and amount to many contrasting line lengths. Every playwright needs at least to experiment with these dialogue styles.

Stress in a polysyllabic word is the vocal force (sometimes pitch and length) given one of the syllables. Stress exists in written words only theoretically; a dictionary discloses what syllables in any word are, by general agreement, to be stressed. But even then, stress marks are merely symbolic indications of vocalization. In English pronunciation, no universal rules govern syllabic stress, and so a writer should tune his ear to notice stress in live speech. The writer should learn to control rhythm in diction by arranging syllables so that the normal stresses in a group of words will match the meaning of that group.

In phrases, clauses, and sentences the words will individually own their proper stress, but for such sequences of words to communicate their meanings fully, *accent* is also necessary. Accent, in this context, means the prominence given to one word within a group. Accent may also give special prominence to stressed syllables, i.e., an accent should seldom fall on an unstressed word. An accent normally involves a change in pitch as well as a change in force; pauses, too, frequently contribute to accent. Accent in verbal rhythm should correspond to sentence climax; thus accented words should assume a climactic position grammatically.

Pauses are as important to rhythm in dialogue as are metrics, stress, and accent. Pauses are, however, more subjective to handle. A lyric poet indicates pauses with punctuation, with line endings, and with verse divisions. A playwright can use verse arrrangement, of course, but in a prose play pauses are equally important. In prose dialogue, pauses are indicated primarily with punctuation (especially commas, periods, and ellipses) and with the word *pause*.

Drama is a time art in that each of its component sounds spans a segment of time. Thus, taken together, a play's sounds make up a sound sequence. Rhythm as patterned sound in a time period is only one of several time factors a playwright should control. Definitions of the other factors should serve to heighten the dramatist's awareness of them. *Tempo* is the overall, but variable, speed of a portion of a play—in speeches, beats, scenes, or acts. It has to do with how rapidly the actions of a play are accomplished in the script and with how rapidly the appropriate vocal and physical activities are carried out by the actors. *Rate* is a more specific term referring to the number of words (sounds) per minute at which the play is presented on stage. *Duration* is the relative length of vowel sounds in individual words. *Timing* refers to the use of pauses of any kind in a script and in its oral performance. The playwright should consciously control all these time factors—remaining constantly alive to contrasts—by efficiently setting them into the diction of a play.

MELODY

A knowledge of stress in sound groups and accent in word groups facilitates a writer's understanding of melody in diction. Rhythm and melody are the two major means to action in diction, and action is the touchstone of drama at every qualitative level. Rhythm comes from changes in stress and accent; melody comes from changes in pitch and contour of tones. Melody is patterned tone, the sequential changes of pitch in a group of sounds.

English is a melodic language as well as a rhythmic one. Speech melodies make possible most of the emotional implications of live verbal communication. In such languages as Chinese or Swahili, pitch changes convey immediate meaning; this is true of all languages in which a given word must stand for a wide variety of things. In English and similar Indo-European lanugages, pitch changes may affect the meaning of some words, but they chiefly serve to provide information about the speaker. Vocal melody obviously makes possible the clarification of attitudes and feelings in word groups such as these:

> You total idiot, shut up!
> Darling, would you kiss me slowly and tenderly?
> I told her to smile, lean over, and then to. . . .
> All right, I'll go if you really want me to.
> Why should he *hurry*?
> The lake was clear, bluish, trembling, and the tiny surface waves were shining like jewels.

A human expresses fears, hopes, questions, commands, compliments, jokes as much in melodies as in words. Even specific physical and psychological characteristics come out in a person's vocal melodies. They reveal a speaker's age, sex, disposition, and emotional state.

Speaking melody and singing melody are similar in many respects, but not exactly the same. A singer prolongs certain sounds—usually vowels—more than a speaker, and a singer makes smoother transitions between pitch points, or notes. Also, a singer tends to direct his vocal process more toward making pleasant sounds than communicating meanings. Nevertheless, there is music in the everyday speech of men and women, and a writer can get it into his verbal creation, provided that he learns to hear speaking melodies and to get them into his art product.

An understanding of the components of vocal melody will help a dramatist control the melodics of his play. The first requisites are *tones* (periodic, regular, rhythmic sounds) and *noises* (aperiodic, irregular, non-rhythmic sounds). A tone is a sound of specific pitch and vibration. *Pitch*, as the second ingredient, is the fundamental frequency of a given tone; the rate of vibration or oscillation of the sound source determines pitch.

Third, *pitch points* are the specific tones of a phrase of sound identified individually; these correspond to musical notes. People commonly use certain pitch points for beginning, continuing, and ending word groups. Fourth, *intonation* is the general rise and fall of vocal pitch. Intonation can be identified as the contours of pitch changes that occur sequentially in phrases or sentences. A *contour,* as the fifth component, is the single melody of one specific phase, clause, or sentence; a contour needs at least two different tones and a change between them. Changes in pitch level that take place without the cessation of tone, the sixth ingredient, are called *inflections,* or slides. A circumflex is a special type of inflection involving one or more alterations in the direction of pitch change. A level inflection—a seemingly contradictory term—is also a special item; it refers to a prolonged phonation with little or no change in pitch level. Seventh, *steps* are changes in pitch between tones; when the voice sounds one note, stops, and then sounds a differently pitched tone, a step (skip, shift) has been accomplished. Vocal transitions, then, are either inflections or steps. Eighth and last, a *sound group* consists of a series of words which comprise a sense group for meaning. A sound group in speech melody corrseponds to a musical phrase. Just as this sort of phrase amounts to a musical "idea," so a sound group is a vocal "idea" that supports an intellectual idea symbolized by words. Such word groupings are normally phrases, clauses, and sentences. A playwright can learn to utilize each of these eight ingredients of melody, and through them he can command the auditory effects of his play.

A few other principles about words and their effects may also be useful. Variety in pitch and contrast in contour are qualities of sound patterns that help maintain interest. A playwright can set them in dialogue by choosing and arranging the words in such a way to require their presence in delivery. For example, if a character says, "There were pies in Grandmaw's window," the melodic contour is quite different from this arrangement: "In Grandmaw's window there were pies." The first sentence requires a downward inflection at the end; the second requires an upward pitch change. Another principle is that excessive repetition of one consonant sound tends to be irritating or ridiculous. The following sentence, for example, contains too many sibilant sounds: "The spring's waters seemed suddenly to spurt from the sod." Further, whenever a writer has a choice between an easily pronounced word and one difficult to articulate, he should always choose the former. Words with mostly consonant sounds are hard to say, those with a balance of vowels and consonants easier, and those with mostly vowels easiest. Note the melodic differences between words within the two following sets of words: penitence, contrition, repentance, remorse, regret; delectation, enjoyment, zest, glee, joy. When selecting the right word, a writer should think not only of meaning but also of melody. This means that how words blend together is important too. Every writer will undoubtedly wish to establish his own criteria about matching end sounds with succeeding initial sounds. But the guiding principle is that, unless words are to be run together, no word should begin with the sound which ended the preceding word. The final three words in the following phrase run together: "if you don't take care." In general, variety is usually pleasing, but repetition is pleasant only when carefully controlled.

A dramatist, like a lyric poet, should also know the major melodic devices of diction, especially rhyme, assonance, consonance, alliteration, onomatopoeia, sound suggestivity, euphony, and cacophony.

Everyone supposedly knows what *rhyme* is, but evidently not every writer knows all its major forms and functions. In general, rhyme is the identity or repetition of sounds in two or more words set in auditory proximity. Commonly, it is thought of as identity in the terminal sounds—an accented vowel plus any following phonemes—of two or more words. *Hot* rhymes with *cot*, *huddle* with *puddle*, and *tenderly* with *slenderly*. Identical rhyme, however, is not true rhyme; hence, the consonants preceding an accented vowel should be different. *Alight* and *delight* make identical rhyme; *slight* and *right* make true rhyme. Rhyme does more than merely titillate the ear. Its most important function is to give a group of words coherence. It helps create unity in sound. A playwright will find many occasions to use rhyme, even in a play composed of contemporary prose. But most good prose writers employ rhyme imaginatively and sparingly. In a prose play, internal rhyme is more useful than end rhyme; rhymes within a sentence or in the middle of two succeeding sentences are better than rhymes at the ends of two or more sentences.

Assonance is a device closely related to true rhyme, and a device of greater utility to a dramatist. Assonance is the identity of two or more vowel sounds in different words. When vowel sounds are repeated without the accompanying repetition of consonants, the effect is pleasing and more subtle than that of rhyme. Assonance is sometimes called vowel rhyme and is best when focused on accented rather than unaccented vowels. The following sets of words, for example, are related by assonance:

> scream, tiki, please, meet, steal
> father, blotter, hop, honesty, Tom
> pie, item, bribe, bicycle, fright

Assonance binds together the sounds of sentences such as this: "Sn*o*wy evenings are best for telling st*o*ries." The [o] vowel in the first and last words of the sentence provide auditory coherence.

Consonance is a device similar to assonance; both are types of rhyme because they involve sound resemblances. Consonance requires the repetition of consonant sounds, especially at the ends of two or more stressed syllables when the accompanying vowels are different. The words *posts* and *frosts* are related in consonance. It is more difficult to employ than assonance and therefore more rare. When skillfully used, it makes coherence, too, but its major function is consonant harmony.

Alliteration, also a consonant device, is more common in prose. Sometimes called head rhyme, it involves the repetition of initial consonant sounds in two or more words. It occurs frequently in everyday speech, and most people enjoy using and hearing it. When third-grade children tease by chanting "sis-silly-sissy," or when adults tell someone to "drop dead," they are using common alliteration. Some parents even choose alliterative names for their children: e.g., Margaret Mead, James Jones, Stephen Spender. Most everyone uses many timeworn alliterative phrases everyday: e.g., *f*irst and *f*oremost, *h*ouse and *h*ome, *l*ast but not *l*east. Alliteration, then, is a natural device, one easily controlled by the writer. If overused, however, it can be pretentious, monotonous, and sometimes comic. Shakespeare often utilized the device for humor. The following illustration comes from the prologue to the rustics' play near the end of *A Midsummer Night's Dream:*

> Whereat, with blade, with bloody blameful blade,
> He bravely broacht his boiling bloody breast;
> And Thisbe, tarrying in mulberry shade,
> His dagger drew and died.

Except in special circumstances, alliteration should not call attention to itself, but rather it should work as a binder in a series of words or as an emphatic quality of a series. A reasonable use of rhyme, in all its forms, promotes melodic richness.

Besides being able to repeat sounds, the poet—and this includes playwrights and novelists—should also be adroit with other aspects of verbal melody. Some words possess their own music, especially the words whose sounds describe or suggest their meaning. Some words imitate the sounds they represent, for example *clank* and *wheeze*. Such words possess *onomatopoeia*. Certain other words with this quality are: sizzle, whirr, hiss, honk, crackle, bang, fizz, murmur, whisper, roar. Each of these represents and imitates a specific life sound, and only such words are truly onomatopoeic. It is important for the writer to understand the rather strict meaning of this quality. True onomatopoeia occurs only in words that denote a sound.

Many words, however, in proper context or because their meaning is easily grasped, possess *sound suggestivity*. Such words provide an impression of what something might sound like, but they do not precisely denote that thing's sound. Their auditory effects imaginatively suggest the meaning, or their meaning enhances the impact of their melody. Typical sound suggestive words are: scissors, ripple, merrily, comfort, brat, feather. In order to create this effect with any frequency, a writer should think of it in relation to word combinations. With sound suggestivity, both sound and meaning can be made to flow from word to word. Part of a writer's job is to take common words and give them fresh impact by setting them in imaginative and unusual contexts, and one way he can do so is by using their melodic qualities. For example, to say that "the horse is shod with steel," may be more effective than to say "the horse wears a horseshoe." Or to speak of a bird's "whistling wings" makes a different auditory impression than to describe its "whirring wings." John Keats employed sound suggestivity in this line from his poem "To Autumn": "Thy hair soft-lifted by the winnowing wind." The combination of phonemes in "soft-lifted" and in "winnowing wind" help to suggest the image Keats wished to create. Shakespeare was, of course, a master of sound; there are many instances of both onomatopoeia and sound suggestivity in this speech of King Lear's at the opening of Act III, scene 2:

> Blow, winds, and crack your cheeks! rage! blow!
> You cataracts and hurricanoes, spout
> Till you have drencht our steeples, drown'd the cocks!
> You sulphurous and thought-executing fires,
> Vaunt-couriers to oak-cleaving thunderbolts,
> Singe my white head! And thou, all-shaking thunder,
> Strike flat the thick rotundity o' the world!
> Crack nature's moulds, all germens spill at once,
> That make ingrateful man!

The sound effects in words can function well in contemporary plays too. Tom Stoppard used sound as a significant qualitative element in *Rosencrantz and Guildenstern Are Dead*, as did William Hanley in *Slow Dance on the Killing Ground* and John Arden in *Serjeant Musgrave's Dance*. Sound suggestivity—and sometimes onomatopoeia—is a major means for the composition of dramatic dialogue.

Euphony and cacophony are also auditory qualities of diction; they too can be important to a dramatist. When applied generally, euphony means a pleasing sequence of sounds, and cacophony means a dissonant sequence. But in order for them to be useful to a writer, they need to be defined more specifically. Euphony can signify a harmonious relationship in a series of vowels, and it can refer to a series of easily pronounced consonants. A playwright can test his dialogue for euphony by reading it aloud and noticing whether or not it is easy to articulate. Cacophony, although it involves disharmony or harshness, is not necessarily a negative quality. Often in a play, as in life, an individual needs to make harsh sounds. Functionally defined, cacophony is a consonant grouping that causes a forced pause in pronunciation and a slowing of articulatory rate. Both euphony and cacophony should represent specific qualities to a writer, and he had best learn to use them appropriately. Both can be either distracting or vivifying qualities, and neither should be used indiscriminately.

The melodies in a play's diction should spring from its writer's imagination and from his store of technical knowledge. The kinds of melodies a writer composes will depend somewhat upon his personal habits of employing vocal melody and his habits of hearing melodies in the speech of others. And the melodies in a given play will be harmonious in proportion to how much conscious effort the writer uses to structure them.

Above all, a playwright should remember that the melody of words should be an organic factor. It will be effective in a play only insofar as it functions with other qualitative components. Only well combined means will stimulate desired effects. Only when melody merges with rhythm and meaning can it be felicitous.

THE ACTOR'S VOICE

The voices of actors are the physical instruments that actually project the tones, noises, and melodies of drama. Just as a composer must know the capabilities of musical instruments, so a playwright should have more than common knowledge about voice. The actor's voice is an instrumental factor in playwriting. However indirectly, a playwright controls the voices of all the actors who ever perform his play. This section, therefore, presents information to help a playwright better understand the human voice as a melodic instrument in speech.

The human voice is the most complex and the most flexible of musical instruments. In fact, all constructed instruments are limited indeed by comparison. It is misleading, then, to discuss voice production as something akin to sound production with orchestral instruments. There are some similarities, and these have prompted many authors to use instruments as analogies. The production of sound for speech, however, is so complex in a human being that some simplifications must be made in order to compose a brief, yet comprehensive description.

The human vocal mechanism consists of four major segments—a motor, a vibrator, three resonators, and a group of articulators. Four processes—respiration, phonation, resonation, and articulation—together create vocal production.

Respiration, or breathing, provides the energy that activates the entire mechanism. Although breathing is regulated most of the time by internal stimuli and occurs involuntarily, it can be consciously controlled. In fact, a man can regulate his respiratory mechanism so precisely that he can vary his volume of exhalation from one fraction of a second to another. The actor, then, must be able voluntarily to control respiration. It not only makes possible all the other phases of vocal production, but also it is the chief determinant of loudness, or intensity. A playwright should indicate—

mostly by contextual references rather than stage directions—how loudly or softly a character is speaking. He should also keep in mind that tension and excitement cause respiratory rhythms to change from regular patterns to irregular ones and from a slow rate to one more rapid.

Phonation is the vibration of vocal bands in the larynx. When the airstream leaves the lungs, it passes through the two bronchial tubes into the trachea, and it then enters the larynx, or voice box. The larynx is the first point of obstruction to the outgoing airflow, the beginning of the vocal tract. Within the larynx are six pairs of muscles with supporting cartilages, and these combine to make a valve. The vocal folds are the muscles that form the actual closure. When the airstream builds up sufficient pressure, it forces these folds apart, and the air escapes in tiny puffs. This released air causes the cartilage-like edges (vocal bands) of the vocal folds to vibrate. In this manner, voice occurs as tone. As the muscles comprising the vocal folds change position, they vibrate at different rates; hence, a variety of vocal tones of different pitches are possible. One thing should be understood about "voice placement." The human voice, strictly speaking, is always located in the larynx and can never be "placed" anywhere else. Although singers talk about "high" or "front" placement as though the voice can be moved around, actually "placement" has more to do with resonation and overall relaxation than with voice, which is phonated tone. Phonation, thus, is the process of tone production, and an actor uses it to set the basic music of each vowel series.

Resonation is the echoing of tone in the hollow chambers of the vocal tract. After phonation occurs in the larynx, the sound and the released air move upward into the pharynx, or upper throat. This cavity, approximately five inches long, is irregularly shaped, and its volume can be voluntarily altered through the use of three sets of muscles. The pharynx is the first and most important of the three resonators. Within it the voice gains its main expressive qualities. Most of what is termed emotional expressiveness results from how a person handles the pharynx during vocalization. When the sound and the airstream leave the pharynx, they enter the oral cavity, the mouth. Although this is primarily an area filled with the articulators, it also acts spatially as a resonator. With each change in the size and shape of the oral cavity, the sound is altered in reverberant quality. When actors talk with an unusual amount of mouth movement, they not only articulate more clearly but also create a wider range of oral resonation. In English speech, only two of the three resonators are used at one time. When the sound and airstream leave the pharynx, they continue into the mouth *or* into the nose. The nasal cavities, then, are the third locus of resonation. Only three English sounds are properly resonated there, the final sounds in the words *sum, sun,* and *sung.* If a playwright selects words that contain a variety of sounds and if he chooses emotionally colorful words, he will better utilize the resonating capabilities of the actors' voices.

The fourth and final step in the process of vocal production is articulation. It is the final formation of vocal sounds and of the airstream into speech as a combination of individually distinguishable phonemes. All the articulators are located in and around the oral cavity. Of the seven main articulators, three are movable—lower jaw, lips, and tongue; and four are stationary—teeth, alveolar (gum) ridge, hard palate, and soft palate. The moving articulators work in combination with those that are fixed to make vowels, voiced consonants, and unvoiced consonants. The articulators are capable of minute variations and extremely rapid movement. A writer, however, can put sounds together in such a manner as to make articulation difficult. Tongue twisters are examples of such combinations. Every play, except those written by dramatists with a heightened awareness of sound, contains some tongue twisting phrases. Every playwright certainly should attempt to write only easily articulated sentences. Smooth sentences create more pleasing melodies, and they make meanings more clear.

From this brief description of the process of human vocal production, it should be obvious that an actor's voice is capable of infinite modulation. Also, everything mental—emotions and thoughts—must, when expressed in words, have a physical manifestation. The vocal mechanism *is* the physical action of word formation. The actor's vocal task is complex and difficult; thus, the playwright should help the actor by providing colorful and well-ordered sound sequences.

The human voice is capable of auditory variety of four principal sorts —intensity, pitch, color, and time. Although each of these factors mechanically relates to the others, they are best understood separately. *Intensity* is loudness, or magnitude of sound, and it most directly depends on respiration. An actor's voice projects more or less loudly as he varies the cubic centimeters of air per second that he exhales from his lungs. Intensity results from airstream pressure. *Pitch* is the frequency of phonated tones, the highness or lowness of a sound. The vocal folds and bands in the larynx are its determinants. An actor creates the basic melody of a play with pitch changes. *Color* is the reverberant quality of each sound or word that helps make it individually identifiable. It is related somewhat to timbre, which is a special set of traits that differentiate one person's voice from others. Color also refers to emotional quality, mainly determined in the pharynx but also affected by the oral and nasal cavities. *Time* can be divided into many factors. In relation to vocal sound, it is controlled primarily by the articulators. Other considerations of time are the duration of each sound, the rate of sounds per second, and the auditory clarity of each sound. These are four characteristics of the actor's voice which a playwright should understand.

Regarding the actor's voice, one other matter demands a playwright's attention. Although an actor is—in a musical sense—an instrument for the playwright, the actor is also a human being. Thus, the playwright needs an understanding of how an experienced actor approaches vocal change. When considering the actor's work in any regard, one must remember that most of his work on a role takes place during the rehearsal period, and the results of that work appear in performance. An actor may have a few emotional impressions of a play from an early reading, or from the play's reputation. But when he begins rehearsals, he first learns about it intellectually. Under the tutelage of his director, he analyzes the script. If he is competent, his study includes the matter of sound and melody. He is likely to approach a play's diction, however, not as a combination of individual sounds but, rightly, as an aspect of character. He sees the reading of lines as a character's expression of ideas and emotions. And this, he knows, requires of him an interrelation of mind and body. He tries to understand the character in terms of action; then, he discerns the character's thoughts; and finally he gets at the precise meaning of the speeches. Character, thought, and diction lead him—and this is the crucial circumstance—to think of the *subtext* of every sentence and speech. Subtext for the actor is what a given sentence actually means. The subtext of "I love you!" might be an expression of sexual or parental love, or it might even mean "I hate you." The meaning depends upon how the actor "reads" the line—*how he gives a line meaning through the use of melody.*

For an actor, melody is the subtext of a play. The subtext is the meaning of the diction; thus, melody communicates meaning. An actor tries to know every sentence of his role as a memorized sequence of words, as an intellectual idea, and as an emotional expression. The actor must combine these three and communicate them theatrically as sound and as physical movement. Sound is physical both as it requires physical movement of the actor's vocal mechanism and as it exists as the motion of those air particles between the actor's mouth and an audience member's ears. Although an actor's work of making melody, i.e., of saying the words, can be logically discussed, the actor himself makes melody more by inspiration than by conscious technique. His work with melody has more to do with his creative intuition than with his intellectual acumen. A playwright, therefore, should endeavor to build controls into the play. He should write sentences not simply that *may* be read correctly but that *must* be read in a certain appropriate way. Furthermore, he should consider the subtext of every sentence as thought and also as a melody expressive of that thought. Only by writing in such a manner can a dramatist have any certainty about controlling his play's melody and hence its meaning.

THE POTENTIAL OF NON-HUMAN SOUNDS

The diction of a play consists of phonemic sounds, but other, non-human sounds may also be necessary. This section not only deals with how such sounds contribute to a play, but also it presents a challenge to every contemporary playwright. Few playwrights, ancient or modern, have utilized many of life's sounds. Anton Chekhov and Tennessee Williams are among the small number of modern dramatists who have orchestrated plays with non-dialogue sounds. Even fewer widely recognized playwrights evidently possess much knowledge about the potential of the new sound equipment available to theatres today. Although most dramatists know that music can serve various purposes, few use it as effectively as most motion picture directors. The challenge to contemporary playwrights regarding sound is this: How can a playwright best express the auditory milieu of life, and how can he fully utilize modern sound reproduction systems? Originality of every sort is one aspect of creativity, and the use of non-human sounds is a significant potential source of originality.

Let it be understood immediately that the use of non-literary sounds can be dramatic! Too many theatrical commentators and literary agitators claim that playwriting has only to do with drama composed of words. Auditory and visual stimuli of all sorts stand as possible materials for drama. The inclusion of sounds of various sorts in a play is not necessarily a matter of superficial theatrical effect. All sounds in a drama should be organically necessary; they should be integral structural units.

Because so few playwrights indicate the sound milieu of a play and because so many permit some "incidental" sound or music to become a part of a production, non-verbal sound is grossly underdeveloped in the modern theatre. Playwrights do not attempt to perfect it as material for drama. Directors and production personnel do not employ it very skillfully. Theatre managers do not bother to discover and buy satisfactory sound systems. All these circumstances simply emphasize the potentials of sound. Even the term *sound effects* is misleading. Life sounds and abstract sounds could be much more important in plays as organic items than as mere embellishments for charming audiences.

But what about non-human sound in musical comedy and in opera? Have not playwrights and composers collaborated to create some works of quality? Such questions are, in this context, both misguided and misleading. They focus upon special forms somewhat related to, or dependent upon, drama. These forms, however, possess a special nature and aim at different purposes than drama itself. Both, in whatever degree, attempt to combine the art of music with the art of drama and to balance them. Creative accomplishment with musical comedy or with opera may indeed be of artistic value. This discussion, however, is directed to the use of sound—music, abstract sound, and life sounds—as material rather than a formal part of drama. That is to say that sound should operate dramatically in a play, whether or not it functions musically. Good musical comedy and opera as special forms are as much musical as dramatic, and in many instances the music is more important than the drama. The challenge stated above is not directed toward the musico-dramatic collaborators, but rather it is aimed at playwrights.

In order for a playwright to learn about the dramatic potential of sound, he should begin by thinking of it as integral rather than incidental. Incidental music or sound effects are, in fact, impossible. Every sound produced during a presentation of a play contributes to that play; all sounds add to the complex of auditory and visual stimuli that comprise the play. What most people call incidental music is often distracting music, and it is usually music added by the director rather than demanded by the writer. Truly incidental sound is accidental sound. What playwrights should be concerned with—directors too—is integral sound, sound as inherent to the action of a play.

Sound in drama is *environmental.* For the non-verbal sounds in any play, this is a major principle. Every sound that truly contributes to the creation of a given drama is a unit of environment. Hence, dramatic sound is environmental sound. This does not imply that all sounds must be representational, realistic, or illusory. The chirp of crickets, the roar of a jet, or a gunshot might be necessary in one sort of play. But perhaps audible but unidentifiable sounds—such as electronic noises—might be crucial to another. And music would perhaps be an important, though intermittent, accompaniment to a third. Environmental sound, then, can be illusory or non-illusory, realistic or abstractionist. It can be continual or periodic. As such, it may serve to establish locale, time, emotional tone, or even psychological attitude. The potential functions of environmental sound are as infinite in a play as are the melodies of human speech.

The chief criterion for determining the inclusion of a certain sound, or sequence of sounds, is whether or not it operates organically in the action. When sound is genuinely contributive to a play, it can occur simultaneously with dialogue, or it can occasionally be the sole auditory stimulus. When sounds disrupt the art work as a whole and call attention only to themselves, they are seldom appropriate.

Non-verbal sound can, of course, be produced live or recorded. It includes music, identifiable "sound effects," and non-realistic abstract sounds. Most commonly today sound is produced live during a performance or broadcast from tape recordings. Another aspect of sound, one seldom used by dramatists or producers, except in musicals, is reinforced live sound. The potential uses of electronic reproduction systems are now so great that any sounds the playwright might require can be accomplished.

Playwrights need to learn something about acoustics and about sound systems. They also need to develop an increased awareness of sounds in life and then put those sounds in their plays. All day long people are deluged with noises—hums, whirs, snaps, and rumbles. When one listens for a few seconds, sounds are always apparent. All buildings have some noise, at least from the heating or air conditioning systems. Offices have the snapping of typewriters. The supposedly quiet outside is filled with the sounds of wind, rustling, and chirping. There is a difference between the sounds of rain and snow, of a sports audience and a concert audience, of a group of men and a group of children. The variety is endless. Even more important, the sounds a person hears profoundly affect his physical and psychological existence. The circumstances of sound in the contemporary world behoove a writer, if he is fully to capture the truths of life, to make sounds an integral part of any environment he creates.

Sounds of all sorts comprise the qualitative part of drama that Aristotle called melody, or music. The primary sounds are phonemes; they are the individual units of sound that make up words. The alignment of such sounds in sequence by a writer and the physical production of such sequences by an actor's voice create melody. And melody becomes meaning. Without meanings as items of thought, there can be no characters, and unless there are characters no plot can come into being. In this fashion, sound stands as material to the other organic parts of a drama. Sounds also contribute to drama as non-human elements of the environment of a play. These affect —support or disturb—the verbal melodies. In order to formulate a play, a dramatist should consciously structure its sounds as auditory action.

Reading a few plays whose authors adroitly arranged sounds and melodies will be helpful to any dramatist. As he reads such plays, he can profitably note the major principles treated in this chapter. The following plays skillfully employ the accompanying concept: easily spoken word sequences, *The Knack* by Ann Jellicoe; melodic effect of words, *Under Milk Wood* by Dylan Thomas; use of balanced consonants and vowels, *J. B.* by Archibald MacLeish; verse rhythm, *The Old Glory* by Robert Lowell; prose rhythm, *The Brig* by Kenneth Brown; modulation of melody, *Serjeant Musgrave's Dance* by John Arden; poetic devices, *A Sleep of Prisoners* by Christopher Fry; utilization of the actor's voice, *A Man for All Seasons* by Robert Bolt; non-human realistic sounds, *Cat on a Hot Tin Roof* by Tennessee Williams; formal instrumental and vocal music, *Mother Courage* by Bertolt Brecht; non-human abstract and electronically reproduced sound, *America Hurrah* by Jean-Claude van Italie. Among the new generation of playwrights working in the United States, Jean-Claude van Itallie is one of the most imaginative in the total employment of sounds, both human and non-human, in drama. A study of his plays and others, such as those mentioned above, should serve to increase a playwright's awareness of the melodic potentials of contemporary drama in English.

In order to place this chapter's information about sounds properly among all the facts and principles of dramatic composition, a playwright should consider the extent of human physical involvement in producing sounds. To suggest that a dramatist consciously control tone, rhythm, and melody is not to indicate that he should write as a mere technician of sounds. Nearly half the human body is directly involved in the speech process, and all of it is somewhat affected. Vocal production requires more than just the operation of a person's lungs, throat, and mouth. As the activator of the speech process, the brain and the entire nervous system responds to a stimulus and then initiates physical action. The auditory system is involved, too, as is the upper segment of the digestive system (and indirectly all of it). The respiratory system acts not only as the vocal motor, but simultaneously it also furnishes the body with certain life-extending elements. Thus, speech is crucially allied to several major organic functions of the body. It can occur only within the limits of the body's biological demands. And all this says nothing of the telling psychological effects of talking and listening. Both verbal and non-verbal sounds are, therefore, a major means of involving total human activity in the art of drama. When a playwright controls the sounds of his play's diction and the noises of its environment, he thoroughly controls and affects the lives of his characters, the actors who will portray them, and the audiences who will hear them.

A final reminder for the playwright is that the sounds of drama should be organic. Seldom, if ever, should verbal melodies or environmental sounds be the goal of dramatic construction. Sounds are contributive as material items in the formulation of a word, a thought, an emotion, or an auditory setting. Although a writer should frequently think technically about sound as he constructs a play, his management of technical factors should be for the sake of an overall aesthetic goal. While manipulating sounds, he might well ask: Do these material tiems and technical devices modally contribute to the whole? If a play's sounds are functionally adapted to the total intention, the auditory techniques will support other factors, and they will engender specific effects. Along with other factors, such as vocabulary and imagery, the sounds of a play are greatly responsible for its texture. Diction and sounds together make *dramatic texture.*

But do skilled playwrights, as they write, consciously think of all the considerations about sounds? How can a dramatist think of so many things at once? The creation of an art object requires both spontaneity and craftsmanship. Craftsmanship eliminates the awkward and the unnecessary, and it heightens whatever comes from the imagination. In truth, the writer's craft stimulates his imagination; craftsmanship and imagination are interdependent. Both are essential. Yes, playwrights do think of such minute items as how two consonants fit together. Of course, an experienced writer has formed certain habits of creation—good ones, he hopes. Often, a novice must labor more to control the same elements. A writer also acts as a critic of his own work as he composes and revises it. The work of good writers is more conscious than most people realize. Like anyone else, though, a writer cannot think of everything at once; thus, he learns to work intelligently and patiently. Only if he does so can he expect to be a master of the sounds of drama.

8: *SPECTACLE*

The life of an artisan of the theatre, filled as it is to a great extent with sheer execution, is incomplete and unfertile unless backed up by the creator of the theatre, the WRITER, *the author whose rôle it is to provide the theatre-material . . . The author is the Father.*

JEAN-LOUIS BARRAULT
Reflections on the Theatre

The quotation above by one of the great actor-directors of the century reveals how the best artists of the theatre regard the playwright and his dramas. Drama is for a playwright to conceive and for theatre artists to deliver. That the art of drama has to do with the craft of playwriting is revealed in the foregoing discussions of structure, character, thought, diction, and sounds. Although the treatment of sounds in drama introduced some connections with acting, the previous chapters primarily deal with literary matters. But despite the assertions of many poetic theorists from Aristotle to T. S. Eliot, drama is not only a literary art. In drama, the poetic composition is for the sake of something else—a performance. In order to provide the core for the most intense sort of performance, a play must above all be an image. It is an image proceeding from a writer's vision of life. It is a life image first and a verbal construction second. A play, although it exists as words on pages and although it can be read silently or aloud, is meant to be consummated in a live production. *An enactment is necessary for a play's ultimate being.* The author's image vitalizes the performance, and the performance effectuates the image. All this should remain uppermost in a playwright's mind. Because of these circumstances, writers who adroitly handle other literary types often fail at dramaturgy. Precisely in the study of spectacle and with the conception of a drama as spectacle, a writer contacts the unique nature of drama. Spectacle is the element that sets drama apart from all other poetic forms.

ABOUT MAN, BY MAN, FOR MAN

This chapter begins with a reference to an actor-director, and all its sections refer to theatre artists, because a dramatist should write for these people as much as he writes for an audience. The artists of the theatre are an extension of the dramatist's creative self. They are facets of his own creative personality, not as a human individual but as the central creator responsible for the art object that comes to exist in time and space on a stage. The actors, the director, the designers, and the technicians—all are inherent factors in every play. Only through them can a play truly become a drama. What far too few playwrights and almost no critics have understood is that the other theatre artists are tools of the playwright—not mere tools, but creatively contributive ones. (Many young playwrights, such as Megan Terry and Jean-Claude van Itallie, have discovered this to their benefit and to the benefit of the American theatre.) Actors, directors, and designers as effectuators are integral materials to the playwright, and he must use them *in* a script, stimulate them with his script, and depend upon them for the enactment. Only when these contributive circumstances are fully understood will writers and producers truly realize the necessity of making playwrights centripital to all theatre companies. Only when a writer fully understands this point can he really become a dramatist—an image-making, verbal artist of dramatic art.

The theatre is a seeing place. This is a simple enough statement. The production of a play obviously is meant to be seen and heard. The "seeing" that goes on in a theatre, however, should be more than mere visual observation. The Greek word *theatron*, from which the English word *theatre* derives, implies more than that. *Theatron* is related to two other Greek words that make the full implications of *theatre* more complete. *Thea* is the act of seeing; and *theoria* means both spectacle and speculation. A theatre, as these words connote, is a place where people are involved in the human activity of seeing a spectacle and speculating about it. Seeing drama and thinking about it are thoroughly intertwined for an audience. The actors and the other workers make possible the seeing, and the playwright furnishes the material. He conceives what is to be seen and thought about. That the theatre is a seeing place implies that people are themselves involved in personal action and must see in order to speculate.

A playwright is as much responsible for creating the spectacle of his drama as he is for its characters and thoughts. The formulation of a play's spectacle is another organic element in the total creation; without it the play is merely a verbal poem, not a drama. A dramatist conceives the spectacle—the acting, setting, costuming, etc.—not as stage embellishments, but as integral elements of the total image which is the play. A play is a visual image as well as an imaginative and verbal one.

For too long theorists have attempted to separate what they call "the art of playwriting" from "the art of the theatre." When these are separated in a playwright's mind, his plays will fail to be dramatic. Drama as an art demands that spectacle be an organic element, one properly integrated with all the others. A playwright, therefore, is more than a constructor of word groups and a handler of time sequences; he is also a sculptor of space and a choreographer of movement. As an image maker, he structures words and time, space and movement. Without usurping the function of the director as the manipulator of a play's rendering, a dramatist conceives spectacle as the physicalization of sounds, words, thoughts, characters, and actions.

A drama is as significantly visual as it is audible. In life, many of everyone's most intense experiences are lived without words, and so it is in drama. A kiss is not verbal, or a smile, nor even a realization. A verbal motivation may cause such actions, or a verbal reaction may follow them. Hamlet does not kill Claudius with words; Oedipus does not wear words; Willy Loman's home is not composed of words. Drama depends largely on the physical fact that human beings prefer to see things and events rather than to hear about them. Visual treatment in drama generally provides a more heightened credibility than does verbal treatment. Cyrano and his friends can talk about his expertise with a sword, but it becomes believable only when he actually uses one in a duel. "Seeing is believing," although a trite aphorism, represents a cardinal principle of spectacle. Insofar as a play is visual, it takes on the special immediacy of belief. Thus, drama is present action; it possesses spontaneity, the illusion of the first time. Drama lives as a visual art of the present tense by virtue of its actualized spectacle; all other forms of poetic art are usually past tense, because they appear on a printed page.

The performance of action—a simple definition of spectacle—strikes the mind, by means of sight, directly. The verbalization of action requires mental translation. A sequence of words only symbolizes a thought or an emotion. Mental impressions are most intense when aroused by physicalization. Sight is quicker than hearing; light moves faster than sound. Drama at its best, of course, utilizes both—sound and sight, word and deed. Any playwright who fails to consider spectacle seriously and comprehensively cannot expect to create a complete dramatic image.

Drama is a synthesis of the verbal and the physical, the auditory and the visual. It involves a rarer experience than reading words or watching activity. A playwright as a maker of dramatic images should not only master verbal construction and control mimic rendering, but also he should blend them inextricably to create direct experience for theatre artists and audiences, together. The brilliance of a work of dramatic art depends not just upon literary skill, philosophic penetration, or superb acting, but upon all three. The art of drama is not a singular one, not just literary nor merely theatrical; it is plural. Although sayings and doings are essential ingredients, the intuitive and direct experience of a living image is the thing. The total work surpasses all the parts. Thus, a playwright should be more than a writer of words. Working with others, he creates a total, complete, and live image of human existence. And by doing so, he makes present life more intense. Spectacle, then, is not a matter the dramatist should leave for execution by actors, directors, and designers. The playwright conceives a totality and thus provides all the others the basis for their cooperative creativity. In this manner, a playwright can best and most functionally serve the ensemble art of drama.

THE WORLD OF THE PLAY

To explain the world of a play, Aristotle's terms *imitation* and *artificial* are helpful. To understand a play's world, one must realize that it encompasses more than merely a stage setting. To comprehend dramatic spectacle, a playwright should be able to distinguish between milieu, place, and setting. And he should realize how to effectuate each.

Every play establishes its own world as a total *milieu*. This world is an imitation of the world its author (and his co-artists) has experienced. A play's world is imitative, not in the sense of a photographic replica nor in the sense of being a copy of some other play, but as a world selected, delimited, and organized by an author and shown in all the elements of the play. The world of a play is a creatively constructed world. It is an artificial world, not because it is phony, but because it is man-controlled. A play's world is one strictly of man's own making; it could not come into being naturally, not without the endeavor of some sort of playwright. Every event, character, thought, word, sound, and action in a play describes and delimits the world of that particular play. Thus, a playwright communicates his imitative vision of what the natural world is like, or what the natural world means. The more consciously he applies his vision, the more comprehensible is the dramatic world he establishes. And certainly the more penetrating a vision he possesses, the more the work is likely to have value for others. Everything that an author (and his director) admits into the play contributes to the total milieu, and only what is *in* the play is contributive. Therefore, a playwright needs to sift all the materials through the screen of his selectivity, knowing all the while that he is making a dramatic world, a total milieu.

A play's milieu, or world, is totally inclusive of everything in the play, and all the materials of a play, taken together, project the milieu. The nature of the characters, their relationships, the kind of events, the environmental factors, the logical (or illogical) progression of the action, the types of sentences, the costumes—all such things circumscribe the milieu. The world of Hamlet's Denmark is quite different from the world of *Waiting for Godot*, especially in the depiction of character motivation and expression. The decadently sensual milieu in *A Streetcar Named Desire* by Tennessee Williams contrasts with the ordered universe established by Sophocles in his *Electra*. Thus, with regard to spectacle, a playwright's first problem is to decide upon the nature of the entire logical, social, and physical milieu in which the characters will carry out the play's action.

Usually after this decision, but sometimes simultaneously, a playwright then discerns the play's locale, or *place*. Every play has a place, an area where the action occurs. Within the broad realm of a certain kind of milieu, a play's action occurs in some locale or series of places. In many modern plays, the place is specific, no matter whether the style is realistic or abstract. Anton Chekhov placed *The Cherry Orchard* on the estate of Madame Ranevsky. The place of Sean O'Casey's *Juno and the Paycock* is the tenement apartment in Dublin of Jack and Juno Boyle. In the nonrealistic world of *Six Characters in Search of an Author*, Luigi Pirandello placed the action on "the stage of a theatre."

In a playscript, the author normally names the place in one or more sentences before the dialogue begins. For *Look Back in Anger*, John Osborne explained that the action of the play occurs in the one-room flat of the Porters' in the Midlands of England. Edward Albee identified the place of *Who's Afraid of Virginia Woolf?* as a living room in a house on the campus of a small college in New England. Harold Pinter named "an old house in North London" as the place for *The Homecoming*. The statement about where a drama takes place is no mere conventional identification. Whether realistically specific as in Shaw's *Arms and the Man* or purposefully generalized as in *The Chinese Wall* by Max Frisch, place is the environment necessary for the play's action. It is the locale that concretely represents the milieu.

Milieu in drama can functionally be considered as the abstract limits of a play's total image. Place is the location in space and time of the image. And the concrete items on a stage that bring both into the realm of reality make up the setting.

The *setting*, physicalized as the stage set, is what an audience actually sees. A play's setting should visually establish its total milieu and identify the specific place. As the physical environment of the action, it should affect all characters that enter it. Because as a whole and in its parts a setting is an active environment, a playwright can best conceive one according to the following principles: (1) as a specific representation of a place, (2) as a physical image of the whole drama, (3) as an environment that stimulates certain kinds of human relationships and activities, and (4) as the only environment where the unique events could possibly happen.

Choosing a setting for a play is like selecting a sailing craft for a voyage. The length, beam, draught, displacement tonnage, sail area, spatial layout, age, condition, and specific equipment—all these things and more are scrutinized by the prospective owner. He considers every item separately and then develops a feeling or judgment about the craft as a whole. He realizes that it must function in certain ways if he is to accomplish the voyage, and he knows that his life will to some degree be affected by every item. So it is with a stage setting. It is the physical craft in which the characters will live. Its features will affect them at least as much as those of a ship affect a seaman.

A play's setting, like all its other elements, should be organic with the whole. All the other parts of a drama, together, require a certain setting. Hence, a playwright does not select a setting so much as he discerns what setting the play demands. Yet there is no mathematical formula that will reveal what setting is necessary, and few playwrights conceive a setting after determining all the other parts. The choice of a setting, like everything else, relates directly to the dramatic image which vitalizes the whole. Once a playwright has begun to conceptualize that image, the choice of a setting will occur naturally, appropriately, and with relative ease.

When a skilled scenic designer reads a play with the intention of creating a stage setting, he tries to discover a *scenic metaphor*. He looks for indications *in the play*, rather than only in the stage directions, about what the visual image should be. He reads the play for visual, physical, spatial, and temporal requirements and relationships. All these he combines in one metaphor. A playwright, therefore, should also first think of setting as a scenic extension of the overall dramatic metaphor which is the play. If he thinks of a scenic metaphor, it need not be specified in the script. It should simply help him imagine the physical details comprising the environment of the characters. He can then—when he writes the dialogue—permit the characters to refer to specific items, and he can conceive scenes with some sort of spatial and temporal relationships in mind. For the playwright to think carefully about a scenic metaphor, a visual projection of the central image, is much more helpful to the designer than is a long, detailed description of a more or less conventional stage set. The former will stimulate the designer, and the latter will inhibit him. A scenic image is likely to be dramatic, but a set description will probably be not only literary but also merely imitative of a setting for some other play the dramatist has seen. To establish a setting, the playwright should first arrive at a scenic metaphor.

A stage setting is a dynamic image of theatrical poetry comprising drama. The scenic images in *Waiting for Godot* by Samuel Beckett, *Winterset* by Maxwell Anderson, and *Ring Round the Moon* by Jean Anouilh, although quite different in conception and purpose, are excellent examples of settings that permeate the actions they house.

Examples of scenic metaphors are not easy to discover, because one cannot be sure of exactly what metaphor was in an author's mind as he minimally described the set and wrote the dialogue. The metaphor itself may never be mentioned in the script, and although it may work actively in an author's mind, he may not even be fully aware of its existence. Most playwrights, however, consciously think of a visual image; and some write about their scenic metaphors after their play is completed. Stage directions, when well written, sometimes reveal the metaphor used by the author to arrive at the specific setting. Thornton Wilder evidently thought of a town as the image for *Our Town*. For *The Blind Man*, Friedrich Duerrenmatt imagined heaven and infinity in the guise of the nebula in Andromeda. According to Arthur Miller, the image that enlivened *Death of a Salesman* was the inside of a man's head; he first thought of a face, as high as the proscenium arch, that would open up and reveal the inner reality of a man where past and present are one.

If a playwright thinks first of a play's set as a series of items that merely describe a place, the setting will likely remain incidental to the action. If however he originally envisions a scenic metaphor that visually represents the overall dramatic image, then the setting will suggest, intensify, and sometimes compel action.

What about unity of place? For most contemporary playwrights, this is seldom a problem, or even an interest. And the scripts of the great Greek and Elizabethean playwrights attest to the fact that unity of place did not concern them much, if at all. Some theorists and writers, from the seventeenth through the nineteenth centuries, endeavored to promote unity of place in plays. But it is not necessary for the best kind of drama. Unity of place is often a worry of some Broadway producers because they want simple, inexpensive settings. But when a dramatist fully conceives a play, the scenic image will naturally be singular throughout, even if a number of stage sets are necessary. But they seldom are!

The best contemporary theatre companies—and even some of the worst—no longer have serious difficulty handling plays with multiple sets. Working with flexible stages, abstract scenery, and easily controlled lighting, today's imaginative designers are masters of suggestive settings. They rightly tend not to bother with realistically detailed box sets but attempt to build a visual environment that is active, unified, easily altered, and within the limits of the production budget. Perhaps modern technology has helped, perhaps the de-emphasis of pictorial realism, perhaps the realization that audiences have imaginations too. Certainly, a playwright should realize that he composes a drama for a stage rather than for a motion picture camera or for the printed page. But a writer need only read Brecht's *Mother Courage,* Ionesco's *Rhinoceros,* or Megan Terry's *Viet Rock* to comprehend the scenic flexibility possible in contemporary drama. Or he need only see a production by Tyrone Guthrie, Peter Brook, William Ball, or Jerzy Grotowski to understand theatrical flexibility. A playwright need not be an expert of scenic execution. He need not know the difference between fragmentary and cut-down scenery—although such knowledge will not harm him—but he does need to develop for himself an idea of what sort of theatrically visual experience he wishes to make with his play.

A play not only takes a place, but also it takes a time. A setting represents time—sometimes non-time, circular time, or fragmented time—as both duration and location. The action spans a length of time, during which a setting should *change.* A fully dramatic setting is alterable and constantly altering. In life, chairs get moved, milk spilled, walls painted, houses burned, and buildings bombed. An unchanging setting is a deadly setting. The essence of the dramatic is action—change. A play also occurs, however abstractly, at some single or multiple location in time, some century, some decade, some year, some time of day. Even if the playwright fails to designate the time placement of the play's action, the designers will to some degree establish it. Time location affects the costumes, stage set, hand properties, and lighting. The designers must make choices about these things, and the dramatist needs to understand his obligation to conceive, if only suggestively, these things in both stage directions and dialogue. Time is not only an important factor in a play's structural organization, but also it prescribes the features of the visual representation.

In the foregoing, this discussion has treated a setting for a play as a specific portion of a place, as an environment represented by physical items. There are, however, three other equally important features of a setting—space, light, and temperature.

Space, as a general condition of life, is an infinite three-dimensional extent of distance, area, and volume. In it objects exist, events occur, and movements take place, each having a relative position and direction. In the spatial sense, a play's milieu is a broad and somewhat abstract portion of space; place is a more delimited space; and setting is a specific space. Or to think of space in a slightly different context, dramatic space is the abstract and ever-changing image of space suggested by the play and imagined by the audience. Localized space is the comprehensive spatial extent of the play's activity, not all of which is visible to the audience. And stage space is the three-dimensional area housing the immediate and visible action. A playwright's dramatic image should include some creative conceptualization of space. It is, after all, a continuous condition of human life. The physical items of the setting must exist in space. And the characters, as represented by actors, must move in space. The most effective stage settings are not those that form pleasing visual pictures, but rather those that treat space sculpturally. For the best dramatic utilization of space, a playwright should conceive a setting that includes imaginative spatial conditions and possibilities.

Light is a stimulus to most living things. It is a physical necessity to plants, and for most human beings it is a psychological necessity. As an electromagnetic radiation, it makes vision possible. Within a circumscribed cubic area, it provides for the perception of space as distance, area, and volume. Man has learned to create and control it. Light, as another means available to the playwright, should also be a factor in his total dramatic image.

As twentieth-century physicists and psychologists have demonstrated, space, light, and time are interrelated both in man's physical world and in his emotional world. Man realizes his life's duration with time measurements, and he comprehends his spatial boundaries by means of light. He is strangely disturbed by the infinities of time and space. He reacts emotionally and physically to light. From childhood onward, darkness implies loneness, fear, and the unknown. Light suggests happiness, identity, and security. Color, as one characteristic of light, also affects man's emotive life. Cool colors make him feel differently than warm ones. Contrasting light, whether slowly or rapidly changing, is a strong influence on everyone's sensory and psychological reactions. Most people react strongly, for example, when a bright light is switched on them just as they are awakening from a night's sleep.

In a drama as in everyday life, light serves at least five functions. It furnishes illumination, its most essential function. It reveals form as dimension. In other words, it has to do with spatial plasticity because it illuminates extent as perspectives of height, width, and depth. Third, it helps establish psychological moods for the characters, actors, and audience. Further, light commands attention; thus in drama, it can be useful in visual composition, or what some designers call selective visibility. Finally, light functions as representation. It can create illusions, both natural and abstract.

The three controllable properties of light are intensity, color, and distribution. Each play should incorporate stage directions and dialogue references specifying the employment of each property as affective of the entire action. A playwright can best indicate degrees of brightness, the kind of color, and the spatial arrangement, not by orders to the designers, but by suggestive references in the dialogue. Every play demands certain effects; how these are executed is the province of the lighting and scenic designers. The writer conceives the demands for light.

The most frequent effects called for by dramatic references are the following. Light can suggest the time location and the time span of an action. It can provide indications of season, weather, and time of day. A second effect of lighting is the establishment of emotional atmosphere.

Through the varied application of intensity, color, and distribution, light can emotionalize the environment of characters, actors, and audience. For example, theatre artists have known for centuries that a brightly lighted stage stimulates an audience to laugh more quickly and more often than a dim one. Also, the *magic* of theatre usually has to do with the emotional toning of light and with the way light reveals and conceals. Light saturated with one color makes a very different impact than with another color. Most people have relatively little control of light in everyday life. They cannot control the sun, and they control the light in their homes only by means of on-off switches. Thus, when light is modulated during a drama, most people have some feeling of mystery because the changes are so different from the changes they know. Lighting changes affect people physically and emotionally.

A third effect of lighting in drama is rhythm. When one, or more, of the three properties changes, the contrast can contribute to visual rhythm. Naturally, if the changes tend to be regular and recurrent, the rhythmic effect will be increasingly strong. Lighting changes ought to be orchestrated—through hints from the playwright as well as decisions by the designer—to emphasize the other rhythms of the whole drama. The communicative and emotive potentials of light are only in the earliest stage of development. Technicians and even designers can only innovate to a limited degree in drama, unless playwrights make more and better use of lighting as a strong physical and emotional factor in their creations. A play's dramatic image needs illumination of a certain brightness, saturation, and modulation.

How strange it is that dramatists, actors, and directors take so little notice of *temperature*. It at least matches in importance other conditions of human life such as time, space, and light. With few exceptions, dramatists fail to mention heat and cold. Insofar as characterizations are concerned, actors usually take temperature for granted. And most directors only think of temperature as a condition in the theatre building which may adversely affect performers and audience members. In contrast, how often is temperature mentioned by most people every day? When they discuss weather in general and temperature in particular, it is not a particularly intellectual topic of conversation, but it is nonetheless a frequent and important one. The physical comfort and well-being of humans, as warm-blooded mammals, depends on an even body temperature. Because external temperature and internal metabolism affect human life physically and mentally, a playwright might well give temperature some attention in the establishment of a total setting, in the alignment of activities, and in topics of conversation of his dramas.

It should be emphasized again and again that stage directions are not the best means available to a writer for describing the physical environment in a play. In fact, stage directions and introductory descriptions should be a last resort. The most organic—i.e., functional, meaningful, and coherent—way to handle setting is to establish the environment through the words of the characters. After all, in everyday life, people continually voice their attitudes and enact physical responses to the physical conditions surrounding them. When a character expresses feelings about milieu, place, setting, or when he acts in a certain way because of time, space, light, and temperature, then each of these conditions becomes integral to the action. And the character himself gains in verisimilitude. Although a play's production designers need some hints about scenery, lighting, properties, costumes, and makeup, the playwright had best handle references to those items as he does references to all else in drama—by putting them into the dialogue.

At this point, an explanation of *atmosphere* is appropriate because it also has to do with environment. In drama, however, atmosphere is coordinate with *tone* and *mood*. As explained in relation to sound in Chapter 7, tone means a vibrated, regular sound, but in the context of spectacle it has to do with emotive state. Tone is often used, in relation to various art objects, as a term identifying a general emotional quality. In drama, it can be used to mean the emotional relationship between one actor and his particular role. Precisely then, dramatic tone is the emotional intensity of a character as performed, live, by an actor. Mood is pervasive and compelling emotion. In drama, mood is shared emotion. It is the emotive connection between characters in a play and between actors onstage. But even more significantly, mood is the shared emotional state of two or more characters in the living performance of a drama. Tone can thus be seen as a necessary condition for mood. Both tone and mood are conditions necessary in a script and hence in a performance for the establishment of atmosphere. Atmosphere is the total emotional condition of characters in a particular affective environment. It can also mean the shared emotion between actors and their characters, between the actors themselves, and between the living characters and an audience. Atmosphere, most broadly defined, means the overall aesthetic environment of a drama in performance.

Each of these terms—tone, mood, and atmosphere—should be used with caution. They are abstract, and the playwright and his associates should work out clear meanings for them. They all have to do with the relationships between characters and environments, and it is in this regard that they are most important to the writer. As such, they are factors of a drama that he should control.

As a writer selects environmental features from life and begins establishing them as a part of the spectacle of a play, he should be aware of the overall dramatic image he wishes to project. The placement of an action in time and space depends as much upon the emotional intent as it does upon the visual intent of the total image. Questions about the influences of environment on the characters should outweigh all others. The playwright should consider the unity of the visualization of the dramatic image, and in all his formulation he should concern himself with unity of action. But partly because environment is a constantly changing set of conditions and partly because stage design is now so imaginatively fluid, the playwright should not worry about the relatively inconsequential unities of time and place. This fact does not, however, excuse the incompetent writer who puts together a "realistic" play of the sort popular from the late nineteenth century through the 1940's and who calls for an impossible setting or a multitude of three-walled boxes. Various styles of drama undoubtedly need different settings. If the writer's purpose is to compose a small, realistic comedy that will be relatively cheap to produce, then he had better think of a single conventional setting. If his purpose is to create a dramatic image and provide it with structure and action, then he can disregard the cost of a play's setting, and concentrate on how imaginative the setting is, how it fits the image and the action, and finally how much it may stimulate the designers.

ACTION, ACTING, AND INTERACTION

The essential difference between a character in a novel and a character in a drama is that the dramatic character actually comes to life on a stage. A novel narrates, and a drama enacts. A character in a play is a personage not meant to be read about, but to be seen. One major aspect of spectacle, therefore, is acting. The major action of spectacle is, in fact, the motion of an actor in space and light as he portrays a character. In drama, the noun *action* and the verb *to act* are related not only grammatically but also aesthetically. A fully realized object of dramatic art must be acted.

As the enactment of characters by actors, acting is, like all the other aspects of spectacle, an inherent and qualitative part of drama. Although a playwright does not "write" the acting, he suggests and controls it by the way he conceives the action, constructs the characterizations, and writes the dialogue. Acting is necessary in drama in order for the sounds and sights of a drama to be communicated. A person can read the words of a play, but he cannot contact a drama as a completed object unless he sees and hears actors performing the characters. Playwriting is the art of structuring action, and this includes the acting as well as all the other elements in a drama. Thus, playwriting deals with the structure of acting as much as with the structure of action.

To develop a fully creative attitude toward spectacle in drama, a playwright needs to consider the relationships between himself and actors, directors, and designers. These relationships must exist on at least two levels, social and aesthetic. First, the other theatre artists stand as coworkers. Regarding such external relationships, Chapter 9 introduces some key considerations. As human craftsmen working together to make a special sort of object—a drama—a more or less normal set of human social relationships is likely to arise.

A second level of relationships, however, prevails in a dramatic ensemble. These are aesthetic relationships, and they are far more crucial to the inner nature of the play. Aesthetically, a playwright *plus* all the other theatre artists make up a single human unit. This creative ensemble is *the artist* that formulates a drama as an art object. The artist, or effectuator, in drama is multiple. The efficient cause of a drama is a group. Writers trained in fiction or some other genre often find this circumstance disturbing. Some never understand it, and others cannot accept it. Many commentators about theatre fail to comprehend the nature of the creative relationships in drama. Even most theatre artists—perhaps because they are so deeply concerned with a single creative job—fail to discern the essential connections. Part of the confusion arises, perhaps, because not every playwright is a member of a theatre company and because most theatre companies do not have member playwrights. A playwright, however, who works with a particular company well understands the potential and actual contributions that actors, directors, and designers can make during the creation of a drama and certainly during its final public performance. The ancient argument about whether actors are creative or interpretive is as ridiculous as it is dull. They, like other artists, are involved in creating an art object. Although actors *act*, they too are producing a drama; their knowledge must be as "productive" as a playwright's. The aesthetic relationships between a playwright and other theatre artists can be epitomized in this statement: All members of the theatre ensemble act as necessary creative facets of the playwright himself. The playwright should compose drama as though actors, directors, and designers are extensions of his human and creative self. The play should demand their contributions, and it should control the nature of their activity.

A character in a novel is connotative and suggestive; a character in a drama is doubly so. A fictional character, represented only by words on paper, comes into being in the imagination of a reader. A dramatic character, however, comes into being first in the imagination of an actor, then in the stage activity of that actor, and finally because of the actor in the imaginations of a group of audience members. Thus, an actor's imagination and creative work extend a playwright's character-conception. Actors present a play not as words on paper, but as sounds and sights on a stage. Because a play can be brought to life by a group of actors in New York, by another group in Berlin, by another in Kansas City, or by a group anywhere, dramatic characters can be judged as changeable and changing. *The characters in a play differ in different productions of that play.* And rightly so! This circumstance is basic to the aesthetic nature of drama. This appears to be a confusing, and often infuriating, fact to critics and academicians. There is no *one* ideal production of a play. Hence, there can be no definitive critique or explication of a dramatic text. A play is finished only in performance, and different performers "finish" it in different ways.

A playwright, thus, depends upon actors. Some are his close associates, but more often they are people he never meets. They are the only ones, however, who can complete his creative act. The more he is able to build basic requirements into his characters and simultaneously to set them as imaginatively suggestive to *actors,* then the more he has fulfilled his aesthetic function in a dramatic ensemble.

None of the points mentioned above demean the work of the playwright or of the actor. They are intended only to plant in a playwright's mind the nature of dramatic creation regarding the essential element of spectacle. A playwright should learn all he can about acting and actors, and he should write plays for actors.

Chapter 7 indicates the connections of acting and sound. It shows how an actor's voice expresses melody as subtext. Subtext is also an important matter to spectacle. The thoughts and feelings supporting the words of a play are expressed in the diction as verbal symbols, in the sounds as melodic symbols, and in movements as physical symbols. An actor sets his jaw, turns and stalks across the stage, bangs his fist into the back of a sofa—he thus represents, whether he is speaking or not, a certain subtext of the play. The ideal, though impossible, visualization of a play would be for the actors to physicalize all the thoughts and feelings so perfectly that words would be unnecessary. Hence, mime is a crucial factor in the spectacle of a play insofar as the playwright formulates the requirements of movement and visualization.

To perform a role appropriately, an actor must first understand and feel, then vocalize and move. An actor must comprehend the thoughts of his character and of the play itself. Next, he must perceive the feelings of the character he portrays. He must be able to speak the words clearly and appropriately. And he must conceive and enact meaningful movement. The actor assumes what he can understand about the thoughts, emotions, words, and actions of a character, and then he puts these together with his own ideas, feelings, and motions. Through this highly complex act, he creates an immediate, live, and dramatic character. He aligns the details of his own personality with those of the character and thus completes the work the playwright began.

Actors by necessity are the most emotional and the most physical people of the dramatic ensemble. Their creative contribution depends largely upon their being so. Well trained, experienced actors usually possess startling emotional imagination. It is the core of their art. The tangle of emotions they must handle, reveal and stimulate is enough to make them absolutely unique among all humans. Their emotional complex includes understood, remembered, felt, communicated, and stimulated emotion. They must discern as well as possible the emotion of the playwright as he wrote, the emotions that appear in the character, and the emotions demanded by the director. They must combine those emotional factors with their personal emotions—remembered, simulated, and felt—during rehearsals and performances. Actors must then communicate the entire complex to an audience by making noises and motions. Finally, they must be aware of and react to the emotions of an audience during each performance and adjust, however slightly, their entire enactment of emotion each time through. It is no wonder that they are constantly concerned with such matters as how to think of emotions, whether or not to feel emotion during a performance, and how to project emotion vocally and physically.

The more clarity, variety, and intensity of emotion a playwright puts into his play, the better he will stimulate the actors who will perform it. An actor works in the present tense. As the coordinating artist who brings the play to life, an actor lives in performance solely for the pleasure or pain of the moment. Actors provide immediacy of response, and thus to use Kierkegaard's phrase about aesthetic man, actors are "the immediate ones." To construct a drama of high quality, the dramatist must write it for actors. It should be supremely actable. Even though Aristotle evidently did not understand acting very well, he recognized how essential it is to drama and how inherently it is a qualitative part of drama. One of the four segments of his definition of tragedy explains that every play comes into being through acting rather than narration. Whether or not a play is actable and to what degree it is so, therefore, helps determine its quality, its beauty, and its accomplishment.

What makes a play actable? The answers to this question appear in the foregoing discussions of diction and sounds and in this treatment of the actors' creation of spectacle in drama. Clear, appropriate, and easily articulated diction is important. Rhythmic, varied, and melodic sounds contribute. And the clarity, variety, and intensity of emotions in the play stimulate the physical actions so crucial to acting. Nearly any written material is performable; actors have effectively performed sections from novels, epic poems, and even telephone directories. But dramas formulated especially for actors are the most potently actable. Although it is not essential that a playwright be an actor himself, acting experience probably is advantageous. Undoubtedly, part of the reason the plays of Shakespeare and of Harold Pinter are so actable is that they themselves knew drama as actors. Sophocles and Brecht were not only great playwrights but also excellent directors; they knew and cared about actors and acting. The actability of *Hamlet*, *The Homecoming*, *Oedipus the King*, and *The Good Woman of Setzuan* is one of the overwhelming virtues of each. Of course actability alone is not enough to make great drama, as the plays of David

Garrick or Emlyn Williams demonstrate, just as a good story or colorful verse is not enough. But every drama needs to be well formulated as spectacle, as a piece to be acted. Hence, the contemporary practice of some experimental and improvisational companies to include one or more playwrights in their number cannot help but encourage plays of at least an actable sort. A playwright who works more than just a few weeks with an acting company has the great advantage of being in touch with actors, of learning about their work, and of permitting them to assist in the origination and testing of material.

The actor is the dramatic artist as integrator. He takes what in the play is abstractly implied and verbally formulated and makes it immediate. And regardless of what goes on inside the actor as he rehearses and performs, he brings the play to life as *mime*. An actor's physical actions "tell" the play. Emotion is the actor's inner work, and his outer work is movement. He must glean from the script intellectual and emotional motivations for the kind, quantity, and style of his external activities. The visual action of a play is the pantomime of the actors. With mime, the actors bring together what is ideational, emotional, and physical. In rehearsal, actors can to some degree discuss the intellectual, the emotional, and the physical aspects of a drama, but in a performance they can never really separate these within themselves. Actors are human beings as artists and as instruments. Ideas, emotions, and movements occur simultaneously within them, as within people in life situations.

A play's actability is not so much a matter of a writer worrying about what pantomimic techniques should be used. He will make a play actable mostly through conceiving a vivid dramatic image, by structuring the necessary action, and by verbalizing both. But he can help insure the proper spectacle in his drama by weaving requisites for movement into the script. Shakespeare implied the appropriate physicalization in his plays, and not with stage directions either. As Shakespeare did, a contemporary playwright should put into the dialogue as many implicit and explicit suggestions for physical movement as the specific dramatic action demands. The more he visualizes the mime of the characters while he writes, the more his words will imply movement. And the more his words imply movement, the more the author will control and stimulate the actors. In such a manner, a playwright concretely establishes the play's spectacle, its final enactment, its ultimate coming-into-being. Even though all plays of approximately equal length contain about the same number of words, some are labeled as "talky" or "wordy." When a play truly deserves such a label, its author has failed to conceive the spectacle, failed to impel the expression of human emotion in movement.

Four terms may be useful in describing a play's physicalization through actors. Three of these terms, as commonly used in most theatre companies, suggest the general categories of movement: blocking, business, and peripheral movement. *Blocking* is the movement of actors from place to place in the stage space. This kind of movement includes actors' entrances, walking patterns, locations, and exits. Blocking is most significantly the actors' use of three-dimensional space. *Business* is the specific, smaller activity that makes a character life-like. It includes most of the pantomimic actions an actor performs in conjunction with the blocking. Business usually involves gestures or the handling of stage properties. All the apparently meaningful gestures of hands and arms, head and face, shoulders and spine, legs and feet belong in this category. Examples of business are such actions as dealing cards, lighting a cigarette, opening a book, drawing a sword, and hugging a lover. Even the manner in which an actor executes movement, a bow for example, is stage business. Most pieces of stage business can be described as symbolic, descriptive, or emphatic. All such moves should naturally rise from the overall characterization and be appropriate to it. Stage business encompasses all the immediately meaningful activity by an actor for giving verisimilitude to a character. *Peripheral movement* is marginal movement, the small details of physical activity that are not directly meaningful but are always suggestive. Business can be described as denotative activity and peripheral movement as connotative activity. No single piece of peripheral movement is absolutely essential to a character, but each may be so for the actor playing the character. In everyday life, this sort of movement is usually called involuntary or subconscious. On stage, it is sometimes planned and sometimes spontaneous. Examples of peripheral movement might be when an actor blinks his eyes, grits his teeth, scratches his ear, lets one shoulder sag, or drums his fingers. A dramatist can write in such a way to stimulate all three of these kinds of movement, and he can even call for them with dialogue references or stage directions. All three of these categories of movement are necessary in drama, and characters achieve lifelikeness through them.

Actors are also involved in another sort of physicalization. Technically, this is *pictorial composition*. It is an element of spectacle usually controlled by the director. Stated simply, pictorial composition is the location of a character in the stage space. The spatial relationships between actors visually indicate the social and emotional relationships between the characters they portray. The overall arrangement of scenery, furniture, properties, and actors makes a visual impact on an audience in a manner similar to the arrangement of forms in a painting or a piece of sculpture. At the very least, a playwright can attempt to conceive dramatic scenes with compositional potential. He can utilize interesting locales with implicit spatial variety, such as those in Brecht's *The Good Woman of Setzuan*. And he can establish scenes with both divergent and convergent character relationships, thus requiring changing compositions of heightened importance; such relationships occur, for example, in Shakespeare's *The Comedy of Errors*.

Costuming, as another significant aspect of spectacle, can be discussed from several viewpoints. For a playwright, however, there are two especially important considerations. Costumes affect the physical movement of the actors and visually communicate some aspects of character. For writers who have had little acting experience, it may be strange to hear that movement control is the most important principle of costuming. A person's movement is affected, for instance, by the sort of shoes he wears; the heel height, the weight, the overall flexibility of footwear dictates the manner of a character's walk. Other items of apparel similarly affect how a person stands, sits, and gestures. Some stage properties also affect a character's movement and should be considered as items of costume. Swords, knives, capes, handkerchiefs, purses, and flasks are items that people handle, and such items affect physical movement. The other major principle of costuming, having to do with what a character looks like, is more obvious. Few playwrights are likely to fail to realize that costumes can suggest period, social status, income, and even disposition for a character. A competent playwright should, thus, properly employ the basic principles of actor movement, pictorial composition, and costuming as aspects of spectacle written into his play.

In the process of creating a drama, both the playwright and the actor are essential. Their work is coordinate, not separate. The improvisation necessary for the total formulation of a drama is not a matter of a playwright making up what a character says and an actor making up what that character does. When an actor improvises without a script, he becomes in fact a playwright. Conversely, as a playwright composes a script, he must be an actor. A playwright is an actor on paper, a director in manuscript. Theoretically and practically, the playwright and the actor should improvise together for the sake of the character. Playwrights and actors are today exploring together the many potentials of their creative relationship.

The persistent problem of the playwright-actor relationship is whether an actor should adapt a character to his personality or adapt his personality to the character. Because of their contacts with a variety of actors, most playwrights have struggled with the problem, and many—Yeats, Shaw, and Brecht among them—have written about it. William Butler Yeats wanted a theatre of poetry, of intellect, and of words. He called for simplified acting in which speech is more important than gesture and in which the actor is totally subsumed to the play. George Bernard Shaw realized that a demand for actors to submerge their personalities in a play's characters was no guarantee they would do so. Hence, he tried to write actor-proof plays, ones that would compel actors to perform the characters as he wished. And he recommended that the playwright should manage his own theatre and direct his own plays to insure their proper enactment. Bertolt Brecht, however, realized more fully the potential value of the actors' creative contribution. It is not so important that Brecht directed some of his own plays as it is that he constantly drew upon the imaginations of his actors and accepted some of their improvisations and suggestions. Brecht was not just a playwright, but a man of the theatre. And as such, he has been one of the forces in twentieth-century drama impelling writers and companies to co-create.

The possibilities of *interaction between playwright and character and actor* have never been fully explored, in theory or in practice. This is, however, one of the areas of genuine excitement and innovation in contemporary theatre. It is a stimulating concern of the best playwrights who have worked with experimental companies. The influences of Bertolt Brecht and Antonin Artaud have vitalized some of the current experiments involving actor-author relationships. But every playwright should realize that he must learn for himself. He can never afford simply to read about experiments. He can learn a little from reading the scripts of others and from reading theoretical works. But reading is not the whole answer. No artist can afford to be merely derivative. No playwright can afford to be only literary. Every dramatist needs live contact with a theatre company that will permit him to join their ranks, not just for a few rehearsals but for an experimental span of time.

Granting that the success of co-creative experiments of playwrights with actors depends upon the specific individuals involved, there are still some working principles worth identification. Playwriting has always required improvisation on the part of an author, but only recently has improvisation become a widely practiced means for actor development and creation. The improvisational innovations of companies in the United States, as well as in England and Poland, Germany and Russia, are among the most significant developments in dramatic art in the second half of this century. The aesthetic interaction possible in improvisational sessions can involve a playwright in various ways. Each may be useful or useless to a given writer.

The creative interaction between playwright and actors can possibly occur in any one of the following stages during the formation of a play:

Conception
Development
Structuring
Rehearsal and revision
Performance

Since the actors must do the performing, the final stage is theirs, and undoubtedly the work of structuring is rightly that of the playwright. Hence, the first, second, and fourth stages of development are the ones in which co-creativity can best occur.

For the first stage, conception, when a writer is working with a company, he can bring one or more germinal ideas to improvisational sessions, or he can simply attend the sessions and pick ideas from the work of the actors. The choice of the germinal idea on which the company will work is most often made by the company director. If he has sufficient insight and sensitivity, he will pay attention to the opinions of the writer and actors, but more importantly he will see which idea holds the greatest creative potential for his particular company. Although the writer thus surrenders some of his power of decision, he gains the combined judgment and creative power of the ensemble. Most writers no doubt prefer to make their own choices about the germinal idea, and the best company directors, realizing the ability of writers in selectivity, heed them attentively. Many groups surmount this problem by putting the playwright in charge of the conception sessions.

The second stage of ensemble dramatic formulation is development. It would be a rare dramatist who would not benefit from the improvisational work of a skilled group in this matter. An ensemble arrangement should be established that stimulates free exchange of ideas between writer, actors, and director. They should naturally and freely stimulate each other. At this stage, the work depends also upon the suggestive power of the germinal idea and of the participating writer himself. By comparison with the number of sessions spent on germinal ideas, there should be many more devoted to development. Ideally, the writer should be suggesting many ideas, the actors exploring their possibilities, and the director prodding everyone. During such a period, the playwright needs to prepare in writing many ideas for the ensemble sessions, and he needs to record the best materials he sees and hears during the actor explorations. Improvisational development sessions can help a playwright with his collection of basic dramatic materials, and they can even provide him with scenario elements. Sessions of this sort permit the writer to test ideas, characters, conflict scenes, bits of physical business, and even dialogue sequences. Developmental periods should be fluid enough to permit the writer to try whatever he wishes so long as he stimulates the rest of the ensemble.

No group work can fully substitute for a writer's meticulous and lonely work of structuring the action and drafting the dialogue. Thus, the third stage is his. The fourth stage in a play's formulation, however, can involve an entire company. This is the rehearsal period for the actors and the revision period for the writer. Although a writer can benefit from attending the rehearsals of nearly any sort of company, he will benefit most from working in rehearsals with actors he knows, and preferably with the actors who helped him develop the dramatic material. The salient notion for a playwright working through a full rehearsal period with an improvisational company is that the play should not become frozen nor the author protective.

The final stage in a play's development is performance—not publication. A playwright and actors work together to make drama, not literature. This is true of any production given by any company of any play. And it is an even more significant truth when a playwright and actors co-create. It is a fact of contemporary drama that the renewed vision of drama *as drama* is driving the theatre's best new writers. Hence, even though performance is normally the province of the actors, the playwright still participates. The play can and should continue to change. Performance does not mean that the play is finally and absolutely rigidified. In a performance, a play can go on evolving.

At least two volumes of plays serve as examples of the results of co-creation between playwrights and an improvisational company. The first is *Four Plays* by Megan Terry, and the second *America Hurrah* by Jean-Claude van Itallie.[1] These two playwrights worked closely with The Open Theatre, an experimental acting troupe involving such directors as Joseph Chaikin, Peter Feldman, Jacques Levy, Michael Kahn, and Tom O'Horgan. The seven plays in the two books indicate a wide imaginative range of actor-writer collaboration. And the introductions to the two volumes—by Richard Schechner for Terry's plays and by Robert Brustein for van Itallie's—are enlightening. Schechner's piece is especially informative about new forms and about how a playwright can truly be a "wrighter."

The foregoing description of some of the possibilities of creative cooperation between writer and actor is appropriately the core of this discussion of spectacle. Acting is part of the spectacle of drama—spectacle as an integral and essential element of any play. Too long have critics separated what they call "the dramatic" from "the theatrical." A playwright is as much a poet of visual and auditory activity as he is of verbal arrangements. A dramatist formulates the acting of a play in the sense that he structures the actions, feelings, and associations of characters. Actors can help him to take risks, to avoid old solutions, to penetrate the unknown. He must learn to find actor stimuli as well as audience stimuli. From actors, he can learn new values of contact, discovery, and confrontation. With them, he can more rapidly test his material and find elements that help him get beyond the individual to the personal.

HOW MUCH STAGECRAFT FOR THE PLAYWRIGHT?

The traditional advice to playwrights is for them to get firsthand experience in all the practical aspects of theatre. This implies that sometime a playwright should work in nearly every job in the theatre. According to the counselors, he should "get his hands dirty," "sell a few tickets," and "develop a feel for the business." If a writer heeds such advice, he will certainly learn—how to avoid writing. He will waste his time!

A writer does not need to know how to build a flat, cut a costume, or apply makeup. But he must know how to verbalize a thought, construct a beat, and make a verbal melody. He probably will not be mentally injured by practical experience as a theatrical artisan, but he probably will not be

[1] Megan Terry, *Four Plays* (New York: Simon and Schuster, 1966, 1967); Jean-Claude van Itallie, *America Hurrah* (New York: Coward-McCann, Inc., 1966).

helped either. This is not to say that the writer should be ignorant of stagecraft. He should know it theoretically and productively, but not necessarily practically. He needs intellectual knowledge about the potentials of theatrical materials, forms, styles, and functions. He needs a productive understanding of theatres, scenery, costumes, and makeup as stimuli to his imagination. All the mechanics and physical materials of the theatre—and of life—represent means for the construction of a play. If learning about them requires pounding nails and sewing hems, then he should do so. But never at the expense of writing time! A playwright needs to know stagecraft not as a carpenter but as a playwright. Ideas and concepts about it are essential to him. Hours in a scene shop are not. His concern involves everything visual and auditory, but not everything manual. A playwright best studies theatrical practice not to learn "how to" but to understand "for what purpose."

Most university drama programs perpetuate the myth of the necessity of practical experience in technical theatre for the playwright. Most university theatre curricula stand as examples of the misdirected advice about stagecraft that a playwright is likely to encounter throughout the theatrical world. Too many playwriting teachers are willing to encourage the novice to spend his time not dreaming metaphors, learning phonetics, or studying dramatic theory but rather manning the scene shop, treading a sewing machine, or pushing tickets. The playwright needs to study theatre design, scene design, and audience psychology. Yes! But too many university theatre faculties are more interested in getting man-hours from their students for scenery construction than in devising courses and workshops that will encourage them to practice their art.

The essential areas of technical knowledge for playwriting are theatre design, scenic design, costume design, and makeup. A study of each will assist the writer's theoretic understanding of theatre and impel him to innovate more meaningfully. A playwright should be, at least imaginatively, an originator of design concepts. Every dramatist, if he is to construct thoroughly dramatic plays, needs to know the potentialities and the limitations of each phase of theatre art.

As performer-spectator spaces, all theatres can be placed in a small number of architectural categories. Although such categories generally characterize the spatial arrangement of individual theatre buildings, they do not reveal the surprising and unique personality of each individual theatre. The following kinds of stages typify the main categories: proscenium, arena, thrust, flexible, and environmental. Since this book concentrates on the formulative aspects of drama, it explains only the types. A writer should turn to other sources for descriptions of specific theatres.

A proscenium stage is an acting space in front of a seating area with a separating plane containing a frame. The chief function of such a stage is to promote pictorial illusion. It gives an audience an omniscient feeling of being a god-like observer of other people's lives. The proscenium stage dominated American theatre until the second half of this century. Now that drama is less realistic and more imaginative, the planar division between performance and audience no longer suffices. The proscenium stage is rapidly becoming outmoded, at least to innovative directors and playwrights. Most of the "Broadway" playhouses in New York are of the proscenium type.

An arena stage is an acting space surrounded by seating area. Operating psychologically as an almost primitive magic circle, an arena functions to enclose the action. It draws a magic line around a play. It promotes an audience feeling of control and superiority. Despite its inherent simplicity, it is a stimulant of ritual. Scenic potential is severely limited in an arena, but such a stage emphasizes the performed play more than does a proscenium stage. Although some critics maintain that arena stages are merely voguish and that the vogue has passed, a few American companies—for example Arena Stage in Washington, D. C.—are exploring with fascinating results the potentialities of an enclosed arrangement.

A thrust stage is an acting space that projects into a seating area. At best, it is three-fourths enclosed by the audience. Functionally, it emphasizes three-dimensional space and yet permits a scenic background. It provides both performer and spectator with a heightened awareness of life sounds and movements. Visually, it makes an important contrast with the camera media, cinema and television. A thrust stage tends to arouse in an audience a feeling of personal participation in an intensely human event. An increasing number of new theatres—in contrast to public and school auditoriums—are being constructed in thrust stage arrangements. In the best of these, the audience surrounds the stage from 180° to 210°. Many new theatres, however, possess a modified thrust stage. Most of these are actually proscenium theatres with large and slightly projecting apron-stage areas. Although these theatres partially avoid the planar illusionism of the proscenium stage, they do not engender the liveliness or imaginative potential of the full thrust stage. The major advantages of a thrust stage have to do with its emphasis on three-dimensionality, its immediacy, and its staging fluidity. It is particularly well suited to free-flowing, multi-scene, abstract contemporary plays. Its major disadvantages, at present, have to do with the lack of experience of most directors and designers in staging plays on a thrust. The variety of possible production styles is much greater with a thrust stage. Thus, of all the stage arrangements, it probably offers the best stimulus to the "new" playwright. The Tyrone Guthrie Theatre in Minneapolis and the Vivian Beaumont Theatre in Lincoln Center are examples of workable thrust stages.

Besides the three major theatre types, there are a number of less often used spatial arrangements. A few theatres feature multiple stages. This type usually consists of three proscenium stages fanned across and partly enclosing a forward audience section. Another type is the open stage. Characteristically, its performance and audience areas share the same architectural cube, and lighting rather than scenery creates the performance environment. Some special variations of the open type stage are the corner stage, the central stage (audience on only two opposing sides of the acting area), the diagonal stage, and so on. A few theatres own what might be called a mixed performance-spectator plan, in which miscellaneous acting spaces surround and penetrate the seating area. Amphitheatre and stadium stages are sometimes used for drama, but they are not primarily built for that purpose. Also, wagon or portable stages still exist. The modern version consists of one or more semitrailer trucks that unfold to make a temporary stage.

A fourth major type is the flexible theatre. It can be transformed into two or more of any of the arrangements so far discussed. The differences between individual flexible stage theatres is great. Some are simple and consist of a large room with folding chairs. Some are complex and costly to build and maintain; these may possess movable audience sections, portable walls, and hydraulic or revolving platforms. With each transformation, however, a flexible theatre is likely to take on the appearance and aesthetic implications of one of the other three major theatre types.

The fifth type of stage is not really a stage at all, but rather an acting location. An environmental theatre is simply a place—anyplace—where a dramatic event might be made to occur. It is simply a theatre locale appropriate for happenings, theatre games, and other improvisational or semi-improvisational events. Most environmental stages resemble, in performance-audience relationship, one of the more conventional theatre arrangements.

Each of the stage types possesses one or more unique features of spatial arrangement and of emotive stimulation. And each theatre, regardless of its physical organization, projects an individual architectural personality. This is another reason why every play is altered in production. As the architectural milieu changes, a play must be creatively rendered in a different manner. If a playwright studies the various types of theatres, he will likely understand that to write for the modern theatre is to write for a variety of spatial arrangements. Fluidity, variety, and flexibility of spatial conception should be the architectonic guides for the contemporary playwright.

Although the practical study of stagecraft as construction methods is not essential for the dramatist, a study of scene design is crucial. He should be aware that a scenic designer must visually capture the play's central image. Scenery and stage properties are the specific means a designer uses to complete a play's visual spectacle. Rather than exhaustively to explain scenery and properties, it is appropriate here only to note the two key criteria for the selection of physical elements for a stage setting: (1) to make the actors move in certain patterns and perform certain activities, and (2) to present appropriate and affective visual stimuli to the audience.

There are four major divisions of scenic items. First, floor pieces include ground cloths, rugs, traps, platforms, ramps, furniture, walls, doors, and the like. Second, backing items are such things as curtains, flat walls, painted drops, cycloramas, and three-dimensional constructs. Third, overhead pieces may be ceilings, balconies, chandeliers, draperies or other cloth pieces, sculpted items, symbolic pieces, and lines. Fourth are portable pieces: movable furniture, hand props, rolling items, small segments of rooms, and so on to an infinite variety. For each of these four divisions of scenery, both the designer and the playwright can think of style too. To what degree each stage piece is illusory or abstract, realistic or symbolic, descriptive or suggestive affects the overall presentation of spectacle.

Costumes and makeup are actor-scenery. As such, both are functional, descriptive, and decorative. Costumes should direct and aid the actors in their movements. Because various pieces of clothing cause differing physical sensations, costumes even help the actors to generate and communicate feelings. As descriptive items, costumes reveal information about the characters. To say that costumes are decorative implies not that they are unimportant but that they provide visual interest and variety. Costumes are scenery in action. A writer should consider costumes as another means in the realm of spectacle for structuring his play's action.

Makeup is a mask. A mask is makeup. Both transform an actor into a character. An actor's bare face and a complete head mask are the two extremes of makeup. The appropriate makeup for a given character is usually somewhere between the two extremes. A playwright should put some hints about degree and style of makeup in the script. He need not, and probably should not, describe the makeups in stage directions, but he will naturally indicate, as he envisions and verbalizes the characters, the relative degree of illusion of the drama. It is this degree of illusion that controls the makeup.

Makeup is more or less necessary depending on the competency of the actor, the verisimilitude of the play, and the physical circumstances of the production. The better the actor, the more skillfully and totally he will use makeup to transform himself into a character. If the play tries to present the illusion of everyday life, then the makeups will be realistic. The more abstract or symbolic the play, the more presentational and expressive the makeup should be. The larger the theatre and the more intense the stage light, the heavier and the more coarse the makeup.

Makeup is functional, decorative, and symbolic. It functions in two realms. First, makeup complements a character's costume, emphasizes facial features, makes a face discernable at a distance, sets the color of a character, and combats the flattening effect of intense light. Second, makeup functions as an aid to the actor. Most actors begin serious concentration on their role before each performance as they apply makeup. Also, the feeling of makeup on one's face is a constant reminder about the character. Makeup is decorative in a manner similar to costuming. It becomes increasingly decorative—*and meaningful*—as it avoids illusion. That makeup as a mask is symbolic, and even mystic, is readily apparent in the rituals of primitive peoples. This "facial magic" can be operative in the modern theatre of (presumably) civilized man as well. From the street makeup of a pretty girl, to the whiteface of a clown, to the celastic full-face mask of the actor in a Greek tragedy, a makeup mask is a significant, though often ignored, means of dramatic imagery.

STYLES OF PRODUCTION

A selected few stylistic categories are explained in this section in one paragraph apiece. Each description is short not because the style for a play's production is unimportant, but because a writer ought to create his own style in the realm of spectacle. He should perhaps know about stylistic categories, if only to avoid them. The style of a play's spectacle is best named *after* it is performed. But every writer should control his play's visual style.

The style of a drama is most apparent in its spectacle, that is in all its visual, temporal, and spatial aspects. For purposes of critical analysis, a play's verbal or poetic style can be differentiated from its production style. And in some respects, these two aspects of a play can be separated in the finished drama. For example, a realistic play can be produced in a nonrealistic manner. But such unharmonious execution usually fails to make the best possible work of dramatic art out of the particular play. At best, a play's verbal style and its spatial style are organically connected. When the director, actors, and designers perform their artistic work well, the play's spectacle will be an organic concomitant of all the other qualitative parts (plot, character, thought, diction, and sounds). The responsibility of the writer is as great as that of the other theatre artists in creating a play's production style. If he structures the spectacle *within the play itself*, then he will establish an ample basis for the production.

Style in drama, no matter which part of drama is being considered, proceeds from a playwright's vision of life. His conception of the world, as reflected in the play, controls the manner in which the drama is carried out in mime as well as in words, in scenery as in characterizations. A dramatist always establishes some sort of probability in the play, and this suffuses its production. The style of a play's verbal organization is coordinate with the style of its performance. And the styles of no two plays are precisely alike. Although several categories of style are briefly explained below, each category merely provides a general name for certain groups of plays. Actually, every play's style is unique. The isms are dangerous generalizations, and for the most part they are the easy nicknames used by popular and academic journalists. The following explanations do not include long historical descriptions, but rather some ideas for the playwright's active use. This is not a comprehensive list, but a functional one.

Realism, the dominant dramatic style from the 1880's to the 1950's, is an illusory style. The world of a play onstage represents the everyday world of ordinary men and women. Realistic drama usually proceeds from the idea that common experience and ordinary sensory perceptions reveal objective reality, and that objective reality is ultimate reality. In realism, the appearance of life supposedly represents what is most true about life. The writer of realism observes, selects, and reports life as anyone can experience it. He writes about what is familiar to him. Verisimilitude, or life-likeness, is the goal of the realist. For the audience, realism provides feelings of omniscience and empathy. Examples of realistic plays are *Hedda Gabler* by Henrik Ibsen, *Candida* by George Bernard Shaw, *Golden Boy* by Clifford Odets, and *A Streetcar Named Desire* by Tennessee Williams. For realism, a playwright strives faithfully to depict the sensory world. Direct knowledge of the actual world, according to the realist, comes best from a report of objective realities of life. Realism stresses the universal nature of the particular.

Naturalism became a significant aesthetic style in the latter half of the nineteenth century. Emile Zola, among others, brought it to prominence in literature and drama. Naturalism is a style closely bound to realism in intent and result. It requires, however, a more extreme objectivity from the artist. The naturalist attempts to employ the scientific method of observation and recording. He trusts only his five senses and tries to eliminate his personal imagination by substituting objective knowledge. Naturalist dramas tend to show that all human behavior is chiefly a result of a person's environment and heredity, rather than his will. Many naturalistic playwrights have dramatized scientific principles or put case studies into dialogue. Naturalism requires more detachment on the part of the dramatist and demands that he not permit his own attitudes to interfere with the objective, recordable truth of life. Naturalism, like realism, is an illusory style insofar as dramatic spectacle is concerned. A naturalistic production attempts a scrupulously faithful representation of sensory reality. It is more a photographic reproduction than a realistic production. The audience is stimulated to feel not so much like a god-observer as a scientist-observer, and naturalistic works stimulate less empathy. Naturalism requires the elimination of decorative and expressive elements. Some examples of naturalistic dramas are *Thérèse Raquin* by Emile Zola, *The Vultures,* by Henri Becque, *The Lower Depths* by Maxim Gorki, and *Beyond the Horizon* by Eugene O'Neill. Although the scientific orientation of these early naturalists is no longer in vogue, some more recent playwrights are using a naturalistic manner to make didactic drama, for example *The Connection* by Jack Gelber and

The Brig by Kenneth Brown. The goal of the naturalist is to report and thereby illuminate what actually happens in life.

Romanticism is another major stylistic category in drama. Like realism it continually reappears in modern theatre, but unlike realism it stresses idealization rather than objectivity. The dramas of the various periods of romanticism differ in rendering, but they all tend to represent a similar view of man in life. The romantic plays of Elizabethan England, for example, are different than those of nineteenth-century France or the twentieth-century musical comedies of the United States. Whereas realism stresses the actual, romanticism attempts to show the ideal and the beautiful. Naturalism emphasizes the objective nature of man, but romanticism represents the felt qualities of experience. Naturalism depicts man as victim; realism displays man's everyday life; and romanticism demonstrates man's potential. Writers of romantic dramas usually interest themselves in the struggles of an unique individual to achieve his own potential, to knock down conventions, and to dominate his environment. Romanticism at its best tends to be more individualistic than sentimental. Ideas such as the perfectability of man, the truth of beauty, and the interconnection of all things are germinal in romantic plays. The romantic playwright is a lyric writer, if not in his dialogue at least in his conception of man. Some descriptive examples of romantic plays are *Faust* by Goethe, *Cyrano de Bergerac* by Edmond Rostand, *Liliom* by Ferenc Molnár, and *Green Grow the Lilacs* by Lynn Riggs (which became the musical comedy *Oklahoma!*). Barrie Stavis is a contemporary playwright whose works, such as *Lamp at Midnight* and *Harper's Ferry*, display some of the best features of romanticism. In general, romanticism is an interpretation of the goodness, beauty, and purpose of life. Romantic dramatists attempt to seize natural phenomena in a direct, immediate, and unconventional manner. They assert human values of sincerity, spontaneity, and passion.

Symbolism is another style important in drama. Although somewhat related in attitude to romanticism, it was one of the earliest denials of realism, and as such it is thus paradoxically related to it. Whereas the realist presents in art the illusion of actual life, the symbolist usually tries to make the illusion of the reality under the surface of actual life. Symbolists deny the evidence of life furnished by the five senses. They assert that intuition is more important to the artist than detached observation. For them, the logic of science is antithetical to creativity. Truth about life, they say, is better suggested by symbolic images, actions, and objects. A symbolist drama, when fully effective, is meant to evoke feelings in an audience that correspond to the emotive reality experienced by the artist. Symbolist playwrights concentrate on verbal beauty, on the inner spirit of man and things, and on the affective atmosphere of nature. They try to capture the mysteries of life. Symbolism was aesthetically one of the first impulses in modern art toward abstraction. Its practitioners attempt to represent spiritual values by means of abstract signs. Some notable examples of symbolist dramas are: *Pelléas and Mélisande* by Maurice Maeterlinck, *Purgatory* by William Butler Yeats, and *Pantagleize* by Michel de Ghelderode. Although most major symbolist dramas influential in the second half of the twentieth century are not so romantic as those of Maeterlinck and Yeats, they are nonetheless intuitive and probing in a similar manner. The works of Jean Genêt and John Arden show a special kinship, even in their treatment of alienation, with earlier symbolist drama.

Expressionism is another stylistic departure from realism. From its rise to popularity about the time of World War I, expressionism has been a vital trend in modern drama. An expressionist play presents not the external world of sensory life, but the subjective and imaginative inner world of the artist's consciousness. The expressionist attempts to convey the subjective truth imbedded in the human mind. He creates as an expressive, not a reportorial, act. He goes through a process of manifesting his personal memories, emotions, intuitions, absurdities, and improvisations. Expressionist art is a record of felt experience. For the expressionist, sincerity, passion, and originality are the major aesthetic criteria. The audience is likely to think of expressionist works as distortions of life, and at best they receive new insight into the nature of human experience. August Strindberg, in such plays as *The Ghost Sonata* and *The Dream Play*, initiated expressionist drama. Some of the best early examples are *From Morn to Midnight* by Georg Kaiser, *Man and the Masses* by Ernst Toller, *The Adding Machine* by Elmer Rice, and *The Hairy Ape* by Eugene O'Neill. The influence of expressionism with its imaginative freedom, its variegated technique, and its abstractionism is still dynamic in contemporary drama, even in such widely divergent plays as those of Bertolt Brecht, Arthur Miller, and Eugène Ionesco.

There are many other stylistic categories of drama which might be of interest to the critic or the scholar—impressionism, surrealism, constructivism, plus epic theatre, absurdist theatre, and total theatre. All these, however, are less widely practiced styles, and each has some relationship to one or more of the broad types explained in the foregoing. One could say that there are only two basic styles—realistic and idealistic. The most influential contemporary styles are generally explained in the discussion of abstractionism in drama which follows.

No creative playwright slavishly follows the technical precepts of any one stylistic school, and no single play perfectly fits any critically contrived category. A playwright should, nevertheless, consciously develop his own theories about style, especially in relation to the spectacle of his dramas. And considering style broadly, he might beneficially recognize that his work is essentially either representational or presentational.

Representational style in drama is an attempt to present on stage an illusion of real life. It closely approximates the reality of everyday existence, especially its surface, whether objectively or subjectively realized. Representational dramas imitate life. Of the styles described, realism, naturalism, and romanticism tend to be the most representational. Representational art stimulates sympathetic emotions in an audience. The spectator becomes involved in an action through identification with one or more characters. The dramas of Ibsen, Shaw, Odets, Miller, and Williams are mostly representational. At best, representationalism in drama is a manner of putting materials together not for the purpose of making an easily recognizable story for an audience, but for destroying the impersonality of life. Representational works of quality usually attack abstractions, destroy sentimentality, and deny the infinite. The representational playwrights depict the world as it is.

Presentational style in drama is an attempt to create on stage an intensified experience. It is non-illusory, even anti-illusory. By means of exaggeration, distortion, fragmentation, and imagination, it surmounts everyday reality. Presentational art denies surface reality in order to examine its substance. Presentational dramas dominate, rather than imitate, life. Of the styles described, symbolism and expressionism are the most presentational. Presentational art stimulates personalized emotions in an audience. Presentational drama is an objective portrayal that generates subjective mass response; it is a subjective offering that initiates objective individual realizations. With the presentational, the spectator's involvement is with the work itself rather than with a single character in the work. The later dramas of Strindberg plus the works of Andreyev, Brecht, Frisch, and Peter Weiss are mostly presentational. Presentationalism in drama is at best a manner of arranging materials to provoke insights about the human condition. Although presentational works contain dislocation of forms, distortions, and non-logical manipulations, the authors of such works seldom intend obscurity for its own sake. The best contemporary dramas of this general sort are meant to arouse controversy, consciousness, and personal autonomy in life. Presentationalism emphasizes that art can be more than an objective contrivance by a conscious will. The presentational playwrights depict what the world is becoming.

Unquestionably, the most intense and imaginative plays of the second half of this century are more often presentational than representational. Most of the new playwrights—whether they realize it or not—are employing a new abstractionism. They consider realism to be a limited manner of rendering life. They deny that objectivity and reason, science and logic can provide a full picture of man's condition, or that mere observation can offer a set of solutions for human survival. The new abstractionism provokes its audience and accuses mass civilization. The abstract art of our time—the new drama included—is somewhat related to the older presentational styles, such as symbolism and expressionism. But no one category is now sufficiently broad to include all the new species of art. Thus, for the purpose of this discussion, it can simply be labeled abstractionism.

The new abstractionism takes on many different forms, treats fresh materials, and serves new functions. In the realm of style, however, it is the most unique. Many contemporary plays feature bleakness, negativism, and shocking effects. They question the established ideals of society many of which in our time have turned out to be hollow abstractions. Hence, the new drama is paradoxical. It eschews the abstract rationality of mass technological society and replaces it with an emphasis on the individual, the subjective, and the human. But it employs purposeful abstraction in form and style. It negates abstractions in content by means of abstractions in execution. Such plays as *No Exit* by Jean-Paul Sartre, *Waiting for Godot* by Samuel Beckett, *The Chairs* and *The Lesson* by Eugène Ionesco, and *Mother Courage* by Bertolt Brecht—although these are vastly differing kinds of plays—serve to identify the change to the new drama of abstractionism, of personal responsibility, of human dilemma. These playwrights attempted to say the unthinkable, the uncertain, and the contradictory. They pointed to the spiritual poverty of modern society and of all too many individuals within it.

Although the new abstractionism is a movement to destroy old forms, it simultaneously represents an attempt to expand the possibilities of art. And dramatists such as Albert Camus, Edward Albee, and those mentioned above have broken many barriers of convention and thus made an even richer creativity possible for the new dramatic artists who are now rising.

The abstract drama, like other kinds of abstract art, shows the artist turning away from a logical concern with things to a subjective treatment of the inner spirit of his own self. It tends to be a drama of introversion rather than extroversion. And as such, it ridicules such formulas as the one that says a drama should depict a conscious will striving toward a goal. The new dramatist is much more likely to be concerned with the subconscious non-will of his characters as they struggle to remain human. Because the world is inexplicable, time and space are flattened, fragmented, or dislocated; climax is no longer the crowning achievement, the moment of release; communication is opaque and sometimes unintelligible; and sequence is irrational. The artist now, when creating most qualitatively, refuses the superficiality of restricted classical constructions, and he attempts a more spontaneous revelation of the sort of which perhaps art alone is capable. The danger for the new abstractionism is that it often turns out to be merely disordered and inconsequential. Its potential, however, is that it may help man to recapture his identity as he daily confronts the void of nothingness. The new drama is a drama of image and intuition, and its style is abstract, or presentational, but the works themselves are redemptively personal.

Style of spectacle in any play should simply be the most appropriate rendering of the organization and meaning of all the other parts of that play. Production style should be as organic a part of the entire drama as any other factor. And dramatic art will benefit if in our time playwrights and theatre artists will concentrate together on the coordination of style in spectacle with style in the other facets of the play. Lastly, the artist should always be concerned with style as the manner of marshaling materials into a desirable form and never as the slavish imitation of critically voguish categories. A dramatist should construct a play and not worry about its categorical placement. Style in drama is the verbal-auditory, spatial-visual execution of structured action. A playwright's senses help determine his dramatic style.

SPECTACLE IN STAGE DIRECTIONS

Stage directions in a play are a verbal means of rendering its spectacle. When a playwright wishes to specify an element of spectacle, such as a character's movement or a scenic item, he can do so in one of two ways. First he can put the specification in dialogue. If one character says to another: "Why are you standing there, staring out of that window so sadly?" Then, it is clear that the room has at least one window, that the referential character is standing by it, and that he appears to be sad. The second way to get a specification into a play is with a stage direction. Although this is a less effective indication of spectacle, it is sometimes necessary. And the only stage directions a playwright should ever admit to his script are the unavoidable ones.

The purposes of stage directions as integral segments of a play are primarily to place the action, to qualify the sayings and doings, and to specify other non-verbal elements. Stage directions should be written for the theatre artists who will read the play with production in mind. Such directions are not primarily meant for the reading public, because a play is aimed first at performance and only second at publication. A playscript is an organic creation that serves a production, and thus the manuscript is a performance version. Nevertheless, stage directions of whatever type need to be clear and interesting.

There are also several functions that stage directions are not meant to serve. They should not impose upon the production people ideas, interpretations, and limitations which the author could not weave into the dialogue. Although a playwright should conceive his play for acting, he is in fact not usually the actor, or the director, or even the designer. Stage directions should reveal *what* the theatre artists might do to activate the play's spectacle, but such directions should not tell these co-creators how or why. It is appropriate for the writer to describe a physical activity of a character as required for the progression of the play's action, but he should not try to suggest how to execute it or what secondary activities might accompany it. He can rightfully indicate the components of the stage setting, but he should resist the temptation to specify non-essential details of placement and decor. He should reveal what characters are on stage, but he should defer their pictorial arrangement to the director. In short, a playwright will compose functional stage directions by remembering that his actors, director, and designers will be—or should be—as expert in their areas of creativity as he in his.

The subject matter of stage directions normally has to do with spectacle alone. And it encompasses all the visual and auditory specifications that do not melt into dialogue. Also, whatever appears in a play's speeches need not be repeated in its stage directions.

There are three main kinds of directions: introductory, environmental, and character. The introductory specifications preceding the beginning of the dialogue—the character list, the indication of time and place, plus the description of the setting—if they are to be included at all, are a part of the play. They help to establish the play's given circumstances. Environmental references, some of which appear in the introductory material, may occur throughout the play. They include all specifications about time, space, light, temperature, and physical objects. Character references are those that state or qualify what the characters do and say.

The criteria for stage directions are necessity and clarity. All non-dialogue directions should be held to an economic minimum. They should not appear at all unless positively necessary, and their length should be as abbreviated as clarity will permit. Simple declarative sentences, or their fragments, are the most functional. And the diction in them is most expressive when nouns and verbs predominate. The trite is nearly as bad in stage directions as is the fancy. The verb tense in the sentences should be the simple present. The past and the future tenses are as ridiculous as the perfects are messy. All the sentences should be lean, not fat. Their melody and rhythm need to be as carefully wrought as the dialogue's. Stage directions can be sentence fragments implying complete constructions. But each should be the best sentence for the purpose the writer can possibly make.

A few informative items about stage directions may be useful too. Although these may seem to be technicalities, each item will aid clarity and consistency. Each sentence, whether fragmentary or complete, should begin with a capital letter and end with a period. Lone adjectives and adverbs reveal a writer's ineptitude with both dialogue and stage directions. For example, if a character is to display unhappiness in a speech, the speech should be unhappy; an adverbial qualifier—e.g., (*Unhappily.*)—would reveal the writer's laziness or stupidity. Contractions, though essential in realistic dialogue, are inappropriate in stage directions because they reduce clarity. Likewise, slang and ready-made phrases are inept devices because they are always overused items with diffuse meaning. "And so forth," "and the like," "etc."—these are meaningless. If a character is supposed to laugh, a stage direction—such as (*He laughs.*)—is better than the words "Ha, ha" in the dialogue. *Continued* need never be placed at the bottom of a playscript page. All entrances and exits made by every character should be noted in a stage direction. These are a few of the specific practices that will help a play's stage directions operate functionally and clearly. Chapter 11 explains the preferred typing format for a playscript.

This treatment of stage directions is meant to free the playwright; with this information in mind he should be more able to devise and control the spectacle of his drama. Each writer should naturally identify his own purposes and perfect his own style for stage directions.

VISUAL IMAGINATION

Eyes are windows for the mind. They are sensors, and they are imagination accelerators. Sight leads to insight, the eyes to the mind. Most people see only fragmentarily. They focus on details. An artist creates according to the integrative quality of his vision. He is an image-maker. A dramatic artist must develop creative vision, both as sight and as idea. Spectacle is drama's unique way to make a play immediate and pertinent. A drama is a visual revelation.

As Schopenhauer pointed out, each person believes the limits of his own field of vision to be the limits of the world. Hence, the more inclusive the artist's vision the more likely he is to depict what the world is. Furthermore, his creativity will be enhanced if he begins to understand the vital interdependence of eye and mind. Both are necessary for vision. And vision is a form of awareness. Since awareness occurs only in the present, it is a peculiar necessity for the writer of plays. Drama is present action. It is art as the *now* of human life. Eyes do not remember, but they are the sensors of the *now*, the makers of awareness. Seeing leads to feeling, thinking, and acting—on stage and in life.

These ideas can lead a playwright to the realization that drama is far more than a literary art, though a dramatist needs to be marvelously literate. Drama is at best a perfect joining of words and pictures, of sounds and sights, of poetry and motion. These are the components of dramatic action.

No one can understand a drama simply by reading it. Every drama is an object that depends upon verbal and optical communication. Thus, a playwright cannot hope to create a play with words alone. He searches for words that generate spectacle. His words must impel actors, directors, and designers to create—with him—that spectacle. It follows, then, that every dramatist needs to develop his visual imagination.

Plot is the organization of a play, the structure of the action. Spectacle is the representation of a play, the revelation of the action. Characters, thoughts, words, and sounds provide the content for both. Playwriting as image-making depends upon the author's skill with all six elements of drama. It also depends upon his understanding of the social nature of dramatic creation. Only if he sees, thinks, and feels can he come to make a play. Only if he structures action in such a way that it can be communicated orally and visually to an audience can he bring a drama into being. With the six parts of drama, a writer can create a play, but he must vitalize them with vision. A playwright is both craftsman and seer, artist and visionary.

PART III: *PROBLEMS OF PRODUCTION*

Contemporary art is sensual, has an insatiable appetite for pleasure, is replete with restless urgings toward an ill-understood aristocracy manifesting itself as an ideal of voluptuousness, or of violence and cruelty.

BENEDETTO CROCE
Guide to Aesthetics

9: *WORKING RELATIONSHIPS AND CONTEMPORARY THEATRE*

. . . although in theory the talents who undertake to recreate his [the playwright's] *world onstage by other means, flesh, movement, color are embodiers and deliverers of his vision, they all in fact deliver two substitutes: their vision of his vision, and the totality of themselves.*

WILLIAM GIBSON
The Seesaw Log

A great play, dully performed, can be a great bore. A trifle, greatly performed, can be a tremendous experience.

TYRONE GUTHRIE
In Various Directions

When a playwright finishes his solitary work of creating a play and when the play as a concrete literary object exists as a whole, the collaborative travail begins. This book would be incomplete without an explanation of what a playwright can expect from theatre artists, associated business people, and audiences. He needs to know something about the ideal relationships that might exist—but so seldom do—in the production situation. He needs to learn to study audiences and their reactions. He needs a sense of the strongest forces at work in contemporary theatre. And he needs some information about the essentials of marketing a play.

After a playwright "finishes" his play, he rises from his desk and turns to face a strange and segmented audience. It consists of his friends who first read the play, of those who might buy it with production in mind, of the theatre artists who may collaborate with him in its final creation, of the people who might pay money to see it onstage, and of those who perhaps will criticize it in periodicals. This section, then, treats the production of a play in the specific matters of relationships, audiences, and business procedures.

Discussing all these matters, this section, like the others in this book, is concerned with dramatic art in general and at its best. The discussion is not limited to the production procedures of the New York commercial theatre. More theatre companies exist outside of New York than inside it. The ideas that follow are general in the sense that they apply as much, if not more, to production relationships and business procedures in resident theatres, university theatres, and even foreign theatre organizations than in Broadway production companies. The emphasis is on ideals, not as unrealistic goals or as commercial ones, but as creative and artistic objectives.

If a playwright wishes to know about the Broadway situation, about the agony of writing one play and seeing a poorer but more pleasing version produced, about the convolutions of a group of intelligent and dedicated people working to revise a play so that it will win the Manhattan show audience, then he should read *The Seesaw Log* by William Gibson.[1] There is no other document like it for demonstrating that professional theatre in this country is often—one hopes not exclusively—money oriented, not a place to be serious, and destructive of creativity. In Gibson's *Log*, he damned the collaborative nature of theatrical production by stating that the more a company works on a play, the better show it becomes but the worse it grows as a work of art. Perhaps this is true in all commodity theatre companies whose objective is first and foremost to repay investors and make a profit. But a playwright's collaboration with theatre artists—despite Gibson's New York-Washington-Philadelphia experience—does not have to be anti-creative. There are directors and actors, producers and designers whose primary concern is not monetary success but rather the fulfillment of the writer's vision. For a writer who is used to creating only in solitude and to reaching only solitary readers, the co-creative situation necessary to bringing a drama into being can be

[1] William Gibson, *The Seesaw Log: A Chronicle of the Stage Production, with the Text of Two for the Seesaw* (New York: Bantam Books, 1962).

shocking. This very fact, however, is part of the *raison d'être* of this book; the theatre needs writers who are trained as playwrights.

CO-CREATION WITH THEATRE ARTISTS

Many a playwright, after working alone for years, suddenly finds himself smack in the middle of a company producing his play. Because of the recent explosion of new playwrights, many theatre artists are today for the first time discovering blood-real dramatists as fellow workers. Such situations are so new to so many that problems about functional and acceptable relationships often burst the creative bubble. Is the playwright a co-creator or a pest? What should he contribute during a rehearsal period? Is the director the playwright's ruler or slave, his benefactor or tormentor? Should the playwright ever direct his own play? What can the playwright expect from the other theatre artists—actors, designers, and crew members? When and how can he critique them? Or they his play? What, *ideally*, are the working relationships a playwright should expect and attempt to establish?

Knowing what to expect from other theatre artists in functions, skills, relationships, and attitudes is important to every playwright. The explanations which follow describe not only what the relationships inside the theatre should be but also note some frequent departures from the ideals.

Because dramatic art requires a group of contributing artists, people call it a social art, but the artists themselves more functionally speak of drama as an ensemble art. The necessity of depending on directors, actors, and designers does not belittle the playwright. These people give a drama more immediacy, complexity, and impact than any other poetic composition can ever have. Together, the artists of the theatre—the playwright but one among them—can provide an unforgettable experience for other human beings. The ensemble, which this special group constitutes, works best as an organic unit to produce a single art object.

Playwrights have seldom discussed the centuries old question about whether the director, actors, and designers are creative or interpretative. Nobody argues whether or not a serious playwright is creative. He obviously uses productive knowledge to give selected materials some form. He thereby constructs a play as an object. But what about the others in the theatre? Experienced playwrights know that their plays are incomplete except during performance. In printed form, plays lie as merely seeds for growing final art products. The other theatre artists take the play and bring it into full existence by investing it with life. They add their imaginative skill to that of the playwright, and thus they assist in producing the final art object. True, they are involved in an activity, but its end is a product which exists only through their physicalization and lasts only for a limited time. All the artists working with drama, therefore, put together specific materials in order to formulate a time product. Certainly interpretation is a part of the activity of the theatre people, and certainly the playwright is the central creator. But without the volition and activity of the others, the play remains a seed in the playwright's hand. The playwright and the theatre artists, then, comprise an ensemble of creators making a drama.

To understand the most functional relationships with theatre people, a playwright ought to realize—and for many this is a fist in the ego—that in production the play, not the author, is absolutely and solely centric. The play forms the one essential nerve center for an entire production. As a concrete but only partially realized object, it assumes far more importance than the playwright himself. The people staging the play must know what is in *it;* and although it originated within the writer, what is in *him* matters little to the production. Whenever he has an opportunity to work with a company through a rehearsal period, he suddenly becomes merely one member of that company, and not an essential one at that. In fact, if rehearsals proceed well, the others will normally begin to think of the play as *theirs;* they will feel it belongs to them as much as to the playwright. In a sense, they are correct. They, not the author, complete the art work. The theatre artists whom a playwright needs to understand are the director, the actors, the designers, and on some occasions the composer.

The director totally commands the artistic production of a play. Although in the past, he assumed varying degrees of dominance, in the twentieth century, he stands second only to the play itself. Sometimes, always unfortunately, a director may even override a play. At best he plans, coordinates, and controls all artistic elements of the production. Anything that goes wrong is his fault. What is less widely recognized is that everything good about the production in some way proceeds from his creativity. He functions in an overwhelming number of capacities. For most companies, he selects the play. Then, he edits it, or works through it with the playwright. His analysis of the play sets the interpretation for the ensemble. He leads and instructs the scenic, costume, and lighting designers. He teaches, coaches, or directs the actors in every movement and sound they make. He amasses the thousands of details—intellectual, emotional, and sensory—to give the drama life. The epitome of his action is to discover how best to make a given drama properly visual and auditory.

A director's knowledge must extend over an unusual range of skills. He should know dramatic theory and structure, and he needs a thorough acquaintance with ancient and contemporary dramatic literature. For example, if he cannot identify the given traits of any character or if he does not understand the objectives of a specific playwright, he cannot fulfill his function. Also, he must know acting and actors. His knowledge of acting should include everything from the mechanics of articulation to the use of tempo-rhythm. His success depends, too, upon how well he understands and handles the personalities of individual actors. Further, the director employs graphic knowledge as he presents ideas and requirements to the designers and aids them in perfecting their conceptions. He applies principles of visual design to his own work of establishing balance and contrast with line, mass, and color in all the staging. And, finally, he must be an expert at manipulating people. Above all, his chief quality must be volition. He is one who thrusts himself into action and who elicits productive action from others. The director, not the playwright, is the leader of the ensemble of theatre artists.

Normally, the playwright's most crucial theatre relationship is with the director. The director is the one person with whom the playwright can freely talk about the production and the one person who can effect the playwright's suggestions. The director usually wants three kinds of things from the playwright: information, work and cooperation. The desired information ordinarily consists of details about the playwright's intent with the play. The discussions between the director and the writer most often focus on meaning and interpretation. The important work the director expects from the playwright has to do with textual changes to be made before and during the rehearsal period. Sometimes the director wants the playwright to enter discussions with the others, and always the director depends upon him to help stimulate the appropriate rehearsal mood and group morale. The cooperation of the playwright is important to the director in the following: The playwright should not usurp in the open, or in secret, any of the director's work. The playwright should avoid discussing characterizations or interpretations with any members of the company, except at the director's request. He should always uphold the word of the director, except in their private discussions. He must recognize the absolute authority of the director in all decisions about the production and staging of the play, and he should not be irritated, lose his temper, or hold a grudge when the director overrules him. His loyalty to the director ought to be total and firm.

What does a playwright want from the director? Most of all he wants the opportunity to help with the production interpretation. He wants to see the play come alive in a manner at least related to how he imagined it should. The director has an obligation to listen to the writer and to heed him, especially in the matters of meaning and interpretation. He expects the director's cooperation. The latter should be flexible in requiring script changes. There should be no absolute demands regarding the script. Just as the production decisions are solely the director's, the final script decisions must always be the playwright's. Nobody in the production unit should be permitted to cut or alter the script, but anyone may profitably make suggestions. The key to the relationship between these two central theatre artists is learning. Ideally as they work together, the director learns about a play and about dramaturgy; the playwright learns about the intended production and about the means of production.

The ideal director is a playwright's chief co-artist and friend, but the directors with whom the playwright will probably work are not always ideal. The best of them will respect his play and strive to stage it faithfully to the writer's intent, all the while treating the writer himself with attention and courtesy. The best directors inoffensively encourage the playwright to make textual changes that sharpen the theatricality of the play without harming its other values. The worst directors are the opposite. They may be overbearing or uncertain, irritable or servile, puffed up with their experience or frightened for lack of it. If in a specific production a director respects his own ideas above the play's and if he treats the playwright as a servant or a gadfly, he simply proves himself incompetent. The two most frequently encountered flaws in directors are, first, the lack of real knowledge about drama regardless of their experience, and second, the tendency constantly to override the playwright. Good and bad directors exist in New York theatres as well as in resident companies and in universities. Some professional dramatists are amazed occasionally to find an expert director in a campus theatre, and some new playwrights are surprised by the respectful treatment they receive from renowned directors. But all too often playwrights and directors conflict. When that happens, let the director look to his super-confidence and let the playwright look to his egoism. The director's job is to accomodate the playwright, but the playwright should make this easy and pleasant. When these two cooperate fully, dramatic art can reach its greatest creative heights.

The actors, who most directly give living presence to the play, face more difficulties than any other members of the theatre ensemble. As both instrument and instrumentalist, each actor must use himself and transform himself. Although the actors labor through a long period of rehearsal to perfect their enactment of the characters, they cannot see or hear themselves. Of all theatre artists, they need the most freedom, and yet they are the most dependent. Their function is to make the characters live by means of voice and movement.

An actor works with emotion. True, he uses intellect to understand, interpret, and plan; he uses body and voice to reveal thoughts and emotions. But, whether he feels emotion or not as he acts, each actor handles a unique complex of emotions. There are at least five: (1) the basic emotion of the playwright as he composed the play and as expressed in the specific emotive powers in the overall drama, (2) the changing emotions of one character in the play, (3) the emotions called for by the director as a result of the latter's interpretation of the play, (4) the emotion of the actor himself as he portrays his role, and (5) the emotions of the spectators as they watch the performance. No other art or human activity presents such a tangled network of emotional states. The actor must somehow unify these to communicate the play. It is no wonder, then, that actors are mercurial people. They must be emotional—overtly, covertly, or both—to be good in their art. A lot of people, some of whom are ignorant and others temporarily so, have written about how the actor is some sort of exhibitionist. Most true actors are not, in any way. Actors are simply artists who must use themselves as moving, thinking, speaking, and feeling instruments.

The actor's skills are intellectual, physical, vocal, emotive, social, and imaginative. Intellectually, he analyzes the play, interprets, and memorizes lines. Physically, he executes the blocking and invents the business in a life-like style appropriate to the play. Vocally, he strives to make every sound correctly and clearly, and without shouting he drives each utterance into the ears of the audience. Emotionally, he concentrates, remembers, stirs himself, and either imitates or generates the emotion required. Socially, he contacts others in the company, on the stage, and in the audience. If he is not highly sensitive and responsive and if he does not make the requisite contacts quickly and well, he cannot successfully act. His imaginative powers of memory and creativity give credibility and immediacy to the playwright's script.

The relationship between a playwright and actors performing his characters is usually happy. He may not have a lot of individual contact with every actor in the company, but he will undoubtedly become friends with some. Actors want to be liked. If the playwright responds warmly, he will find them friendly and cooperative. A playwright normally has a more informal relationship with the actors than with the director. The playwright must make known, if the director does not, that he will give critiques of the actors' work *only* to the director. But actors want the playwright to discuss the play in general with them. They want him to reveal his feelings toward their interpretation and progress. The writer can best do this with a pleasant manner rather than by inane compliments. Also, actors despise being told how to do their job. The playwright, on the other hand, can reasonably expect the actors to take his play seriously and work on it with discipline and care. In most theatre companies, a good deal of talk occurs regarding how each actor works. The playwright should be forewarned that no actor is totally technical nor purely emotional in approach, no matter what they say. All actors who are more than minimally competent must employ technique in order to communicate anything and must in some way, hidden or apparent, utilize emotion. In fact, if a playwright wants to know an actor, he should pay more attention to that actor's work in rehearsal rather than his claims made over a cup of coffee.

Every playwright will encounter, of course, both excellent and poor actors, normal and abnormal ones. He will come to know the phonies, the incompetent, and the twisted, and some of these may even have experience and skill. But he will also meet among them some of the most fascinating, dedicated, and well balanced people of his entire life. Actors are universally sensitive and sociable.

The designers with whom the playwright may be associated are those responsible for conceiving the play's scenery, stage properties, lighting, and costumes. Although some small theatres may have only one designer, the majority have at least two for these functions. The playwright will have most to do with the designers before the rehearsals begin, although he may be involved in a few design conferences during the rehearsal period. If he joins a company after the planning conferences, he may meet the designers only informally.

The designers' functions are to invent and to supervise the execution of all the technical and physical items for the production. Although each designer analyzes the play for himself, all are directly responsible to the director. His conception controls theirs; his interpretation of the play is as important to them as the play itself. The playwright must, when he wishes to reach any designer, speak through the director. Although the playwright should briefly indicate place in the script, he should leave the stage setting to a designer. Designers are visual experts. Playwrights usually are not. Designers can better work from an imaginative conception of the playwright's than from a detailed description of scenery, lights, costumes, and the rest. Each playwright must learn to depend on the experts of the theatre and not try to usurp their job, especially since they are likely to be so much better at it.

Another significant member of the theatre ensemble is the stage manager. The playwright will get to know him nearly as well as the director. As the director's chief assistant, he commands the production in dress rehearsals and performances. He organizes all the artistic personnel and sees that they perform satisfactorily. Throughout the rehearsal period he attends most planning conferences, and he runs the rehearsals to the director's best advantage. During the performance period he controls everyone's activity backstage. The playwright will do well to understand this man's function and to cultivate his friendship.

A few playwrights enter into an artistic partnership with another writer. The resulting collaboration is a further possible creative relationship. Such collaborations are rare as permanent connections, but the professional writer soon discovers that temporary collaborations are common. This is especially true in the case of Broadway musicals, television dramas, and cinematic writing. In the two photographic media, writing teams are frequently employed, and writing by committee is not unusual. Whether or not the playwright is able to write in any sort of collaborative situation depends upon the clarity and thoroughness of the agreements made between all the involved parties *before* the work of composition begins. The individual playwright must be at ease with his co-author and like working socially. Collaborations are so variable and so many differing working agreements are employed, that much generalization is impossible. Most writers dislike collaborations, and all who enter into such partnerships discover that most of their real work is still done alone.

Some playwrights working closely with a contemporary theatre company collaborate with actors and directors to create a play. The writing occurs as a result of improvisational sessions. This can be, especially for the socially oriented, a stimulating way to work. Most of the resultant dramas, however, have proven to be more theatrical than dramatic. These plays are as strong in visual imagination and dialogue energy as they are weak in form, story, and characterization. But, as Chapter 8 explains, ensemble collaboration is an important lesson that all playwrights should have the opportunity to learn.

The technical director heads the remaining segment of a theatre's artistic staff. Under the general supervision of the director and the exacting supervision of the designers, the technical director sees that each working crew of theatre craftsmen properly produces the items needed for the physical mounting of the play. He controls the following crews: scenery, stage properties, lighting, and costumes. Most companies have a crew head for each of these and a group of artisans. The technical director himself often heads one or more of these crews, and sometimes he also acts as stage manager. In large companies, there are two kinds of crews for each area of production—a construction crew and a running crew. The construction crews build the scenery, costumes, or whatever; the running crews handle these items during each performance. In small companies, one crew usually performs both functions. The technical director handles the construction crews, and the stage manager directs the running crews. In sizable professional companies, the playwright would have little reason to participate in crew work. But in small companies, the writer will find many advantages in occasionally helping the technicians.

When a playwright works with a composer it is unusual. Two kinds of products involve composers in the theatre: incidental music for some nonmusical dramas and the scores for musicals. The composer who creates incidental music normally talks with the director alone before rehearsals begin. The playwright may or may not attend such conferences. This sort of music, though it contributes to the production, must always be secondary to the play itself and not draw undue attention. Competent composers understand this and expect to work with the proper perspective.

If the play is a musical, the playwright and composer are more closely associated. They function as co-artists, collaborating to create the script and score *before* the play is even accepted for production. This book deals primarily with plays and does not purport to treat musicals extensively. Musicals, however, employ many of the principles described in Part II. The relationship between the playwright and composer should be smooth and highly productive. Thus, each such team must establish its unique manner of cooperation. The two artists should always establish the ground rules for collaboration before any work begins. They need to decide whether the script or the score is to predominate. Usually, it is the script.

Sometime every playwright will think of taking on one or more of the other theatre functions himself. He is often motivated to do so either because he knows and loves the other activity or because of an unfortunate experience with an incompetent theatre artist. Most frequently, playwrights have the desire to direct. A few like to act, but seldom do any participate in a theatre ensemble as a designer.

Is it wise for the playwright to direct his own play? Maybe! Both Sophocles and Ibsen, for example, applied their artistic genius to directing their own works. In contemporary theatre, a fair number of playwrights eventually begin staging their own works, for instance dramatists such as Arthur Miller and Albert Camus. No playwright—regardless of his intelligence, writing talent, experience in the theatre, or age—can direct well unless he knows the art of directing. It is not at all the same as the art of playwriting. The only playwrights who stage artistically successful productions are those who have studied directing, gained experience through practice, and possess the requisite skills and talents. The director's craft is at least as complex and demanding as that of the playwright. Also, the playwright who directs his own play must, when he does so, end his existence as the playwright and begin a new life as the director. This requires a different attitude toward the play. In such a circumstance, the person truly assumes dual roles, and highly contrasting ones. When one gifted artist writes a play and another directs it, the production is usually enhanced by the creativity of the second artist. A separate director extends the imaginative circumference of the drama. When a playwright directs his own play, he certainly understands it thoroughly, but he may not be able to fulfill it visually, handle the people concerned, and integrate the details properly. That is the danger. Only if the playwright excels in both activities should he direct. The same is true of acting and designing. The young playwright should never direct his own play. He needs the training, discipline, and extension that even a moderately skilled director will provide.

A playwright's relationships with theatre artists can be personally warm and creatively dynamic. If such conditions exist, the play will benefit, and the quality of its performance will be as fine as that particular ensemble can accomplish. To permit optimum working relationships, the playwright should know the specific functions of each of the other artists. He ought to be able to recognize their skills and their inadequacies. He must know what ideally to expect of every company and also understand what he may find in any company. He should strive for smooth cooperation. More important than himself or anyone else, the play should reign supreme.

CONNECTIONS WITH THEATRE MANAGERS

A playwright should also expect to form important associations with people outside the creative ensemble. These connections will include working relationships with non-artistic members of the theatre staff, business relationships with certain persons outside the theatre company, and social relationships with a variety of people. Although these relationships are not creatively crucial, most influence the playwright's work on the production of a specific play, or they affect his career.

The most significant person on any theatre staff is the producer. In professional theatre, he holds the ultimate financial power and thus controls all factors regarding the artistic production and the related business organization. The director and the business manager are responsible to him for these two areas. Often he will participate actively in both spheres. The playwright will have a formal contract with him, one containing many and varying stipulations. The producer is the organizational leader and has the final say on any question. He can veto anything about the production. He cannot force script changes, but he can exert pressure for them. He often settles the major disputes which may arise within the company or with outsiders, such as unions. Some producers also direct or serve other functions in the company. In many professional and semi-professional theatres, the artistic manager fulfills a role similar to the producer's. In educational theatres a department chairman or head of theatre functions as producer. The playwright should recognize both the broad responsibilities and the rather absolute powers of the producer. He should keep in mind that producers and directors more than anyone else can help the playwright with important theatre contacts. They and their friends are most often those willing and able to produce future dramas by the playwright.

The business manager is the central figure of another segment of the theatre staff. He is usually in charge of all the non-artistic personnel. The playwright's associations with the business manager and his staff probably will have most to do with publicity and promotion. The playwright should get to know all these people and offer to aid them in any way he can. He will undoubtedly be useful by providing photographs and a biography of himself. He can be available for press interviews, and often he can help with the actual composition of news releases. Especially with small production companies, the playwright can help immensely in these areas.

The management staff of a theatre company, then, under the supervision of the producer and the business manager make the production possible. The respective staffs of the business office, the box office, and the house, plus the public relations people, the various consultants, and even the custodians—all contribute to the production of the play, the coming into being of the art work. To comprehend how many people cooperate in the artistic and managerial activities necessary for the production of a play is to understand the social nature of dramatic art.

BUSINESS AND SOCIAL CONTACTS

Although Chapter 11 is devoted to business procedures and practices, the personal relationships between a playwright and other people important to him outside specific theatre organizations are worth mentioning here. These people, at varying distances from a company, frequently exercise influence on the playwright, aid him in his work, or affect the reception of his play. The playwright's business associates usually belong to one or more of four groups: (1) agents and professional consultants, (2) critics, (3) advisors and teachers, and (4) the people of educational institutions.

The factors to consider about when and how to get an agent appear in the marketing chapter. When a playwright arranges a contract with an agent, he should understand that this is a business relationship. The concern of both parties primarily has to do with selling a product. The playwright should deliver the best possible product—the play—and the agent takes over some of the work of selling it. Agents are specialists in marketing literary works, and although they are usually well-read and often capable of astute criticism, they should not be expected to be literary advisors or play doctors. More often than not, correspendence or conferences with an agent about changes waste the agent's time as well as the playwright's. Whatever time the agent spends corresponding with the author cannot be used to sell the author's products. If the association between writer and agent lasts for years and if they become friends, then the relationship may take on further dimensions, but such developments have little to do with the business of selling plays. This relationship, then, should be cooperative and courteous, but more especially it should be businesslike, and it should be economical of the time of both people.

Additional business consultants with whom a playwright may become associated are a lawyer, an accountant, a tax expert, and a union contact. Obviously, his need for such advice as these people can provide will be small until his income grows large or until he has some difficulty. All contracts should, however, be reviewed by a legal advisor, and like everyone else a playwright must pay taxes annually. For the most part, these relationships will be even more businesslike than that with an agent. Many times an agent, or members of his staff, can provide the necessary aid or advice.

To the public and to many beginning writers, all critics loom over the artist like evil genies, all exactly alike. This is, of course, untrue. They differ greatly in ability, benevolence, and function. Although any particular individual may at different times serve multiple critical functions, each person who writes about dramatic art ordinarily exercises one particular critical skill and performs one critical task. Even critics themselves sometimes become confused about their talents.

Critics can be divided into four classes, according to their function. Each class admits variety and permits sliding from one to another. The most widely known type is *the reviewer*. His primary job is to write news stories or informative columns for daily newspapers. Most newspaper staffs include one or more reviewers. At best, they set down the facts about the time, place, and general nature of a given production. Most reviewers add personal impressions and some reflection of their likes and dislikes. Their acquaintance with any production is brief and their writing hurried. But, as everyone knows, reviewers influence audiences. Also, many magazine "critics," especially those serving weekly publications, are actually reviewers, even though they spend a bit more time on their pieces.

The critic is not a reporter, as is the reviewer, but rather an evaluator. Most often, criticism appears in monthly and quarterly publications, or in books. A critic can spend more time considering and writing about a play or a production. He may evaluate the drama—usually his chief concern—or the performance of it. But most critics comment more astutely about writing than about acting. Their most important tasks are to tell the interested reader the relative place of a given work in the art of its kind and time, to compare it to works of the past, and to identify the level of its aesthetic accomplishment. Often a critic is less concerned with a specific production of a drama than with the lasting qualities of that drama or with the theoretic principles of dramatic art generally. Although a critic must deal somewhat with journalistic information and with literary analysis, his primary attention goes to comparisons.

The scholar spends still more time; he works more slowly and carefully. Dramatic scholars involve themselves with historical or analytical investigations about a play or a playwright. The theatre historian collects and records facts. The dramatic analyst carefully studies dramas and explains their structural or textual natures. The writings of scholars regularly appears in professional journals, in books, and in orally presented papers.

The fourth type of "critic" is *the theorist.* He stands at the greatest distance of all from the individual play, and yet his influence may be heavy upon a particular playwright. Some theorists write retrospectively and inductively. Others disregard or attack both the past and present. They address themselves strictly to the future by writing deductively.

After a playwright reads enough criticism from each class to recognize in what manner a certain critic is thinking and writing, he can better comprehend the proper and probable relationship of each to him and his work. The artistic stance is seldom similar to the critical posture. But a playwright should realize that each of these people performs a certain task. Most of them proceed honestly, if not always complimentarily. The playwright can benefit most by leaving them to their work. If they start a personal correspondence, he should try to reply quickly and courteously. If he meets them in person, he should treat them as warmly as his honesty will allow. They can be helpful to him with publicity and sometimes with their constructive commentary.

But what about negative criticism? How should a playwright—or any other artist—react? The best idea is not to react at all. Reviews should not matter. Once an artist realizes that complimentary reviews mean as little to the organic nature of his work as uncomplimentary ones, then he can proceed to ignore them all. At best, reviews and evaluations publicize a work; even negative ones do so. Although in the Broadway theatre, some reviews are likely to be the difference between financial success and failure. In any case, each artist must devise a rationale for himself about all critiques of his work. He will certainly get them, and he must learn how to take them. Probably no counterattack, personal or public, by any playwright on any critic has done any good, except perhaps to make the playwright feel righteous. To create a work of art is to invite criticism of it. Only the foolish believe otherwise. And probably every playwright has, or will have, among his closest and most useful friends some critics from each of the four classes. Not all critics are enemies.

Most experienced writers ask one or more advisors to read their work. This is an especially common practice with manuscripts of magnitude. Nearly all playwrights invite some expert to read their dramas. The advisors are ordinarily friends, associates, or teachers. What the writer usually wants is an objective view of the overall effect of the work, its technical correctness, and its clarity. Such advisors ought to be selected with care. They should be qualified to provide the sort of reaction the writer desires, and they must be willing and able to devote the necessary time. Anyone who offers through advertisement, by letter, or in person to furnish a critique for a fee should be avoided. The writer must warily select only those who will render commentary of value.

For beginning playwrights, their writing teacher is likely to be the most valuable reader. At best, he provided help and direction over a long period, and at worst he probably understands and likes the writer. Despite some current negative opinions, a number of excellent playwriting instructors now teach on college campuses. For example, Nebraska, Iowa, Tufts, San Jose State, Arizona, Immaculate Heart, Trinity, Fresno State, Santa Barbara, Carnegie-Mellon, Brown, Texas, and Yale have now or have had playwriting instructors whose knowledge and encouragement are most valuable to the beginning playwright. Their major problem is not that they cannot help the young dramatist but that they are expected to do so in only one or two courses. Universities would undoubtedly turn out more and better playwrights if the playwriting programs were as extensive, lengthy, and professional as, say, medical training. It is also important to point out that student playwrights usually have not developed a total vision of life or had sufficient practice in setting down their visions. Thus, it would be surprising if there were many student writers ready to step immediately from their undergraduate years into international literary and theatrical fame as playwrights. Most universities with playwriting programs try to provide a foundation for the young playwright rather than attempting to make him a finished professional.

A playwright with any degree of talent will have little trouble finding helpful advisors if he will search thoroughly and choose wisely. He must always remember, however, that even the best advisors may be wrong and certainly have a different viewpoint. Every idea or criticism they provide should be weighed before inclusion. Further, the writer should be most cordial and grateful to any advisor, helpful or not. He may, if he continues to write, wish to call on that person in the future.

Today's theatre is daily increasing its connections with universities. Theatre companies as well as individual professionals are periodically or permanently moving onto campuses. Many of the best young theatre artists are choosing to move back and forth between the professional and educational worlds, and some find the educational milieu preferable. University drama departments are more frequently inviting playwrights to join them for both teaching and production. Large institutions, such as N.Y.U. and Yale, and small ones, such as Immaculate Heart College and the University of Evansville, often commission, employ, or enthrone playwrights. The relationships between the playwright and such institutions are complex and not always predictable. Some things are relatively certain. If the playwright knows his craft and honestly likes people, he will find the students interested and stimulating and the faculties eager and courteous. The playwright's reception depends mostly on his own attitudes. If he is dynamic and friendly, his relationships will probably be happy and beneficial; if he is surly and lazy, he ought to stay away. Serious writers usually find campus associations rejuvenating. The dramatist must remember that while on campus he has new responsibilities. He is there to serve the young, and if he does that unselfishly, his rewards will be as great as theirs.

Among the playwright's business associates one person stands apart from all others. This person is likely to be the most truly useful of all—the secretary. Although beginning writers ordinarily cannot afford a secretary, someone, friend or spouse, often helps with typing and other clerical chores. As the writer rises in his profession and as his output and involvements increase, a secretary becomes a necessity. He should find one who is diligent and dependable, not given to gossip or talkiness. This person's tractability and typing skill are the most important qualities. A good secretary should be an extension of the writer's own working self.

The writer, as he creates, is a lone worker. He must cherish and protect his loneness. Loneliness, however, can injure him. To be lone is to prefer solitude, but to be lonely is to be sad because of separation from others. Dejection inhibits creativity, and belief in self stimulates it. Every writer while learning his art goes through an inner struggle until he comprehends the importance of solitary work and realizes the universal isolation of the contemporary artist. The playwright must, during certain stages in the creation of a play, often operate creatively in a demanding social milieu. Nevertheless, he must defend his loneness.

Despite the importance of working in solitude, nearly every writer needs some significant social relationships. In fact, he will have them whether he wishes them or not. The point here is that the playwright should cultivate the useful relationships, whether functional ones or morale boosting ones, and he should avoid those that turn him from his work or threaten his loneness. All people he likes or dislikes, loves or hates, befriends or opposes should be scrutinized. People are not just social contacts to the writer. They furnish his material, affect his view of the world, make him happy or unhappy, and give him inspiration. Every writer needs a number of friends or acquaintences of each type. The women, men, and children in his life will appear in his art. His political commitments, economic attachments, and social contacts, at least indirectly, affect his work. Even his sexual life matters to his art. He should learn to listen and observe, seriously. But he can just as well ignore the comments of his non-professional friends or enemies when they remark about his work. Only when a large number establishes a community of opinion, or only when someone he loves or hates inspires him will his work rightly and directly be affected. Most writers, before they ever begin, know the secrets of contact between human beings.

THE DRAMATIC SCENE

Every playwright, in his own time, needs to know what is going on. He needs to know himself, his milieu, his society, his country, and his world. Since drama is an ensemble art, he needs also to know what is happening in the theatres of his time. In the United States, this is not easy. Distances are great, and important events occur in many places. Despite the champions of New York City who own most of the mass media loudspeakers, there is no longer a literary or dramatic center. There is no one place for the playwright to go, even though there are more theatres within a small area in New York than anywhere else in the country. And most playwrights learn sooner or later that they must find a private place anyway. Still, to comprehend the drama fully, a playwright needs to devise means to find out what is going on in the best theatres throughout the country. He needs to keep in touch, not so he can adjust his work to conform to the latest vogues but rather so he can avoid doing that. And yet he needs to learn what he can from the best new ideas, and he must at least recognize the relative possibilities of placing a script somewhere in the contemporary dramatic scene.

The intent here is not to survey the latest theatrical crazes but to provide some points of origination from which an individual writer can begin, or renew, his own investigation of the dramatic scene in the second half of this century. This section offers some hints, identifications, reading suggestions, and theories. The new theatre—incidentally the latest term—today will be the out-of-date theatre tomorrow. Thus, a theatre artist needs some leads about what to look for, what to avoid, and how to stay in touch, if he cares to. This section does not exhaustively survey all movements, experiments, nor artists of the current dramatic scene. For such an overview, a playwright should turn to books such as Gerald Weales' *The Jumping-Off Place: American Drama in the 1960's*, Robert Brustein's *The Third Theatre*, or *Stages* by Emory Lewis.[2]

The dramatic scene of today can be epitomized with the term *action*. It is an old term, but its meaning is being freshened, renewed, and extended. It indicates not only the most focal materials in the best recent plays, but also it reveals the movement of artists and audiences spatially and intellectually. It points to the ever more vivid dichotomy between the frozen Broadway business of shows and the lively theatre of campuses and streets, of cafes and clubrooms, of resident theatres and garages.

The new ideas about a drama of action lead the playwright and other theatre artists to re-examine the relationships of art and life. The separation between the two is less than ever and growing narrower each month. The new dramatist has discovered the potential of partnerships with the new directors, actors, designers, and critics. And the new students are the willing audience. The enemies are media commercialism, restrictive institutionalism, administrative Puritanism, playboy eroticism, middle-class Victorianism, and isms themselves. The new dramatic ensemble wants to tumble the taboos and psych out the psychic. They work for free and personal expression. Despite the denials of established journalists, most of the new dramatic artists are committed to painstaking craftsmanship. They are dedicated to explore possibilities, to try the unconventional, to test the individual. The action of the dramatic scene is rebellion and revolt.

But who, specifically, triggered the revolution, and who leads it? What are the key ideas, the fresh subjects, the emerging styles? And where can the isolated playwright find out about all these things?

Soon after the naturalist-realist revolt against romanticism in the late nineteenth century, a few playwrights realized the limitations of the scientific method in art. Realism, however, with its new frankness and its illusory representation of everyday life, predominated and became the standard in dramatic art until the 1950's. The other, less popular genre of presentational drama nevertheless continued unheralded but uninterrupted. Its strongest early practitioners were August Strindberg and Luigi Pirandello. Both were at first masters of realism, and both soon recognized its formal, material, and stylistic limitations in drama. Strindberg began to probe man's personal being, and his consuming imagination, self-flagellation, and psychic intensity permitted him to initiate the first significant dramas of modern presentationalism. Pirandello penetrated the illusory nature of both realism and everyday life, and he injected a new abstractionism into the

[2] Gerald Weales, *The Jumping-Off Place: American Drama in the 1960's* (London: The Macmillan Company, 1969); Robert Brustein, *The Third Theatre* (New York: Alfred A. Knopf, Inc., 1969); Emory Lewis, *Stages: The Fifty-Year Childhood of the American Theatre* (Englewood Cliffs, N. J.: Prentice-Hall, Inc., 1969).

drama and brought relativity of life to the stage. These two men, along with a few others, began the non-illusory genre of modern drama that today has finally pushed realism aside.

In the realm of spectacle—acting, directing, and design—several men stand as important initiators. Constantin Stanislavsky was easily the most influential acting theorist of the modern theatre. He perfected a system of acting that not only permitted actors to perform the realistic plays with verisimilitude, but also he established a tradition of investigation and experiment among actors and directors that is absolutely new in theatre history. But Vsevolod Meyerhold, one of Stanislavsky's associates, is equally important to the contemporary non-illusory theatre. He envisaged an abstractionist mode of production which emphasized ensemble, improvisation, and stylization. Gordon Craig's theoretical writings have influenced several generations of theatre artists, but they are just now coming to some sort of fruition. More interested in productions than in plays, he called for a theatre of visions that appealed directly to the senses. Adolphe Appia's ideas led to many of the dramatic ideals of this century. He stressed artistic unity in dramatic art, and he demonstrated the need for coordination of three-dimensional space, a moving actor, multi-directional lighting with a faithful interpretation of a script. To understand the contemporary dramatic scene, one can begin with the plays of Strindberg and Pirandello and then delve into the theories of Stanislavsky and Meyerhold, Craig and Appia.

Certain later theorists, however, are even more directly responsible for many current practices in contemporary drama, especially Bertolt Brecht, Antonin Artaud, and Jean-Paul Sartre. Although these three began writing before World War II, their influence widened after its close. Brecht's impact on today's dramatists and directors is already immeasurable and still growing. Part of the reason is that he functioned in the triple spheres of playwriting, dramatic theory, and stage direction. The spreading of his ideas occurred with the production of his plays throughout the world, the tours of his Berliner Ensemble, and the translations of his theoretical pieces. His notebook, "A Short Organum for the Theatre," with its discussion of Epic theatre, the alienation effect in acting, and the potential social functions of drama has stimulated others to seek a new immediacy for the theatre.[3] Although few playwrights in Europe and the United States share Brecht's Marxian philosophy, they recognize the strength of his formulative methods. Brecht introduced the twentieth-century theatre to the unlimited potentialities of a new didacticism. He gave drama social relevance and life pertinacity.

Antonin Artaud's theories were published under the title of *The Theatre and Its Double*. Artaud worked in Paris as actor, director, and designer,

[3] Bertolt Brecht, "A Short Organum for the Theatre," *Brecht on Theatre*, ed. and trans. John Willett (New York: Hill and Wang, 1964), pp. 179–205.

but his influence came not from what he performed but what he thought. He called for a theatre of participatory action, of transcendent experience in life, of spectacle.[4] In short, he envisioned a Theatre of Cruelty that penetrates beneath man's protective surface. He demanded action in the theatre that would compel each person to see their own baseness and savagery and that would destroy illusions, reveal powers, and induce a heroic attitude. Artaud conceived ideas about performances full of all the objective surprises the creators could imagine, and he initiated what is now called "Total Theatre." Most significantly, he wrote that drama should present man's "obverse side" and that it should lift the reality of dreams and of imagination to an equal importance with "common" reality of everyday life.

The third member of this trinity of theorists is Jean-Paul Sartre. Like Brecht, Sartre labored in several areas, and his influence arose from his plays, his popular writings, and his philosophic works. As a leader of the existential movement, as a political activist, and as an artist, he has impelled many serious dramatic artists to be concerned with the individual self, inner being, and ultimate fate. He championed the idea of freedom of being, a personal condition made possible through recognition of life's absurdity and then made active through self-engagement. Sartre's book *What Is Literature?* is more the manifesto for late twentieth-century writers than most literary savants, even writers themselves, realize.[5] In it, he explained the necessity for human and aesthetic commitment of a writer, and he suggested that a writer should actively engage himself in life issues and meditatively engage in reflection about his own being. He called for a new theatre of questions and of morality—not moralizing—in which man *chooses*, in which the artist invents himself with his work.

Alienation and commitment, cruelty and absurdity, action and choice, dream and object, self and being—these are more than the popular catchwords of magazines or cocktail parties. They are theoretic concepts that infuse energy into today's dramatic scene. Of course many other ideas and influences are apparent in today's drama. Oriental theories and practices, for example, are significant now, and their effects on today's artist reveal how the "pure" Western tradition is breaking apart.

It is unnecessary to discuss in this context the dramatic works of leading dramatists. A playwright ought naturally to be reading them. But it may be helpful to identify those dramatists whose work most pervades our theatre. And for the sake of the isolated playwright's awareness, it may be helpful to name some of the currently influential dramatists and some of the new generation of playwrights who have taken the experimental stages. In

[4] Antonin Artaud, *The Theatre and Its Double*, trans. Mary Caroline Richards (New York: Grove Press, 1958).

[5] Jean-Paul Sartre, *What Is Literature?*, trans. Bernard Frechtman (New York: Harper & Row, Publishers, 1965).

addition to Strindberg and Pirandello, other earlier dramatists of importance to the new theatre are: Alfred Jarry, Georg Büchner, Frank Wedekind, Leonid Andreyev, Georg Kaiser, Michel de Ghelderode, Sean O'Casey, Federico García Lorca. By far the most influential dramatists of the mid-century were: Brecht, Sartre, Samuel Beckett, Eugène Ionesco, Jean Genêt, Albert Camus, Tennessee Williams, and Arthur Miller. Playwrights now writing whose impact is less but still increasing are: Jean Anouilh, Friedrich Durrenmatt, Max Frisch, Peter Weiss, John Osborne, Harold Pinter, John Arden, Slawomir Mrozek, Edward Albee, Arthur Adamov, Fernando Arrabal, and Günter Grass. Among the most active new American playwrights are: Jean-Claude van Itallie, LeRoi Jones, Megan Terry, Kenneth Brown, William Hanley, Sam Shepard, Ed Bullins, John Guare, Rochelle Owens, Charles Dizenzo, Harvey Perr, Maria Irene Fornes, Frank Moffett Mosier, Lanford Wilson, Ron Cowen, Howard Sackler, Paul Foster, James Lineberger, Rosalyn Drexler, James Schevill, Frank Gagliano, Barbara Garson, Charles L. Mee, Joel Oppenheimer, Terrence McNally, Lee Kalcheim, Leonard Melfi, and David McFadzean. And of course there are more. These new dramatists are innovating with life, not representing it, and a playwright can find stimuli in their works for his own plays. Each writer can learn something from others, if he bothers to read at all—provided he does not read too much.

Writers of other sorts have affected contemporary drama too, especially novelists Franz Kafka, James Joyce, William Faulkner, and Ernest Hemingway. And certainly these philosophers should be included: Friedrich Nietzsche, Sören Kierkegaard, Martin Heidegger, Karl Marx, and Benedetto Croce.

Certain critics have spread the word of the new theatre and have helped form audience opinions about theorists and playwrights. Some of the most well-known critics are: Eric Bentley, Kenneth Tynan, Robert Brustein, Jan Kott, and Martin Esslin. Some of the incisive new voices are: Richard Schechner, Susan Sontag, John Lahr, Simon Trussler, Theodore Hoffman, Ronald Peacock, Robert Pasolli, Michael Smith, Leonard Pronko, and Bernard Dukore. Among the most promising critical movements, structuralism—initiated by such Frenchmen as Claude Lévi-Strauss, Lucien Goldmann, Roland Barthes, and Jacques Lacan—is the most productive of ideas for the creative artist.[6]

A number of directors, too, have contributed significant ideas to the dramatic scene. Some of the directors who especially fostered presentational theatre from the early part of the century to its middle and whose influence is still extant are: Vsevolod Meyerhold, Eugene Vakhtangov, Alexander Tairov, Nikolai Okhlopkov, Leopold Jessner, Charles Dullin, Jean-Louis

[6] For a summary review of structuralism, see: Peter Caws, "What Is Structuralism?", *Partisan Review*, Winter (1968), 75–91.

Barrault, Bertolt Brecht, and Erwin Piscator. Brecht's directorial creativity with his own works and Barrault's continual experimentation and his staging of Andre Gide's dramatization of Kafka's *The Trial* were crucial stimuli to the theatre practices of our time. During this century's middle years, the forties through the sixties, Tyrone Guthrie and Elia Kazan set production standards in the United States. More important than his hyper-realism was Kazan's insistence on the unity and vitality of performance. Guthrie gave a transfusion of craftsmanship and renewed interest in classic plays to an American theatre that had almost totally humbled itself before its own versions of psychorealism and The Method. These men, then, helped set the stage for the new drama.

The next generation of directors, however, was even more innovative, and these are the ones more directly responsible for the best strains of contemporary production. Among the leaders are Roger Blin and Jean Vilar in France; Peter Brook, Peter Hall, and John Dexter in England; and in the United States Alan Schneider, Gene Frankel, Paul Baker, and William Ball. A group of newer American directors are even more interested in the work of non-established playwrights—Tom O'Horgan, Joseph Chaikin, Julian Beck and Judith Malina, Joseph Papp, Gerald Freeman, John Hancock, Andre Gregory, Sydney Schubert Walter, Tunc Yalman, Davey Marlin-Jones, many others.

One of the most vital recent forces in theatrical experiment is Polish director Jerzy Grotowski. His work with the Polish Theatre Laboratory in Wroclaw has furnished many fresh ideas for others, and he has put into actual practice a number of older, theoretical ideas of certain precursors of modern theatre, such as Appia. He has experimented with varying theatre environments, with actor development through exercises relating the physical to the psychic, and with how to find the genuine realm of contact between a play text and the actor. Grotowski represents the new artist-as-director. By his very example, he helps others make discoveries about the nature of theatrical creativity.

Traditionally, the American theatre scene has not furnished many publications worth reading, but perhaps this condition is changing. Today the isolated playwright—whether he lives in New York or Iowa City—can read about the theatre activity in this country and the world. Among the most worthwhile periodicals in the United States are: *The Drama Review, Players Magazine*, and the *Educational Theatre Journal*. The editorial mission of each is unique, ranging from the experimental to the scholarly, and together they can help a playwright maintain a balanced intellectual view of the explorations of his peers. Other publications of value that may better suit a given individual are: *yale/theatre, Drama & Theatre, Modern International Drama, Theatre Today*, and *Theatre Crafts*, Publications not strictly devoted to dramatic art, such as *Evergreen Review, Arts in Society, Partisan Review, Southwest Review*, and *Quarterly Journal of Speech*, also run useful material. And of course many popular magazines and newspapers—especially *Saturday Review, The Reporter*, and the New York *Times*—carry news and reviews. A playwright would do well also to note the drama publications of Grove Press, Hill and Wang, Bobbs-Merrill, and the University of Minnesota Press; these companies are leading the way in setting out new plays. While reading about the dramatic scene, every playwright should remember that the journalists, critics, essayists, and scholars, perhaps necessarily, use captions and categorize artists and their work. Reported styles and pat categories often do not exist as the generalities some of them claim. The two dangers of reading publications such as those mentioned are that the writer will try to follow someone else's prescriptions or that he will get "the critic fear."

The artist, however pessimistic his public pronouncements, is by nature optimistic. His artistic endeavor alone is enough to demonstrate his attitude. Thus, the attitudes in this chapter are mostly positive ones. Nevertheless, all is not happiness and harmony in the theatre. There is still plenty of hokum and hucksterism. Not enough producers or companies are willing to produce new plays by relatively unknown playwrights. Art of high quality is all too rare; there is never enough of it. Segments of American theatre are stagnant, silly, commercial, or pedestrian. Too many members of the national audience possess locked up tastes that resist change. What little subsidy there is, outside the campus, too often goes to the conformist theatres or to the playwright with connections. Too many university theatres are run by men with timorous dispositions and unfocused aesthetic vision. Too much theatre criticism attacks the artist instead of championing him, and too frequently theatre research is unsystematic or spurious.

Good theatre requires more than good intentions. Most novices believe that to enter the theatre requires only desire, that improvement is automatic for those who try hard, and that the crucial element of quality drama is pleasant rapport among the dedicated. Our theatre is predominantly a middle-class entertainment, in middle-class institutions, with middle-class virtues. It lacks social, as well as artistic, courage. Success in the theatre is too often measured only by renown and money. The people who operate, occupy, and attend our theatres forget too easily that art requires talent and intelligence, discipline and craftsmanship, experiment and exploration, audacity and rebelliousness, and perhaps even selfishness and aggressiveness. Most significantly, the theatre needs playwrights with talent and awareness. And both those qualities arise from the playwright's self—what he sees, what he thinks, what he knows. He needs to reflect about what kind of objects he wants to make, and he should set criteria for choosing how to make them. Above all else, the theatre needs playwrights who can write and who can collaborate with other theatre artists to create dramatic art of high order.

Despite the theatre's ailments, the dramatic scene of today is filled with vigorous ideas, fresh subject matter, strange styles, and active forms. A potpourri of these would include: individual dissent, an anti-war position, the blasting of commercial theatre formulas, a revolt against the rational, a recognition of life's absurdity, and assault of the spectator. Today's theatre artists are simultaneously amused and enraged by the collective society they find enclosing them. They are interested in protests, events, and social objectives.

The best of the artists in drama serve ideals of aesthetic order, craftsmanship, and harmony, and they open their works to subliminal, metaphoric, and kinetic imagery. The worst of them are callow, sloppy, and arrogant; some merely disruptive; and many simply conventional in their pretense of being avant-garde. Not all the new dramas have worth. The great enemies are not only the forces of mass society but also anti-intellectualism, sensationalism, violence, and verbal illiteracy. Dramatic art at its best invites each playwright to be free, personal, sensitive, craftsmanlike, imaginative, and intelligent.

Our society wants ideas, and our theatre wants art—as never before. But the day has arrived in our world, our country, and our theatre when the half-man has no place. Those who cannot work in theatre amicably, creatively, and humanly should be rejected and ignored. The irritable, the dictatorial, the insensitive, and the ignorant are enemies. Whenever they appear in the theatre, they destroy it.

In American theatre, one motivating factor causes the greatest upset; it is chiefly responsible for ruining companies, productions, and artists. It is *egotism*. The American economic and social system has contributed greatly to the formation of thousands of egotistic personalities. The theatre has more than its share. Because every artist—actor, playwright, director—must fight so aggressively to make a living with his art, because each must learn to push and praise himself, because the cultural climate permits so few companies and so little regular employment, the dramatic artist often comes to believe that he is an isolated individual surrounded only by those who wish to crush him. The system maintained by the current establishment offers the criteria of money, fame, and quantity of productivity for measuring success. Hence, too many people in the theatre—indeed in all the arts—use most of their energy attempting to promote themselves rather than striving to create works of quality. But artists themselves must assume some of the blame. Too many have learned the techniques of their craft without discovering that self is less important than created objects. Regardless of their incomes, how often they work, or their reputation, too many artists in the theatre fail to make drama worth anyone's attention. Too many directors fail because their goal is to establish tyranny. Too many actors fail, indeed often wreck the few companies in existence, because they are more concerned with the size of their role or the quantity of their publicity than with the artistic adventures of the ensemble and the creation of the drama. Upon joining a company, too many playwrights, after laboring alone for months, jabber only of themselves. They maneuver for the predominance of their playscript rather than endeavoring to perfect a living drama. It is often said that American theatre is its own chief enemy, because there are so many bad productions. More are ruined by egotism than by ineptitude. Every struggle for supremacy within the theatre ensemble simply reveals the ignorance or selfishness of those involved. Although theatre artists must work with their emotions as tools and are thus likely to be emotional, anger with fellow artists is the one inadmissable emotion. Egotism is the one inadmissable quality.

In his relationships, a playwright should take care not to resent the theatricalism of the director and actors, to ignore the business acumen of the theatre managers and his other business associates, or to underestimate the significance of social contacts. All can directly or indirectly affect his plays. The essential components for dramatic art are script, actors, and audience. A playwright is necessary only as an innovator who provides one component. Thus, he owes it to himself and to the companies he may join to submerge himself in theatre. He should learn what he can about it, not in order to usurp the jobs of others but in order to cooperate with them more fully. Increased knowledge of functions and relationships will ultimately help him write better plays.

10: *LIVING AUDIENCES*

"Theatre" consists in this: in making live representations of reported or invented happenings between human beings, and doing so with a view to entertainment.

BERTOLT BRECHT
"A Short Organum for the Theatre"

The word *audience* has a broad meaning, a large field of associations, for a playwright. As a member of the aesthetic trinity—artist, art object, and audience—it demands attention in a book such as this. To write about creativity without considering audience would be to discuss therapy instead of art. A discussion of audience in no way suggests that the artist—as playwright—should write only for effect, but art is by its very nature effect producing. Thus, the artist should learn what he can about audience.

AUDIENCE IDENTITY

Every artist, knowingly or not, selects an audience for each object he makes. The act of creativity contains this choice, however unconscious, as one among the many comprising the efficient cause of each art work. A playwright, too, selects an audience whenever he writes. He makes his choice as much by exclusion as by inclusion, as much by his knowledge of trends as by his appeal to a certain group, as much by his own social milieu as by any wider one, and as much by market as by simple dream. His exclusionary choices are the most simple to discern. Jean-Paul Sartre obviously does not write for an audience of five-year-olds; Harold Pinter does not write for the series television audience; and Peter Weiss does not write for an illiterate audience. Even through the identification of such negatives, a writer delimits—i.e., chooses—his audience.

A playwright who wishes to create as totally as he can, however, cannot write by permitting considerations about audiences to dominate his mind. He, like the actor, must realize an audience, a relatively specific one, is there, but he cannot perform merely for their sake. He creates for himself, because he *must*. As Hemingway and other writers have pointed out, every good writer must write as best he can, and the successful writer is the one who is fortunate enough to find a large audience. No playwright can create well by trying to write plays merely for the sake of pleasing an audience. But neither can a playwright afford totally to shun knowledge of or thoughts about his potential, specific audience.

To study audience characteristics is not to take an audience-centered approach to drama. Learning about the reactions of audiences does not threaten the writer's creative fertility. The considerations which follow amount to an introduction to a study of audiences in the theatre. The discussion is not meant to be exhaustive, but rather to provide some essentials on which each playwright can build his own structure of information.

The first general topic about audience is identity. It has several ramifications. Much confusion arises in discussions about audiences when no recognition is made of the difference between crowds and individuals. Every artist needs to preserve the distinction between the kinds of individuals who will see his play and the responses groups of individuals will have in a theatre. As a member of an audience, the individual is not passive; he sees, hears, and responds. Reacting to drama, or any other art, is never completely relaxed and effortless. It is behavior.

Much of the so-called control that some dramas have over an audience is not so much the manipulatory power of the artists as it is the relationship the art object has to the continuing behavioral patterns and tendencies of the individuals in the audience. To put it another way, drama is one of the arts that lives only in performance before an audience; as an art object performed live, an audience can influence a drama-in-performance—thus, the art object—as much as the drama influences the audience. An audience, through the behavior of its members, can react positively or negatively. Spectators can boo, stomp, riot, and stop a drama; they could suspend it by all walking out. Or with participatory attention, they can spur the performers, on a given night, to make the drama more intense or beautiful than ever before. Most audiences, of course, fall between these extremes of behavior, but most audiences consist of active not passive, behaving not vegetating individuals. To some degree, then, a person sitting in a theatre seat helps determine the nature and the outcome of the dramatic experience. He continues the sequential behavior of his life. And however much he is affected by the drama, he retains a personal identity.

The awareness of identity within the single individuals in a theatre audience naturally varies greatly. So infinite are the possible combinations that no one completely understands the behavior of individuals in an audience. It is certain, however, that their previous experience, what they have learned, affects their behavior. According to communications experts, an individual does not substantially change when he joins an audience. His behavior as a spectator still arises out of his personal complex of traits —his physical, dispositional, motivational, and deliberative facets of personality. Seeing and listening is individual behavior, and as such it is controlled by the same factors that govern other kinds of behavior. An audience member perceives the stimuli of a play and interprets them according to his prior experience. Regardless of what theories one may believe about group psychology, an audience is primarily a collection of behaving individuals with unique identities.

Even the act of identification is not simply passive. An individual must himself have an identity before he can identify with another. The relative degree of self-projection into, or self-removal from, a performance depends on many factors, but neither is possible without self-awareness. A mentally disturbed person, for example, who has no sense of his own identity cannot sympathize with others so fully as a mentally healthy person. To understand empathy, aesthetic distance, and alienation is to realize that all three depend first upon identity within individuals. Most theatre essayists place too much emphasis upon crowd behavior, and most theatre artists likewise err in their narrow view of an audience as merely an emotional collective. Understanding theatre audiences requires a study of human identity and of responsive behavior. The more an artist knows about the specific identities of his immediate audience, the more he will realize the probable nature of their responses, as individuals or as a group.

The word *audience,* as most frequently used among writers and theatre companies, tends to be one of those nonspecific terms that is nearly meaningless. The greatest error occurs when it comes to distinguishing between audiences. Whatever the collective situation, the same principles are operative. Each person in every audience will behave in a certain manner because of his previous experience, the stimuli of the present, and the circumstantial context as he perceives it. As these three variables—identity, stimuli, and situation—change so will the response of the individuals in any audience alter. Audiences, mobs, crowds, groups—all vaguely describe differing collections of *individuals.*

Undeniably, different types of audiences behave in different ways. Indeed, the various audiences witnessing a given drama, although made up of presumably the same general sort of individuals, differ widely. To understand audience variations, the following *audience characteristics* should be noted; a playwright can at least be aware of their existence. The first characteristic is collectivity. When more than one spectator is involved, the others in an audience alter, by their presence and reactions, the context of the individual's response. Second, the size of the audience affects the situation. Third, homogeneity is important because the collective behavior of groups composed of individuals with many common traits is more predictable than groups of persons with little in common. Fourth, conformity is a characteristic referring to the uniformity, orderliness, or regularity of a group in its responses. Fifth, preliminary attitude, or set, has to do with the general state of receptivity in an audience before a performance begins. Sixth, collective response occurs in any audience as its individuals have emotional reactions to the stimuli coming from those around them. The seventh major characteristic of an audience is focus of attention; an art object ought to control the attention of the whole audience. The success of a drama in performance depends to some degree upon its powers to direct audience attention.

Empathy, a translation of the German word *Einfühlung,* literally means "feeling into." Several German psychologists—Theodor Lipps, Karl Groos, and others—brought the term to prominence. They used it to describe certain physiological and psychological responses in a viewer who contacts a work of art. Today, empathy is widely and loosely used to mean audience identification, loss of self, and the dominance of emotion over intellect. The most cogent discussions of empathy, however, relate to the writings of Vernon Lee and stress that it is an uniquely human phenomenon existing in every phase of an individual's mental life. Empathy is a man's capacity to participate in the *feelings and thoughts* of another. By means of a physio-psychological complex of stimuli and associations, a human being merges activities of perception and understanding with the qualities of a foreign object, another person, or even a concept. Empathy depends upon physical sensation as its initiator. But if empathy begins with sensation it ends with preference, as an action of aesthetic contemplation. Empathy is a significant psychological process which requires imagination, sympathy, and judgment. The quality of the empathic response of any individual spectator watching a drama depends upon what he infers about the perceived object, by comparing it with his store of inner experience. Empathy is one important constituent of human *being.* It is man's process of relating his subjective self to the objective void, to the world of things, or to his social milieu.

Undoubtedly, discussions of empathy tend to be opaque and rather personal. And rightly so! The human self is difficult to understand and is, above all, personal. The point to be stressed here is that the idea of empathy is useful for probing the nature of man's response to art. Each artist—every playwright included—should avoid a common, non-definition of the word. He should define it for himself as he formulates his own mental and emotional vision of the connections between his art works and his potential audience. And when reading the statements of others, he should carefully identify their particular interpretation of empathy.

As empathy is an universal human process, it relates to several other terms that describe some aspect of aesthetic response. Empathy encompasses such concepts as aesthetic distance, identification, alienation, sublimation, closure, and catharsis—when they are applied to audience reactions.

Aesthetic distance is a contemporary version of the concept of psychical distance, as developed by British psychologist Edward Bullough. Aesthetic, or psychic, distance refers to the degree of personal involvement of a spectator in a work of art. Assuming that the process of empathy is initiated during a performance of a drama, this question then arises: How much personal identification of each audience member with the drama is desirable? Aesthetic distance does not refer to time or space; it is a matter of attitude and of identity in the viewer. The two extremes of aesthetic distance are obviously undesirable. Complete loss of distance occurs when a spectator loses his own identity through total identification with the art work; his projected feelings destroy his self-awareness and his powers of judgment. By contrast, the other extreme is excessive distance. This circumstance in a viewer means that the art work is so foreign that no involvement occurs. He is psychically so removed from the work that he has no feelings or thoughts about it; he is not involved at all. Most artists hope that their audiences will contact their works to a degree between the two extremes.

Most contemporary playwrights do not wish to destroy the self of each spectator; rather, they wish to make a direct and effective contact with that self. Bullough and others among the early psychologists attempted to establish the premise that there should be a separation between practical and aesthetic concerns, between life and art. Artists today—such as Brecht, Ionesco, and Sartre—tend to deny such a separation. They do not wish disinterested viewers, nor do they want viewers who completely lose themselves. They see art as life intensified. Most artists now attempt to establish a degree of empathic response and a controlled aesthetic distance, both of which are meant to stimulate feelings for the sake of thoughts. Aesthetic perception more and more is coming to mean perception of being, of self, and of life. Art is not meant to encourage a suspension of identity, but to stimulate heightened self-awareness. Art today is not for the sake of art, of escape, or of political programs; it is for the sake of *identity*.

Identity is not only the oneness of an individual human, but also it is an individual's awareness of being. A human's identity, then, is both the unified totality of his being and the degree of autonomy by which he controls his existence. A man is, after all, utterly alone. A sense of loneness —not loneliness—is essential to every person's realization of identity. To realize one's identity is to be aware of one's self.

Alienation effect is a concept popularized among theatre artists by Bertolt Brecht. His German word *Verfremdungseffekt* has no precise counterpart in English. Even the word *alienation* possesses an emotional implication which Brecht did not intend. But regardless of how Brecht's term is translated, his concept is valid for the playwright to consider. Brecht objected to excessive identification on the part of the audience. He wanted his dramas to discourage the spectator from losing personal identity; he preferred that each viewer retain a separate existence, a freedom of critical judgment. Reacting against the popular concept of empathy as vicarious emotionality, he tried to make drama that would impel audience members to think, to judge, to decide. He never denied that drama should entertain, nor that it should stimulate emotion. During a performance, he wanted the spectator to study the dramatized experience, and instead of escaping life become more aware of the socio-historical forces in life. He used the concept of alienation to attack the sentimentality of the drama of illusion, an attack that represents the creative attitude of many contemporary artists. A playwright can advantageously consider a balance of empathy and alienation in relation to his own work.

Psychologists are continually tempted to examine both the creative act and the contemplative response. One of the concepts about art developed by Sigmund Freud is that of *sublimation.* According to him, art represents one of man's attempts to remedy the difficulty, pain, disappointment, and impossibility of daily life. Art can be, for artist or spectator, a substitutive gratification, a diversion of interest, an intoxicating substance. Such are the illusions of art that contrast with or defend against reality. Sublimation involves the use of mental equipment for libido-displacement; a man creates an art work by transferring his instinctual aims into an object in such a way that they cannot be frustrated by outside influences. Through sublimation, the artist finds satisfaction in creation, and the spectator enjoys the pleasure of contemplation. The artist creates a version of his fantasy by elaborating it with aesthetic techniques. By means of aesthetic features, the work then incites identification and enjoyment in others. It becomes a sublimated social fantasy. In the Freudian view, an audience witnessing a drama participates imaginatively in an artist's fantasy. If a drama is aesthetically successful, it to some degree satisfies the unconscious infantile wishes of audience members. Thus, social participation in drama, or any art, requires an "invitation" in the work and "identification" in the audience. The gratification in art for artist and audience is kindred wish fulfillment through art as sublimation.

Many other ideas taken from psychological studies may aid the playwright in his attempt to understand audience. *Funding* and *fusion* are closely related terms originated by John Dewey and refined by Stephen Pepper. Together, funding and fusion indicate the human process of accumulating perceptions of an art work, combining these with memories, and thus forming a whole possessed of a single dominant quality.

Another useful term is *closure*. Developed by Gestalt psychologists, such as Kurt Koffka, closure indicates the human tendency to perceive wholeness of structure and meaning in any configuration. The viewer uses his own psychological energy not to imitate the art work but to seize it as a stable unit. He suffers suspense, discomfort, and tension until he can, in himself, organize his perceptions. With a drama, a spectator integrates aural and visual elements, brings them to a state of completion, and forms a "good *gestalt*."

One other concept about audience behavior needs to be settled in every playwright's mind—*catharsis*. Catharsis has undoubtedly caused more confusion about audience response than any other term. Aristotle introduced it in the *Poetics* as an element in his famous definition of tragedy. Since the fourth century B.C., it has been widely discussed and frequently misunderstood. In the *Poetics*, Aristotle discussed the principles of the formulation of drama as an object. He wrote only about drama as drama, not about drama as philosophy, rhetoric, or psychology. Since he treated audience response not at all, how strange that some critics have chosen catharsis as one word in an entire book which indicates the effect of drama on an audience! If one reads the *Poetics* as the logical discussion of the internal principles of dramatic construction that it is, catharsis obviously has to do with the organization of a play rather than with the emotional purgation of audiences. Catharsis is the "ending" of the emotive powers of a drama. It is, for example, the drawing to a close of fear and the amelioration of pity *in* a tragic play. Such emotive powers must exist in a play before they can affect an audience, and when those emotions, or whatever emotions the play itself possesses as powers, come to a close in a play, then they are ended in the audience too. Quite logically, Aristotle suggested that whatever emotions a play arouses, it should also allay. It is doubtful that many reflective artists have ever thought that their works would purge the emotions of spectators. How silly to think that a person who laughs at a comedy during a performance subsequently walks out of the theatre unable to laugh for hours or days because he has been purged of such emotion! Audience members are more likely to carry the emotive mood of an effective drama as they leave, and most people like to cherish the mood a play stimulated in them. Few playwrights indeed try to create emotional emetics. The interpretation of catharsis as emotional purgation *in an audience* is as illogical as it is ridiculous. When it is considered, however, as an inherent principle of resolution *in a play*, then it can act as a functional tool for the writer.

Few, if any, audience members rush to witness live drama because they want a catharsis of their emotions. But why do people attend? The reasons are nearly as numerous as the individuals in an audience, and most people harbor more than one motive. They attend a drama because it is an interesting experience. It heightens their sense of life. Some people want pleasant social diversion, escape, sensory titillation. Some, perhaps, have a habit of playgoing. Some have discovered that drama presents wish fulfillment or ego satisfaction. A segment goes to theatre because they believe art will enhance them in a mysterious cultural manner. Some attend to see a friend perform, or a friend of a friend, or because a friend asked them to go. Some try to impress acquaintances or business associates by taking them. Some like a secular ritualistic experience. A growing number attend to join a social protest they hear the play makes. And one of the most important motives is to gain knowledge about man. People want to learn about themselves, and they study themselves through the comparison and contrast of a dramatic experience. There are other reasons that impel people to go to the theatre, but these are some key ones.

Every playwright has an obligation to himself and his plays to study such concepts as empathy, distance, sublimation, and closure. He also needs to know why the individuals in his audience come to the theatre. To grasp these as tools for reflection about the social effects of his art works will not spoil his artistry. Without some understanding of spectator identity, perception, and involvement, a playwright will never fully understand what he is doing. None of the concepts mentioned above, however, is the final answer to audience psychology. The terms are not particularly clear; some are approaching verbal senility; and they are severely limited. They are useful as reflective tools, but that is all. Many hours of thought and research need to be spent on audience analysis, and few scholars are now devoting themselves to the subject. A playwright can well afford to pay attention to the recent explorations and discoveries of such innovators as psychologists Eric Berne (*Games People Play*) and Martin Grotjahn (*Beyond Laughter*), sociologists Erving Goffman (*The Presentation of Self in Everyday Life*) and David Riesman (*The Lonely Crowd*), and philosophers Jean-Paul Sartre (*Being and Nothingness*) and William Barrett (*Irrational Man*).

The leading contemporary artists and theorists are examining the connections between art and life, and most are finding new and crucial ones. They are suspicious of the oversimplifications characteristic of all concepts of crowd behavior. An audience is a collective of individual identities, and any judgment about group response can only be a vague generalization that is probably false for most of the individuals in the collective. Identity is, therefore, one of the current topics of audience investigation. The existential psychologists and philosophers—such as Jean-Paul Sartre and William Barrett—are pointing the way. Another promising topic is expectation-obligation network; an investigation of this was begun by sociologist Erving Goffman. Following some ideas of Eric Berne, Richard Schechner has started an exploration of theatrical "transactions" in his writings on environmental theatre. Other innovators, such as Michael Kirby and Allan Kaprow, are contributing fresh ideas about audience-performance relationships through their experimental work with intermedia, happenings, and nonmatrixed art. And certainly Jerzy Grotowski's theories about confrontation have revealed possibilities for new theatrical involvements. All these recent explorations indicate the need for reflection and experiment in the area of audience relationships. Most of these evolving ideas also question the traditional theatrical conventions, a topic the next section briefly examines. The concluding thought for this section is this: A playwright needs to forget the crowd, to identify the spectator-auditors he may be contacting, and to use concepts about audience not as effects but as employable criteria for his selective process of creating drama.

THEATRICAL CONVENTIONS

Every artist who works at drama believes a theatre can, and hopefully will, hold a series of audiences. But how does it hold them? There is more to the holding than mere spatial enclosure and comfortable seats. When a theatrical event occurs, the usual constituent factors are a play, actors and other theatre artists, a building with equipment, created auditory and visual stimuli, business personnel, and an audience. The enumeration of these, however, still does not explain everything about the holding. For every dramatic occurrence, certain conventions are set by the artists and accepted by the audience.

The conventions established by any given play and in any production of that play are more than mere theoretical guidelines for the performers and the spectators. They are the rules of the game. Although flexible from one production to another, they control the quality of the performances and the expectations of the audience. A playwright can simply accept, without much concern, the theatrical conventions most widely used in his time. Or he can study such conventions and by conscious choice alter them to fit his particular dramas. A playwright may well ask, what conventions does my play require? Upon which of the traditional conventions does my drama depend? What new conventions does my play demand, and are any companies prepared or willing to initiate them with me?

Many common theatrical conventions are readily apparent to anyone who wants to think of them for a few minutes. For example, during most plays an audience sits in one architectural block of space and focuses its attention on a single group of performers in another block. But there are a number of conventions that are seldom considered. In fact, many new kinds of dramatic events are being produced in situations with highly divergent convention systems. Because of the new range of possible theatrical occurrences—from demonstrations, to happenings, to street theatre performances, to multi-media shows, to relatively traditional productions—a contemporary playwright needs to review the variables of dramatic conventions to be established and imposed upon audiences.

To comprehend the aesthetic and practical conventions and their potential variety, a writer can consider physical aspects of audience milieu, their sensory perceptors, the reasons people attend plays, and typical play-going behavior. The following enumerations do not exhaust the list of traditional conventions or explore all the possible new ones, but the considerations should aid a dramatist in selecting the conditions for the kind of dramatic experience he wishes to make. The succeeding paragraphs mention the convention factors, the traditional arrangements, and some innovations. With each item, a playwright can envision potential variations in established practices.

The major physical aspects of the audience milieu are space, light, sound, time, and temperature. Traditionally, dramas in this century take place in architecturally enclosed spaces where audiences are seated facing the performance areas. Although the size and shape of the building and the performance space varies, the focus is usually fixed; in few theatres are the seats movable. Further, and even more significant, most dramas are performed in architectural blocks absolutely separated from everyday life space. The conventions of light typically dictate a dark spectator area and a bright performance area. The sources are mostly hidden, and the colors, especially those for the audience, are warm. The power is electricity; seldom are plays illuminated with natural light. Gradual intensity alterations, characteristic of lights dimming up or down, are common. The single most important convention of sound is that a drama is presented in a noise insulated area; extraneous sounds are considered to be distractions. There are many conventions having to do with time, but three are common. Most dramas occur in the evening (some afternoon performances are not unusual), span two to four hours, and are interrupted once or twice for periods of ten to twenty minutes. The "desirable" convention regarding temperature is that it should range from the low sixties to the low seventies Fahrenheit. A majority of the experiments with all these physical conventions are being carried out by small experimental theatre groups. Although some theatres are now being built with flexible seating and with multiple stages, most theatre plants feature these conventions. The guerrilla theatre troupes, the promoters of happenings, and the companies who produce environmental performances are helping other contemporary theatre artists re-examine physical restrictions.

Thinking of the five senses—sight, hearing, taste, smell, and touch—will reveal some further conventions in the majority of theatres. In addition to restrictions of space and light, sight is conventionally unimpeded and uninterrupted. Visual focus is most often directed to singularities rather than multiples. Some new groups are trying flexible, interrupted, and variable focus. The conventions of hearing are not only that the auditorium should be noise free but also that the performers should be audible to everyone. And non-human sounds are considered embellishments of the action. Experimental productions now permit actors to speak to small segments of the audience and utilize sound effects which are ends in themselves.

Except in rituals, primitive and modern, oral taste is seldom considered. Whatever the playgoer has to eat or drink before he comes to the theatre dictates the taste he senses throughout the performance. As any theatregoer knows, taste can become a strong physical concern before an evening is over. Even avant-garde troupes have generally failed to disturb the oral convention. What could happen to affect audience taste sensations before, during, or after a performance?

Regarding the sense of smell, most theorists and producers think only of the occasional abortive attempts to waft perfume, food, or incense odors into an auditorium. But every theatre has a basic odor—of mold, fresh paint, old wood, institutional air conditioning, or something—and it affects the performance. Few producers consider alterations in the extant odor of their theatres, except to be certain that there is sufficient air circulation. It is possible, however, with the wide variety of modern chemicals to control the scent of a performance. Even to consider the control of odor seems almost laughable, but not really. Humans are animals with strong olfactory responses. The traditional convention of smell, then, is usually the accidental.

The conventions having to do with touch and kinesthetic bodily arrangement are numerous. Comfort through physical relaxation is the criteria for most of these, and an illusion of luxury is their goal. Carpeting in lobbies, halls, and aisles deadens sounds and initiates the "soft" physical impression conveyed by many theatres. Next, seats of a generally accepted size and shape are almost universal. The seats themselves are padded and have a shared elbow rest. Most seating areas have individual seating spaces as constricted as fire regulations and human anatomy will allow. Such seats promote minimal physical responses. An audience is packed together, and its members usually have a compressed sensation. This is excused by explanations that as many people must be got in as possible to pay for the performance and that a packed-in audience will respond more readily as a crowd. All such conventions are seldom questioned, except by troupes that cannot afford a fancy playhouse, and perhaps contemporary economics require such conditions. But someone might at least consider possible alterations in seating and increased audience portability. Of course, some theatrical groups who work in "found space" are in fact playing to moving, standing, or ground-seated audiences.

Although most dramatic artists apparently take the established theatre conventions as given conditions, changes are coming. And a playwright who conceives his dramas only for proscenium, audience segregated, soft theatres may delimit his art unnecessarily and reduce his chances for production.

Before leaving the matter of theatrical conventions, the habits of audiences ought to be mentioned. Several typical behavioral activities of American audiences marginally, at least, affect a performance. First, most people wear dress-up clothes to the theatre, and this clothing affects their physical feeling about the experience. Since going to the theatre requires an investment of time and money, people arrive with a heightened sense of an Event. Most audience members are hurrying when they arrive, and many come later than the announced curtain time. Their deportment is generally polite, and it is even more reserved in response to large, expensive productions conventionally conceived and staged. They tend to be rather inert because their experiences of cinema and television watching have conditioned them to be habitual but non-participating observers. A startling percentage of our population stays away from live theatre because they believe they would not know how to behave. And many of those who do attend wonder constantly about what they should or should not do. They are slightly uncomfortable, tense, and watchful. Only after a drama begins to stimulate their emotions do they relax. The most typical intermission conversation has to do with questions about how to respond to the performance, although these are seldom consciously posed. Most people think they should be critical, negatively or positively, but they want to be sure that they generally agree with their peers. At the end, they applaud, sometimes because the performance genuinely stimulates a need to communicate approval, but just as often because they believe it is the thing to do. And finally, they hurry from the theatre to avoid the crowd, to escape the parking lot, to get a meal, or to relieve a baby-sitter. Audience behavior is commonly learned, semi-habitual, and non-participatory. Perhaps the experimental theatres, some campus theatres, and the "new" youth will continue to change these rather lethargic behavior patterns.

A playwright can help to free the prevailing convention network. His knowledge of established conventions will contribute little form or substance to a play, but it will affect the plays he writes through alterations in his logical and imaginative criteria for choice. Conventions, along with many other factors, help determine an artist's powers of selectivity. In our time, a decision to compose a play strictly for a proscenium theatre and a singular audience is deadly. To avoid considering potential changes in the physical arrangement of theatres or to fail to reflect about potential audience behavior is for the playwright to reject total control of his art form. The artist, if he works at capacity, is an innovator. A playwright can be the central innovator in the theatre.

CONTEMPORARY AUDIENCES

Any discussion of audience and the prevailing spectator conventions would be incomplete without some characterization of the audiences of the time. The characterizations that follow are reminders to the aware writer about what kinds of people are likely to meet his play. For the new writer, they can help him begin his own process of audience analysis. Although generalizations about audience composition are difficult, one can at least discern some facts about who makes up given audiences and, from the playwright's viewpoint, what is good or bad about them. There are many different audiences in the United States, but they tend to melt into one another. Although simplified and made to sound perhaps too exclusive, the following audiences are those a playwright should know.

The New York audience is far from being singular. At least three separate segments of it are worth characterizing. The Broadway audience, easily the most often defined and maligned of all, possesses these features: A slight majority of the spectators are women; the average age is in the forties; their basic attitude is self-protection; most are in the upper part of the middle-income group; the educational level, a nearly impenetrable mixture, is short on schooling and long on "experience"; most Broadway audiences want "a show." It is a difficult audience because of its frosty nature and questioning attitude, but it pays more money and reputation than any other. The second segment of the New York audience is for off-Broadway. Since productions there are now more often for entertainment than for experiment, many of the spectators are the most theatre-devoted of the Broadway audience. Compared with the uptown audience, however, the people who see off-Broadway shows are somewhat younger, happier, more responsive, and less affluent. As more shows originate off-Broadway and then move on-Broadway, the qualitative differences in productions are disappearing. And as Broadway fare grows more varied under this influence, the two audiences are mixing, blending, and making commercial what was once provocative. The most stimulating New York audience is that drawn to what is now called off-off-Broadway. These performances usually occur in cafes, non-theatres, and found-spaces. The audience is far younger—early to middle twenties—than the others, and it wants fresh experience at fairly low cost. Its personality is nervous but willing.

Although a large portion of the Broadway audience consists of out-of-towners, the "regional" audiences who attend professional productions have a different character. Many of the contributive facets are the same, but the sophisticate element tends to disappear outside of New York. In regional audiences, the reputation makers are absent, but the weary fifty-year-olds who seek escape and social diversion are not. But despite a certain timorousness, the regional audience is more emotionally responsive than that in New York. The regional audience, too, can be segmented. First, there is an identifiable audience for professional touring companies. It is usually a subscription audience in small cities and an open one in large metropolitan areas. A majority of this audience is female, over fifty, financially at ease. The touring audience tends to be lethargic and says to the performers,

"Please me!" It would be misleading, however, to dismiss this audience so negatively without mention of its percentage of bright, responsive, and fairly young married couples who get professional theatre only in this way. Theirs is the reaction to watch. The second segment of the regional audience for professional drama is the one that attends the resident theatres. In many respects, this is the most sensitive and intelligent audience of all, but it varies from city to city. It is more nearly balanced between males and females; its income level is moderately high (though students lower this factor). These spectators attend more for the stimuli of the art than perhaps any other audience. Usually, the resident theatre audience is lively and quick.

The amateur audience, the third one a playwright should get to know, is an exciting one today, and it will be the professional audience tomorrow. As a whole, it contains segments that are the best and worst, the most intelligent and the youngest, the most imaginative and the most unsophisticated of the country's audiences. First, the campus audience is rapidly becoming the taste barometer for the nation. Composed of college students, faculty families, and educated townspeople, the campus audience is sexually young and intellectually awake. This group does not want art to *be* so much as it wants art to *do*. More than all the crying critiques that preach what campus theatres should be doing, this audience is teaching those theatres to experiment, to provoke, to be audacious, and to try the new drama. This is the audience which supports America's subsidized theatre, and as such it is the first one a starting playwright should get to know and the one the established playwright had better heed. Although the campus audience differs at Brandeis, Yale, Wayne State, or Illinois Wesleyan, it wants action. The second segment of the amateur audience is the high school audience. It is the most honest of all audiences—attentive and responsive to the active but rude and disruptive to the boring. It offers more—if smaller—royalty checks to playwrights than all the others combined. The high school audience is the largest, and though unsophisticated it is wildly imaginative; it so very desperately *wants*. Amateur theatre's third audience segment is for community theatre. This audience is searching for something exciting or even interesting but does not know where to find it. It resembles the audience for touring theatre. At worst, its members think they know everything about drama and wrongly consider themselves full of insight; at best, they are willing to respond fully once a play captures them. The civic audience is superficially a pretender, but underneath an honest innocent.

The other three major segments of the amateur audience are so unique that each deserves separate treatment. The child audience has been almost universally ignored by playwrights. They leave it to the mercy of television cartoons, the animal series, Disney movies, and hack children's theatre companies. The youngest of all audiences, its individuals are setting their major impressions of live theatre—when they have a chance at all—by watching poorly written, unimaginative plays from a startlingly small library. The child audience is far more knowing, interested, and sensitive than theatre artists have ever recognized. The life possibilities are more numerous to a wriggling eleven-year-old than to a sleepy sixty-year-old member of a theatre party. Why should a playwright ignore the child for the sake of the grandparent?

The fifth segment of the amateur audience, already large and still growing, sees its live theatre outdoors. This is the tourist audience which attends the summer pageant productions. Although the average spectator age is about thirty, children and grandparents are there too. It is an audience of people who want something a bit different, for a night. On vacation for two days or two weeks, they want to absorb some local color, to have an experience they will remember all winter. They have seen the summer re-run television shows, and there is little to do in a motel or campsite at night. Typically, this audience perhaps possesses the least formal education, but their understanding is usually great. Its members are often considered to be from a fairly low socio-economic class, but their income is frequently larger than that of most college faculties. It is a challenging and generally unrecognized family audience.

The last amateur audience is often not a theatre audience at all. At least many of its members seldom think of going to a theatre—ever. This is the found-audience of guerilla theatre. The poor and the segregated, the exploited and the protesting, the laborers and the marchers, the Blacks and the Mexicans, the bewildered and the angry—these are the spectators for guerilla troupes. It is potentially a fresh audience. Its needs are urgent, and art is among those needs, though few of its number realize that. This audience is sometimes joyful and sometimes dangerous, but it is always challenging.

The theatre audience in the United States is not singular. No one person or group is typical. Perhaps no single playwright can aesthetically comprehend all these segments. But the best artists always know their audiences, and the luckiest artists are those who stay honest and yet whose works affect a wide variety of people. No one can rightfully claim that it is wrong to write for any one of the specific audiences above, or even for the many other specialized audiences—religious, organizational, centennial, and the like. For the artist with integrity, awareness of audience is crucial, but pandering is personally and creatively destructive. A playwright needs to know audiences not so he can please them but so he can decide how to get at them, to disturb them, to move them, and even how to be clear to them. Studying audiences to learn what *not* to do is deadening. But exploring the possible connections between people and art, between spectators and drama is essential.

11: *SCRIPTS AND MARKETS*

The theatre is the only place in the world, the last general means we still possess of directly affecting the organism and, in periods of neurosis and petty sensuality like the one in which we are immersed, of attacking this sensuality by physical means it cannot withstand.

ANTONIN ARTAUD
The Theatre and Its Double

Handling a manuscript and marketing it are essential activities for a playwright. They are practical aspects of his craft, and without knowledge of them no writer can be counted competent. The work of setting a play in an acceptable typed format and getting the manuscript into the hands of producers and directors has little, if any, effect on the inherent nature of the play itself. These are, nevertheless, matters of practical activity for a writer that help to complete a play and make it a drama. So long as a play, in handwriting or rough typescript, rests on its author's desk, it is not a drama. Only when he puts it in a readable condition, gets it to someone who decides to stage it, and finally when it is performed—only then does the inert play become a motile drama, a completed work of art. Thus in this special sense, setting the format and marketing the script are a part of the work of any playwright as artist.

THE FUNCTIONS OF A FORMAT

A format for a play is the plan of arrangement of the manuscript. It includes the makeup of typed words on various kinds of pages, as well as the shape, size, and length of the whole. Although this matter may seem to be mechanical, format has for a playwright four vital functions.

The paramount function of format is its psychological effect on the writer himself. A playwright first carries an idea for a play in his mind; then he makes notes, composes a scenario, scribbles a first draft, and makes corrections. Eventually, some draft, first or tenth, gets typed in a format he knows to be the correct one. Suddenly, what has always before been a mass of ideas and a welter of words—all this finally *looks* like a play to its author. The playscript itself gives him a sense of achievement, boosts his confidence about it, and gives him a fresh view of it. Each knowledgeable dramatist would express his feelings about this in a different way.

Many persons may fail to comprehend the import of seemingly technical items such as this. Perhaps the attitudes of artists toward such matters are somewhat mystic. But for most playwrights, seeing their play in the proper format gives them an aesthetic stimulus. Ernest Hemingway explained some of the writer's mystical-mechanical attitudes.[1] He wrote about how he liked to write with a pencil because it permitted him to get three sights of the work—in the handwritten version, the typed copy, and the galley proof. Every artist permits himself a complex of emotional attitudes toward his working habits and techniques. For most playwrights, handling format is emotive as well as technical.

The other three functions of script format are more objective. The matter of timing a script, judging its performance length, is next. When a playwright habitually employs one format for all his manuscripts, he is able to judge the time span of each beat, page, scene, or act of any play he writes. And if he uses the generally accepted typing format, professional readers will also be able to judge its length. The format explained in the next section is the widely accepted one. Pages typed according to it will usually occupy from a minute and fifteen seconds to a minute and forty-five seconds, depending on the length of the contained speeches and the number of stage directions. Thus, for a normal short play, twenty-five to forty pages are about right, and for a normal long play, ninety to one hundred and ten pages are desirable.

[1] Ernest Hemingway, "Monologue to the Maestro: A High Seas Letter," in *By-Line: Ernest Hemingway*, ed. William White (New York: Charles Scribner's Sons, 1967), pp. 213–20.

The third function of the format is even simpler, but no less important. It is to give a play a professional appearance. When an experienced reader opens a play and sees that it is presented with the professionally accepted makeup, then the reader has an immediate impression that the play itself will probably be carefully wrought. The format enhances a play's marketability.

The fourth function summarizes the other three. It is readability. Fundamentally, the manuscript format most widely used makes any play easier to read in typed form. This particular format is the result of an evolutionary process to make playscripts more readable and neater in appearance. Not only are many professional playwrights and directors responsible for its development, but also such professional manuscript companies as The Studio Duplicating Service of New York. Clarity on the page is the epitomization of all the functions of format.

AN ACCEPTABLE MANUSCRIPT FORMAT

A typewritten manuscript should assume its own format. It is not a printed, published play, nor should it imitate one. Publishers arrange printed versions variously, according to considerations of spatial arrangement or visual effect. The form accepted among professionals is explained and exemplified in the following pages. If the beginning playwright adopts it as his format for every draft, then it will become a meaningful element of his craft. If the working playwright fails to use the format, or some close approximation of it, he advertises his lack of knowledge, even though his play may be well written.

The subsequent manuscript suggestions appear as a list for easy reference. And following the list of suggestions, a series of manuscript pages—figures 2 through 5 (pp. 270–73)—illustrate all the items.

PAPER

1. Weight: 16 or 20 pound bond; mimeograph paper is acceptable, onion skin is not.
2. Size: 8½″ × 11″.
3. Color: White only.
4. Finish: Non-corrasable, because the print will not smudge.

TYPING

1. Pica or elite type only, although pica is preferred.
2. One side of the paper only.
3. Margins: Top and bottom 1¼″; left 1½″; right 1″.

BINDING

1. A substantial cover, not a manilla folder.
2. Secured at the left side of the manuscript pages.

3. Title and author may appear on front cover.

TITLE, PREFATORY, AND DIVISIONAL PAGES

1. Title page.
   a. Play title in capital letters, underlined, centered on the page, about 4″ down from the top edge.
   b. By-line a triple space below with the author's name in capitals.
   c. Copyright notice in the lower left corner, containing: "Copyright, (author's name), (year date)."
   d. The legal name of the author and his mailing address can optionally appear in the upper left corner.
2. Prefatory pages.
   a. None are numbered, but they should take the following order:
   b. One page headed with these words in capitals: CAST OF CHARACTERS; descriptions of the characters here are optional; if there are only a few characters, time and place of the action can also appear on this page.
   c. The second prefatory page, if included, should carry a synopsis of scenes and a description of setting essentials.
3. Divisional Pages.
   a. Before each act comes an unnumbered divisional page carrying the title of the play, in capitals, underlined, centered, and followed by the act designation.

DIALOGUE PAGES

1. Page numbers.
   a. In the upper right corner.
   b. Act number in Roman numerals.
   c. Scene number follows.
   d. Last is page number, beginning with 1 for each act.
   e. For example: I-1-16; II-3-15; III-1-20.
2. Stage directions.
   a. Indent 4″ to about page center.
   b. Single space.
   c. All in parentheses.
   d. Directions of a word or two appear in normal sequence within the speech unit.
   e. A description of each major character should appear in a stage direction at that character's first entrance, even if the character is described on a prefatory page.
   f. Each entrance and exit of every major character should be noted in a stage direction.
   g. At the end of every scene or act, a designation such as the following should be centered three spaces below the last script line:

END OF ACT I, SCENE 1; this at the conclusion of the play: THE END.

3. Character names.
   a. Indent 4″.
   b. In capitals, wherever they appear except in dialogue.
4. Speech units.
   a. All begin at extreme left margin.
   b. Single spaced.
   c. Dialogue units within a single speech are not paragraphed; an interruptive stage direction can provide the same effect.
   d. Followed by a double space if a speech is next or followed by a single space if a stage direction is next.

Additionally, the problem of multiple copies should be resolved. When a playwright finishes a play and prepares the "final" typescript, he will naturally want a number of copies. The author should always keep at least one copy in case all the others are lost. But he may wish to send copies to friends, critics, and perhaps to several market sources all at one time. Probably the best practice is for him to prepare an original typescript and two carbons. He keeps one of these copies and hands the other two around for reactions. He may even arrange a private reading with these three copies. But he will undoubtedly discover a few alterations to make. Eventually, however, he should prepare multiple copies with either the mimeograph or the off-set (Multilith) process. The spirit carbon (Ditto) process is not satisfactory. Carbon copies and Ditto copies are not suitable for submission to professional markets. Xerox copies, if carefully done on the best paper are all right, but usually too expensive, and there are no stencils or plates to retain. For multiple copies, the mimeograph method is best.

Meticulous proofreading is also a necessity. A playwright should carefully and *slowly* read and correct every draft. But proofing is especially crucial to the preparation of submission copies. Whenever a professional reader finds even one inconsistency or a single error, then he judges the writer to be at least slightly slipshod if not absolutely incompetent.

The neat appearance of the manuscript is the goal of all the considerations in this section. The manuscript is, in the matter of securing a production, the product a playwright tries to sell. Like any other product for sale, it should be attractive to the buyer. Not only should the reproduction process render clear print, but also the general appearance of each script should be fresh. Worn covers, creased corners, smudged pages, yellowed paper—all these detract from a play's value in the eyes of a reader. A manuscript that looks old and shabby shows it has been submitted and *rejected* many times before. A playwright can beneficially take as much pride in his manuscript as in the play itself. And if he does so, his drama is more likely to reach the stage of a theatre and become a completed art object.

Figures 2 through 5 (pp. 270–73) illustrate the acceptable manuscript format for various pages of a play manuscript. The pages included are: (1) title page, (2) cast of characters, (3) synopsis of scenes and setting, and (4) a page of dialogue. Together they show most of the necessary items of format.

COPYRIGHT: WHEN AND HOW TO SECURE IT

A copyright assures an author of certain legal rights to his written work. The United States Copyright Law grants a playwright exclusive rights to copy his work, to sell or distribute copies, to revise the work, and to perform or record it. According to the law, dramatic or dramatic-musical compositions are: "Dramatic works such as the acting versions of plays for the stage, for filming, radio, television, and the like, pantomimes, ballets, operas, operettas, etc." Among the items which cannot be copyrighted are names, titles, ideas, and plans. The author alone, or someone deriving rights through him, can rightfully claim copyright. But no author can secure a blanket copyright for all his works; he must copyright each work separately. If an author writes something as a contracted employee, the employer is then regarded as the author.

Actually, there are two types of copyright protection for unpublished works. *Common Law Copyright* is a type of protection relating to state law, and it occurs when a work is first written, automatically. This sort of copyright may last until the work is published or until a formal copyright is secured. *Statutory Copyright* is protection afforded by federal law, and dramatic compositions, as defined above, are the only written works eligible for this kind of copyright previous to publication. For example, books, short stories, poems, and narrative outlines (scenarios) cannot be copyrighted until published.

Should a playwright secure a copyright for every play he writes? And if so, when? Although the theft of an entire work happens less frequently today than thirty-five years ago—several Hollywood production companies were guilty of it in the 1930's—plagiarism still occurs. The present laws and certain court precedents, however, have made agents, production companies, play-leasing corporations, and contest organizations quite wary. If they are reputable, they will refuse to read non-copyrighted scripts. In the play market, copyright is a protection for both seller and buyer. And any script worthy of professional submissions should be copyrighted. On the other hand, beginning playwrights waste a total of thousands of dollars each year in securing statutory copyrights for plays that do not need such protection.

As soon as some novices complete the first draft of a play, they ship it off to the Library of Congress, wasting their own money and glutting the Library with waste paper. Four simple suggestions explain *when* to copy-

SEVEN TEARS IN EVERY WEEK

by SAM SMILEY

CAST OF CHARACTERS

ELAINE TAYLOR, 15, the daughter of the family

PETE MILLER, 60, a fisherman, META'S brother

META TAYLOR, 52, the head of the family

DAVE TAYLOR, 25, her son who has just returned from a war

BERT TAYLOR, 20, another son, a fisherman

ADELE RHODES, 25, a neighbor girl who loves DAVE

JACK TAYLOR, 16, the third son

LUELLA GUTHRIE, 16, JACK'S girl friend

EVERETT GUTHRIE, 50, her father

The action occurs during a spring evening of the late 1960's at the Taylors' house and pier in the Ten Thousand Islands area of the Florida Gulf coast.

SYNOPSIS OF SCENES

ACT I

The Taylor's house. Late afternoon.

ACT II

The same. A few minutes have passed.

ACT III

The same. Ten minutes later.

SETTING: Both the house and pier are constructed of coarse materials. The pier is long, and the house is built half way out on it, over the water. The visible interior of the house consists of one large room which serves as living room, kitchen, and bedroom. Although somewhat cluttered, it is clean. The family is more interested in fish than in decorations, but their home is comfortable. There is one door to the outside, and two doors lead to other bedrooms. Two large, screened windows ventilate and light the room. Each has a heavy wooden shutter for use during storms and cool nights. One window faces the shore, and the other faces the Gulf. Crates, nets, hand lines, buoys, pieces of cork, and heavy lines litter the visible portion of the pier surrounding the house. Weather obviously conditions and controls the whole environment.

I-5

(PETE'S sea boots lay against the wall near the door. He kicks them into a corner where there are several other pairs. As she speaks, META goes to the pile of boots and takes out a pair. These are older and more worn than the others. She takes them to the center of the room, studying them.)

META
(Speaking slowly.)
There was a fight--he beat those two men--and when he came in at noon, he had blood on his face and hands. But there weren't many fish in his skiff; so he went out again before night.
(She stands the boots in the center of the room. ELAINE sobs.)
A storm came up at dark. I put a light out there on the end of the pier, and I strained my ears for the skiff's motor. It never came. Then he was brought back to me. They carried your father back. (Pause.) "He was wrecked in the storm," they said. "We found him mashed on the rocks near town." That's what their lips said, but not their eyes.

PETE
He fought with the Guthries, and it was them that brought him back here after the storm.
(META slowly sits and weeps. DAVE goes to her and strokes her hair. He then returns to PETE, takes the splice from him, and works it slowly. ELAINE goes to the window, facing the shore.)

DAVE
(To PETE.)
Let's have it straight.

PETE
Don't get riled. Can't you see what it does to her to remember? I'll talk of it when we get out by ourselves.

ELAINE
Dave, here comes Bert.

DAVE
(Not listening.)
I can't believe the Guthries killed him. The families always fought, but it never came to killing.

right a manuscript: (1) only after extensive and careful revision, (2) only after a group reads the script aloud, (3) only when the author thinks he cannot improve the play further, and (4) only just before a submission to a professional or semi-professional market. A playwright's personal friends and his acquaintences in theatre are not going to steal his play. Common law copyright provides ample protection for local submissions. Most significantly, the copyright date on the title page, as required by the statutory copyright law, reveals to a reader the age of the composition. The more recent that copyright date, the more likely the reader will consider the play favorably. It is desirable, therefore, for a playwright to postpone copyright as long as possible.

To secure statutory copyright in an unpublished work, such as a dramatic composition, an author registers a claim in the copyright office. A formal claim should at present include an application form, one copy of the manuscript, and a fee of $6.00. The appropriate form for plays is "Class D Form D," and it can be secured free from:

Register of Copyrights
Copyright Office
Library of Congress
Washington, D. C. 20540

Once an author properly registers his dramatic work, then he must include a *copyright notice* in the manuscript. According to law, the notice should consist of three elements together—the word *Copyright*, the author's name, and the year date. The notice should appear on the title page or the page immediately following.

The duration of a copyright for an unpublished play runs twenty-eight years beginning with the date of registration, and it may be renewed for a second term of twenty-eight years. A new copyright can be made for any play substantially revised. Recently, Congress has been considering a copyright statute more favorable to authors; it would provide legal protection of works for the author's lifetime plus fifty years.

The copyright law protects not only the writer but also those from whom he might quote. If a playwright needs to quote another work—book, play, poem, song, or whatever—he should write for permission to the publisher of that work. Usually publishers grant such permission, but sometimes they require the payment of a fee.

For further information about copyright, one can request from the Copyright Office a copy of Class D Form D and Circular 35, "General Information on Copyright." The following books offer thorough discussions of such legal problems as libel, plagiarism, and copyright: *Rights and Writers* by Harriet Pilpel and Theodora Zavin, *Plagiarism and Originality* by Alexander Lindey, *An Author's Guide to Scholarly Publishing and the Law* by John C. Hogan and Saul Cohen, and *A Manual of Copyright Practice* by

Margaret Nicholson.[2] Securing a copyright is certainly a part of the marketing procedure for a play, and a playwright should know how to handle it properly. This and other legal matters are as much responsibilities of a writer as the construction of a plot or the delineation of a character.

THE BUSINESS PROCESS

The organization of our society perhaps dictates the financial exploitation of artists. And until now, artists themselves have shown their antipathy to the skillful management of their works or services. Business men, rather than artists, have traditionally controlled American theatre companies. But one reason for the change in contemporary theatre is that more playwrights, directors, and actors have proven willing to spend some of their energy on the financial problems of maintaining their art.

Playwrights—whether novices, creative professionals, or commercial craftsmen—more than ever before are recognizing the importance of understanding and controlling the business process of marketing their works. Primarily, this involves judging, circulating, and finally selling their work. After a play is written, the marketing process normally requires several steps: format arrangement, consumer testing, market selection, circulation, selling, and recording. These do not imply that a creative dramatist must dedicate himself to commercialism, but they reveal that a play must—even if to an amateur company—be sold.

For a dramatist, consumer testing is a matter of getting sensitive and intelligent reactions to his work from a few people. The problems in eliciting such reactions are multiple. Most writers experience some difficulty in getting others to read their manuscript. For some, knowledgeable critics are hard to find. And nearly always, those who agree to look at a script are slow at reading, reacting, and returning it. When a writer finds two or three people who are qualified and willing, he should consider himself fortunate. If he finds three, they might ideally be an emotionally close friend whose opinion will boost his feelings about the play, an expert in verbal mechanics and lucidity who will help him correct objective errors, and an intelligent theatre artist who will help him discern dramatic values and theatre markets. Getting such responses should not be a matter of egobuilding. The purpose of seeking reactions is to help the writer perceive his play through the sensibilities of others, to judge it more fairly as a com-

[2] Harriet Pilpel and Theodora Zavin, *Rights and Writers* (New York: E. P. Dutton & Co., 1960); Alexander Lindey, *Plagiarism and Originality* (New York: Harper Brothers, 1962); John C. Hogan and Saul Cohen, *An Author's Guide to Scholarly Publishing and the Law* (Englewood Cliffs, New Jersey: Prentice-Hall, Inc., 1965); Margaret Nicholson, *A Manual of Copyright Practice*, 2nd ed. (New York: Oxford University Press, 1956).

pleted whole, and to discover what segment of the audience responds to it most vividly. For perhaps the first time, the writer comes to judge his drama through others not only as a potential art work but also as a salable object.

The third step is the selection of a market. It is futile for a writer to send a play to an outlet that could not or would not produce it. For a creative playwright to worry about market selection while composing a play is equally foolish. Some writers, however, will wish to write commercially, and they must select a market even before putting together the scenario. In any case, market knowledge will at some point be important to the playwright who wants to see his drama completed—on a stage. A playwright can acquire market knowledge by reading, theatregoing, and gathering information from personal contacts with people in the theatre business. Learning the basic market categories is a matter of connecting audience types, such as those discussed in Chapter 10, with production organizations. To locate a market for a specific play requires that the playwright identify an appropriate audience and that he discover the representatives of production organizations that cater to that audience.

Fourth, a playwright must get his product to the market outlet. He must carry or mail his manuscript to somebody. But he must first discover specific outlets—names of theatres, companies, and producers—and get the name of *the* person in whose hands a play should be placed. Although this can be a time-consuming activity, it is a necessary one. And the more of those individuals a playwright can come to know personally, the more likely he is to get at least a sympathetic reading. The next section of this chapter specifically names various markets and explains how a writer can best approach them.

Fifth, the writer must at some point make a sale. He enters into an agreement with a manager for the production of his play and for the resultant royalty fee. Once the readers of a theatrical organization decide that they like a script and wish to produce it, the sale is easily completed. Because production contracts vary widely, generalizations about their content are not possible. All, however, have a few common characteristics. With amateur groups the contract is a letter of intention, and with Broadway producers the contract is a complicated legal document attuned to the stipulations of the producer, the playwright, an agent, and the Dramatists Guild. If the contract is a legal document, requiring the playwright's signature, for a professional production, the playwright owes himself the advice of a lawyer. The royalty range for staged plays is also wide indeed. A playwright can expect from five to ten dollars a performance for a one-act play produced by an amateur company, from twenty-five to seventy-five dollars a performance for a long play produced by amateurs. With professional companies, the amount differs according to the financial means of the corporation, from amounts such as one hundred dollars a performance from a small resident theatre to ten per cent of the gross ticket sales from a Broadway production company. Amateur and small repertory groups seldom can afford to haggle about royalty, and the playwright himself can settle agreements easily. But financial negotiations with professional companies are usually better handled by an agent.

The sixth step in the business process may seem to be less important than the others. It is the maintenance of business records. And in this age of credit ratings, computer controls, and complicated tax systems, a writer, like every other business man, must keep accurate records. Not only are financial records important but also other kinds as well. The following types of records merely indicate the many different sets kept by professional writers: (1) names and addresses of contacts; (2) correspondence files, including copyrights and contracts; (3) submission record for each manuscript; (4) annual income record for all royalties and services; (5) annual debit record of all expenses incurred as a writer; (6) work records, e.g., a daily log of hours worked plus words written; and (7) files of research sources and materials. None of these is marginal, and each writer will discover many others.

To consider the business process of getting a play from study to stage and receiving money for it is to realize the importance of two things. First, a playwright must be a business man, not just an artist. Every writer hears —at least ten times a day—that if he writes something "really good" it will sell. Well, it will *not* sell unless he markets it appropriately.

Second, a playwright's most important sales device is himself. The personal work of getting to know people who can help is the single most important activity for a writer other than writing itself. Working actors, for example, learn from long and bitter experience that what they call *contacts* are most responsible for their securing employment. Contacts are equally crucial to a writer. The playwright who becomes acquainted with producers, who makes friends of directors, and who stays in touch with all the theatre people he meets—he is the playwright whose plays will be produced. In *The Drama Review*, Robert Pasolli explained the conditions facing a playwright off-Broadway and off-off-Broadway in New York. He wrote that most non-Broadway productions come into being only through the authors' appeals to theatre managers. Many managers decide what plays to produce not by reading many scripts but on the basis of their intuition about, or friendship with, certain playwrights.[3] This is undoubtedly true of the managers of most theatre companies who produce new plays. Favorable contacts are essential in the business of marketing plays.

MARKETS

This section aims to initiate, or amplify, a playwright's collection of play markets. It is not an exhaustive market directory, but it indicates the

[3] Robert Pasolli, "The New Playwrights' Scene of the Sixties," *The Drama Review* (Fall, 1968), 153.

variety of outlets for dramas of quality. The *market segments* are more important than the lists of selected names and addresses. Address lists require constant revision in the files of each writer; market segments, however, remain operational for much longer. A playwright can identify other market outlets in theatrical periodicals, in the writers magazines, and especially in the current *Writer's Market, Simon's Directory of Theatrical Services and Information*, and Manhattan telephone directory.[4]

The basic categories of the drama market and their segments are: (1) theatres and businesses, (2) organizations, (3) contests, (4) publishers, and (5) special markets.

Theatres and Businesses

Theatre companies and associated business organizations, in some cases, are willing to look at new plays. Their characters, of course, vary greatly. Many companies and businesses require recommendations and a signed release form before they will accept an unknown writer's play. *A playwright should always send a letter of inquiry to any market outlet and receive a reply before mailing an unsolicited manuscript.*

The following companies lead their respective realm with records of producing or handling new plays or with the prestige they might lend to productions of a new play.

Professional Resident Companies. There are more than fifty professional resident theatres in this country. Most of them indicate a growing, but not an overwhelming, interest in new plays. Some of those interested in new plays are:

ASOLO—THE STATE THEATRE COMPANY, P. O. Drawer E, Sarasota, Florida 33578

ALLEY THEATRE, 615 Texas Avenue, Houston, Texas 77002

ARENA STAGE, Sixth and M Streets, S. W., Washington, D. C. 20024

CENTER STAGE, 11 East North Avenue, Baltimore, Maryland 21202

CENTER THEATRE GROUP, Mark Taper Forum, 135 North Grand Avenue, Los Angeles, California 90012

DALLAS THEATRE CENTER, 3636 Turtle Creek Boulevard, Dallas, Texas 75219

MILWAUKEE REPERTORY THEATRE, 2842 North Oakland Avenue, Milwaukee, Wisconsin 53211

[4] *Writer's Market '71*, ed. Kirk Polking and Natalie Hagen (Cincinnati: Writer's Digest, 1969) can be ordered for $7.95 from *Writer's Digest*, 22 East Twelfth Street, Cincinnati, Ohio 45210; *Simon's Directory of Theatrical Services and Information*, 4th edition, ed. Bernard Simon (New York: Package Publicity Service, 1969) can be ordered for $5.25 from Package Publicity Service, Inc., 1564 Broadway, New York, New York 10036.

PLAYHOUSE IN THE PARK, Mt. Adams Circle, Eden Park, Cincinnati, Ohio 45202

REPERTORY THEATRE OF LINCOLN CENTER, 150 West 65th Street, New York, New York 10023

Summer Theatres. Over 300 summer theatre companies produce plays each season. Few of them stage new plays; most are interested only in accepted commercial vehicles. Because these groups seem to appear and disappear rapidly and because the addresses of their managers change equally fast, a writer will have difficulty maintaining current files for them. It is better to discover a few who are truly interested and stay in contact with them. Other addresses are available in the American Educational Theatre Association's *Summer Theatre Directory* and in an annual directory published in March by *Show Business,* 136 West 44th Street, New York, New York 10036. Some summer companies of interest are:

ARROW ROCK LYCEUM, Arow Rock, Missouri 65320

PAPER MILL PLAYHOUSE, Millburn, New Jersey

PIONEER PLAYHOUSE, Danville, Kentucky (15 West 46th Street, New York, New York)

THE PROVINCETOWN PLAYHOUSE-ON-THE-WARF, Gosnold Street, Provincetown, Massachusetts 02657

TIMBER LAKE PLAYHOUSE, P. O. Box 29, Mount Carroll, Illinois 61053

VAGABOND PLAYERS, Flat Rock, North Carolina 28731

Non-Broadway Groups in New York. A variety of producing groups in New York—both off-Broadway and off-off-Broadway—are looking for new playwrights. They tend to be more interested in the writer than in seeing individual scripts. The best way to learn about these markets is to visit New York often or simply reside there.

ACTORS PLAYHOUSE, 100 7th Avenue South, New York, New York

CAFE LA MAMA, 122 2nd Avenue, New York, New York

CHELSEA THEATRE CENTER, 30 Lafayette Avenue, Brooklyn, New York 11217

CIRCLE IN THE SQUARE, 159 Bleecker Street, New York, New York 10012

JUDSON POETS THEATRE, 55 Washington Square South, New York, New York 10012

THE OPEN THEATRE, 56 Macdougal Street, New York, New York 10012

PLAYWRIGHTS OPPORTUNITY THEATRE, The 13th Street Theatre, 50 West 13th Street, New York, New York 10011

ST. MARKS PLAYHOUSE, 133 2nd Avenue, New York, New York

Community Theatres. The hundreds of community theatres throughout the country differ greatly in financial means, artistic skills, and production vision. Potentially, they represent an excellent market for unknown writers, but only a few regularly produce new plays. With most of these theatres, personal contact is better than mail submissions. The American Community Theatre Association, a divison of AETA with over two thousand members, can furnish other addresses. The address of ACTA is 815 17th Street, N. W., Suite 842, Washington, D. C. 20006. Examples of strong community theatres are:

DES MOINES COMMUNITY PLAYHOUSE, 831 42nd Street, Des Moines, Iowa 50312

DON JUAN PLAYHOUSE, INC., P. O. Box 777, Los Alamos, New Mexico 87544

MIDLAND COMMUNITY THEATRE, Midland, Texas 79701

SPOKANE CIVIC THEATRE, P. O. Box 692, Spokane, Washington 99210

STOCKTON CIVIC THEATRE, P. O. Box 1701, Stockton, California 95201

THEATRE AMERICANA, P. O. Box 245, Altadena, California

University Theatres. More than six hundred colleges and universities offer degree programs in drama, and thousands of university groups produce plays. These institutions represent by far the largest and most available market for unknown playwrights. Although drama departments are much criticized for their policies of play selection, they nevertheless produce hundreds of new plays each year, probably more than all other theatres combined. Their per capita average is low, but the actual number of original scripts staged is high. The department chairmen and directors are, like all theatre people, overworked. Hence, in-person submissions are better than mailed submissions. Most of the schools listed below also offer, at some educational level, instruction in playwriting. The best comprehensive list of university theatres is available in the annual American Educational Theatre Association *Directory of Members.* The following have often produced new plays:

DRAMA DIVISION, BENNINGTON COLLEGE, Bennington, Vermont 05201

THEATRE ARTS DEPARTMENT, BRANDEIS UNIVERSITY, Waltham, Massachusetts 02154

DRAMA DEPARTMENT, CARNEGIE-MELLON UNIVERSITY, 5000 Forbes Avenue, Pittsburgh, Pennsylvania 15213

DEPARTMENT OF DRAMA, CATAWBA COLLEGE, Salisbury, North Carolina 28144

UNIVERSITY THEATRE, FLORIDA STATE UNIVERSITY, Tallahassee, Florida 32304

THEATRE ARTS DEPARTMENT, IMMACULATE HEART COLLEGE, 2021 North Western Avenue, Los Angeles, California 90027

THEATRE PROGRAM, SCHOOL OF THE ARTS, NEW YORK UNIVERSITY, 111 Second Avenue, New York, New York 10003

DEPARTMENT OF DRAMA, SAN JOSE STATE COLLEGE, San Jose, California 95114

DEPARTMENT OF THEATRE, SOUTHERN ILLINOIS UNIVERSITY, Carbondale, Illinois 62901

UNIVERSITY THEATRE, TUFTS UNIVERSITY, Medford, Massachusetts 02155

DEPARTMENT OF THEATRE ARTS, UNIVERSITY OF CALIFORNIA AT LOS ANGELES, Los Angeles, California 90024

DEPARTMENT OF DRAMA, UNIVERSITY OF EVANSVILLE, Evansville, Indiana 47701

DEPARTMENT OF DRAMA, UNIVERSITY OF GEORGIA, Athens, Georgia 30601

DEPARTMENT OF DRAMA, UNIVERSITY OF ARIZONA, Tucson, Arizona 85721

UNIVERSITY THEATRE, UNIVERSITY OF ILLINOIS, Urbana, Illinois 61822

UNIVERSITY THEATRE, UNIVERSITY OF IOWA, Iowa City, Iowa 52240

DEPARTMENT OF THEATRE ARTS, UNIVERSITY OF MINNESOTA, Minneapolis, Minnesota 55455

UNIVERSITY THEATRE, UNIVERSITY OF MISSOURI, 129 Fine Arts Centre, Columbia, Missouri 65201

UNIVERSITY THEATRE, UNIVERSITY OF NEBRASKA, Lincoln, Nebraska 68508

DEPARTMENT OF DRAMA, UNIVERSITY OF TEXAS, Austin, Texas 78712

SCHOOL OF DRAMA, YALE UNIVERSITY, New Haven, Connecticut 06520

University Resident Theatre Companies. The number of professional and semi-professional theatre companies on university campuses is growing. All have professional staffs; most employ advanced students; and some engage Equity actors. Although few of these theatres have outstanding records of producing new plays, most of the managing directors say they are interested in setting more such plays onstage. These theatres are rapidly gaining financial and artistic strength, and they are a fine potential market for playwrights in the future. During 1969, over fifty of these men formed the University Resident Theatre Association as an affiliate of AETA. Some of the leading companies may be reached through the following managers:

WILLIAM H. ALLISON, Head, Department of Theatre Arts, Pennsylvania State University, University Park, Pennsylvania 16802

KEITH M. ENGAR, Pioneer Memorial Theatre, University of Utah, Salt Lake City, Utah 84112

RICHARD G. FALLON, Chairman, Theatre Department, Florida State University, Tallahassee, Florida 32304

KEITH KENNEDY, Director of Theatre, Memphis State University, Memphis, Tennessee 38111

JOHN KIRK, Chairman, Theatre Area, Illinois State University, Normal, Illinois 61761

LEONARD LEONE, Director, University Theatre, Wayne State University, Detroit, Michigan 48202

RICHARD MOODY, Director, Indiana Touring Company, Indiana University Theatre, Bloomington, Indiana 47401

ROBERT C. SCHNITZER, Executive Director, Professional Theatre Program, University of Michigan, Ann Arbor, Michigan 48104

LAEL J. WOODBURY, Chairman, Speech and Drama Department, Brigham Young University, Provo, Utah 84601

New York Producers and Directors. Most New York producers and directors are unwilling—because of time limitations and danger of plagiarism suits—to accept manuscripts from unknown writers. But they will accept submissions from established dramatists and reputable agents. A few are willing at least to answer the queries of unknowns. The names below are representative. For others, a writer can check the Manhattan telephone directory.

CHARLES BOWDEN, Suite 1504, 230 West 41st Street, New York, New York 10036

DAVID J. COGAN, Empire State Building, New York, New York 10001

JEAN DALRYMPLE, 130 West 56th Street, New York, New York 10019

JOSEPH DI STEFANO, Di Stefano Productions, 1833 East 24th Street, Brooklyn, New York 11229

MICHAEL ELLIS, Suite 501, 850 Seventh Avenue, New York, New York 10019

FEUER AND MARTIN PRODUCTIONS, INC., 505 Park Avenue, New York, New York 10022

FRYER, CARR & HARRIS, INC., 445 Park Avenue, New York, New York 10022

CHARLES HOLLERITH, JR., 18 West 55th Street, New York, New York 10019

SHEPARD TRAUBE, 29 West 56th Street, New York, New York 10019

NORMAN TWAIN, 165 West 46th Street, New York, New York 10036

New York Agents. The ideas in the minds of uninitiated writers about literary agents are often like hallucinations about birds of paradise. Many young writers fail to understand an agent's function, or when and how to secure the services of one. A beginner does not need, and probably cannot get, an agent to handle his work. But he can be sure that when his plays attain a minimal level of quality he can engage one with ease. By means of personal contact or showcase productions, a new writer will come to the attention of agents. And as a writer's business knowledge grows with his craftsmanship, he will have minimal problems in finding interested agents. An agreement with an agent does not, however, absolve a playwright of the responsibility of marketing his work. An agent functions more in the areas of business management, contract negotiation, and subsidiary sales than as a manuscript runner. Agents help a playwright open some doors to professional theatre offices, but the writer himself had best walk through. Most professionals advise that a novice might well send scripts to agents who charge reading fees but never to those who advertise services. A playwright can request a list of reputable literary agents from The Authors Guild of America, 6 East 39th Street, New York, New York 10016. The following are a few well-known agencies that handle plays.

ANN ELMO AGENCY, INC., 52 Vanderbilt Avenue, New York, New York 10017

BERTHA KLAUSNER INTERNATIONAL LITERARY AGENCY, INC., 130 East 40th Street, New York, New York 10016

CREATIVE MANAGEMENT ASSOCIATES, 600 Madison Avenue, New York, New York 10022

HAROLD FREEDMAN BRANDT & BRANDT, DRAMATIC DEPARTMENT, INC., 101 Park Avenue, New York, New York 10017

INTERNATIONAL FAMOUS AGENCY, 1301 Avenue of the Americas, New York, New York 10019

LEAH SALISBURY, INC., 790 Madison Avenue, New York, New York 10021

LUCY KROLL AGENCY, 119 West 57th Street, New York, New York 10019

STERLING LORD AGENCY, 660 Madison Avenue, New York, New York 10021

WILLIAM MORRIS AGENCY, 1350 Avenue of the Americas, New York, New York 10019

Organizations for Playwrights

Because the organizations devoted to helping playwrights are so few yet so important, a list of their names and addresses would be inadequate. The following descriptions introduce eleven organizations every playwright ought to know.

Actors Studio Playwriting Section, a segment of the well-known actor training organization, offers membership to qualified playwrights. In certain Studio sessions, actors perform scenes or whole new plays, and discussions follow. It provides a playwright at least a place to make contacts, if not to receive productions, and some writers would of course benefit from the critiques. Inquiries and applications should go to Actors Studio Playwriting Section, 432 West 44th Street, New York, New York 10036.

The *American Educational Theatre Association*, whose membership consists mainly of theatre teachers, is a large national organization that provides means of communication among non-professional theatres. In addition to its major publication, *Educational Theatre Journal*, it sponsors an annual convention and a number of working projects. AETA's convention, usually held during four days in August in a large city, provides a playwright the opportunity to establish contacts with many of the amateur theatre's professional teachers, scholars, and artists. The Playwrights Program, one of the major projects, attempts to stir interest among university theatre directors in the work of unknown playwrights. The address is: American Educational Theatre Association, 815 17th Street, N. W., Suite 842, Washington, D. C. 20006. A playwright should write an inquiry before submitting a manuscript.

The American Place Theatre came into being in 1964, and its home is St. Clements Episcopal Church on West 46th Street in New York. According to artistic director Wynn Handman, its purposes are to attract writers from other genres to the theatre and to act as a forum for serious and talented but unknown playwrights. It has an established policy of offering a writer various kinds of readings or full production. Because of its unusual record of introducing such writers as Robert Lowell, William Alfred, Ronald Ribman, and Frank Gagliano to the theatre, the American Place has received a number of awards and grants. It is truly outstanding among those organizations in New York willing to help unknown dramatists. The address is 423 West 46th Street, New York, New York 10036.

American Playwrights Theatre fosters connections between dramatists and both university and non-profit civic theatres for the production of new plays. At present, only established playwrights may participate in the program, but the organization is currently showing an interest in some relatively unknown writers. In order for a script by an unrecognized dramatist to be considered, it must be submitted by a theatre person of some repute or by an agent. APT is undoubtedly helpful to some professional playwrights, but its potential for assisting the non-established is still to be realized. The address is American Playwrights Theatre, 205 Derby Hall, 154 North Oval Drive, The Ohio State University, Columbus, Ohio 43210.

The Dramatists Guild, the professional dramatists' union, is an essential organization for playwrights who reach major commercial markets. Along with the Authors Guild, it is a corporate member of The Authors League of America. It functions as an organization to establish professional policies and to provide information to its members. For the latter purpose, it publishes a quarterly journal. Information about membership and services can be secured from: Dramatists Guild, Inc., 234 West 44th Street, New York, New York 10036.

The *Eugene O'Neill Memorial Theatre Foundation's Playwrights Conference*, conceived as a forum for playwrights, now amounts to an outstanding creative workshop. For the annual summer meeting of about a month's duration, a committee of expert readers selects a series of unproduced plays for readings, staged readings, and productions. The Foundation provides authors with room and board plus a modest stipend to encourage them to attend. It also assembles a talented professional company of actors and directors to handle the plays; it attracts intelligent audiences; and it invites other theatre artists and critics to join the group in Waterford, Connecticut, for the presentations and subsequent discussions. The purpose of the whole endeavor is to assist playwrights in the evaluation of their creative explorations. Although the O'Neill Playwrights Conference has turned out to be a leading showcase for new plays, its true purpose is artistic rather than commercial. The addresses are: Eugene O'Neill Memorial Theatre Foundation, P. O. Box 206, Waterford, Connecticut 06385; or Suite 1012, 1860 Broadway, New York, New York 10023.

The HB Playwrights Foundation, Inc., makes available to promising dramatists and to established writers who work with other literary forms a theatre workship. It desires to give such authors a chance to see and hear their works read or performed by professional actors. The staff is interested in original plays and dramatic adaptations from important works of American literature. The program includes weekly seminars, acting workshops, and sessions with directors. The Foundation's work is proving helpful to a growing number of writers. It is associated with the Herbert Berghof Studio. An inquiry should precede submissions to: The HB Playwrights Foundation, Inc., 122 Bank Street, New York, New York 10014.

The *New Dramatists Committee Inc.*, an organization sponsored by the Dramatists Guild, tries to encourage talented young writers and to help them gather knowledge about the professional theatre. The Committee's program provides for its member playwrights: (1) theatre tickets, (2) craft discussions with professionals, (3) rehearsal observation of a major New York production, (4) workshop reading or staging of the member's play, (5) publicity within the business for finished plays. The Committee is interested in writers nearing professional maturity, not in beginners. Although membership is by invitation only, any playwright may query the Committee and express his interest. The membership is limited to forty new playwrights per year. The address is: New Dramatists Committee Inc., 130 West 56th Street, New York, New York 10019.

The *Office for Advanced Drama Research* at the University of Minnesota began with a grant from the Rockefeller Foundation. Devoting its resources solely to new playwrights, it is one of the most helpful of these organizations. Its goals are to assist the playwright himself, to provide him time through financial aid, and to help him solve problems of creativity and experimentation. The OADR staff reads all previously unproduced manuscripts submitted to them; they attempt to identify writers of talent rather than merely slickly written plays. They select about twenty scripts a year and submit these individually to associated theatres in the Minneapolis area. If a theatre indicates interest in the writer, the Office then arranges an experimental rehearsal period and, perhaps, performances. In association with the University of Minnesota Press, OADR publishes the best of the resultant plays to encourage other productions of them. Its reluctance to encourage showcase performances, its interest in the writer as artist, and its record of intelligent selectivity have made OADR a leader among playwriting organizations. It provides an example that major universities in other sections of the country might well emulate. A playwright may send inquiries to: Office for Advanced Drama Research, 102 Shevlin Hall, University of Minnesota, Minneapolis, Minnesota 55455.

In 1964, Richard Barr, Clinton Wilder, and Edward Albee formed an off-Broadway production company, and established an experimental playwrights program in connection with it. *The Playwrights Unit*, as the organization is called, permits a selected dramatist to choose director, actors, audience, and discussion plan for a staged version of one of his plays or scenes. The program offers a writer financial means for the production, assistance, freedom, and visibility. It is certainly accessible, at least to talented writers residing in New York. The address is: The Playwrights Unit, Village South Theatre, 15 Vandam Street, New York, New York 10013.

Theatre Communications Group, sponsored by the Ford Foundation, acts as an advisory and liaison organization for about fifty resident professional theatres, most of which are outside of New York City. Although it does not offer any direct services to theatres or writers in relation to the production of new plays, it at least owns a current list of addresses of the best professional theatres in the country. And if a non-professional playwright makes a living with other theatre skills, TCG offers a personnel service for placement in resident theatres. During the 1967–1968 season 6 per cent of major productions by professional resident theatres were of new plays, and in the 1968–1969 season the number rose to 13 per cent. For information, one can write to: Theatre Communications Group, 20 West 43rd Street, New York, New York 10036.

There are, of course, other organizations interested in new plays and new writers. The eleven identified above, however, are the most outstanding. These groups offer the kind of help playwrights most need.

Contests

The number of playwriting contests in the United States ranges from about thirty to one hundred per year. An individual playwright would have difficulty discovering all of them, and even the AETA editors who occasionally publish a contest directory have trouble maintaining a current listing. Certainly, fewer competitions are active now than a few years ago. In 1964 for example, about eighty-six contests were running, but in 1969 only thirty-three of these remained. Contests are sponsored chiefly by universities, community groups, and small theatre organizations. Many of these become discouraged about the lack of good scripts, the labor of reading, or the pain of raising money. Some find that open, year-round invitations to writers of skill is the best way to secure plays of quality. A few competitions, however, stand out above the rest either because of their staying power, because of the size of their awards, or both. Each contest has its own submission requirements and deadlines, and thus only the names and addresses of the major contests appear.

AMERICAN NATIONAL THEATRE AND ACADEMY, Playwriting Department, 245 West 52nd Street, New York, New York 10019

BROADCAST MUSIC, INC., 589 Fifth Avenue, New York, New York 10017

BROADWAY THEATRE LEAGUE OF EVANSVILLE, 618 S. E. Riverside Drive, Evansville, Indiana 47713

EAST-WEST PLAYERS AUXILIARY, 1906 Redcliff Street, Los Angeles, California

THE MCKNIGHT FOUNDATION, W-2762 First National Bank Building, St. Paul, Minnesota 55101

SAMUEL FRENCH, INC., National Collegiate Playwriting Contest, 25 West 45th Street, New York, New York 10036

SIDNEY HILLMAN FOUNDATION, INC., 15 Union Square, New York, New York 10003

STANLEY DRAMA AWARD, Wagner College, Staten Island, New York 10301

UNIVERSITY OF CHICAGO THEATRE, The Charles H. Sergel Drama Competition, 5706 South University Avenue, Chicago, Illinois 60637

PUBLISHERS

A number of companies publish plays. Some issue only scripts that have been produced on Broadway, but a growing number are willing to look at manuscripts that have not been professionally produced. Playwrights should recognize two categories of publishers and understand their goals. First, book publishers issue reading versions of new plays, and their goal is the sale of books. Second, play-leasing companies market acting versions of plays, and their intent is to make money both from the sale of scripts and from production royalty fees. Both kinds of publishers can bring a playwright income, recognition, and productions. An uninformed writer needs to be cautioned, however, about sales to some publishers who will require him to sell his play outright for an immediate single payment rather than for a percentage of every playscript sold and of every performance royalty paid the company. To play-leasing companies, he should sell only the nonprofessional and subsidiary rights. A playwright should see that all copyrights are made in his own name. And when selling a play to a book company, an uninitiated author should seek expert legal and literary advice before signing a contract. Publishing companies are, for the most part, willing to treat playwrights fairly, and they can provide the unknown dramatist with a wide production market. The following list is representative of publishing companies with a proven interest in new plays.

BOOK PUBLISHERS

A. S. BARNES & CO., Box 421, Cranbury, New Jersey 08512

THE BOBBS-MERRILL CO., 3 West 57th Street, New York, New York 10019

COWARD-MCCANN, INC., 200 Madison Avenue, New York, New York 10016

DELL PUBLISHING CO., 750 Third Avenue, New York, New York 10017

GROVE PRESS, INC., 80 University Place, New York, New York 10003

HARPER & ROW PUBLISHERS, 49 East 33rd Street, New York, New York 10016

HILL AND WANG, INC., 141 Fifth Avenue, New York, New York 10010

DAVID MCKAY CO., 750 Third Avenue, New York, New York 10017

RANDOM HOUSE, INC., 457 Madison Avenue, New York, New York 10022

SIMON & SCHUSTER, INC., 630 Fifth Avenue, New York, New York 10020

UNIVERSITY OF MINNESOTA PRESS, 2037 University Avenue, S. E., Minneapolis, Minnesota 55455

THE VIKING PRESS, INC., PUBLISHERS, 625 Madison Avenue, New York, New York 10022

PLAY-LEASING COMPANIES

BAKER'S PLAYS, 100 Summer Street, Boston, Massachusetts 02110

TOBY COLE, INC., 234 West 44th Street, New York, New York 10036

THE DRAMA SHOP, 109 14th Street, N. W., Mason City, Iowa

THE DRAMATIC PUBLISHING CO., 86 East Randolph Street, Chicago, Illinois 60601

DRAMATISTS PLAY SERVICE, 440 Park Avenue, South, New York, New York 10016

ELDRIDGE PUBLISHING CO., Franklin, Ohio; or 1247 Curtis, Denver, Colorado

EVANS PLAYS, 500 East 77th Street, New York, New York 10021

SAMUEL FRENCH, INC., 25 West 45th Street, New York, New York 10036

THE HEUER PUBLISHING CO., Cedar Rapids, Iowa 52406

PIONEER DRAMA SERVICE, Cody, Wyoming

Special Markets

In addition to the so-called standard market outlets, a dramatist can look for more specialized markets. The largest of these is the mass entertainment industries of cinema, television, and radio. Since this book aims primarily to discuss playwriting for the theatre, a directory of the media markets is outside its scope. Subsidiary sales of plays to motion picture or television production companies, of course, can be most lucrative to a playwright. Agents are usually responsible for sales to these media, and in fact such sales are nearly impossible otherwise.

The publication of plays in periodicals is another possibility. Most periodicals that print plays are interested, however, only in a specialized sort of drama. An author should acquaint himself with such publications and not submit manuscripts unless he knows the play matches the outlet. The following selected list of periodicals includes a note for each about its particular bent.

Chelsea, Box 242, Old Chelsea Station, New York, New York 10011; a literary quarterly that sometimes uses new plays.

Chicago Review, Faculty Exchange, University of Chicago, Chicago, Illinois 60637; a literary quarterly interested in short experimental drama.

The Church School, 201 Eighth Avenue, South, Nashville, Tennessee 37203; a monthly religious publication that uses short plays for holidays.

Dasein, The Quarterly Review, P. O. Box 2121, New York, New York 10001; a literary magazine that uses some short drama.

Drama & Theatre, Department of English-Speech, State University College, Fredonia, New York 14063; a quarterly devoted to publishing mature new plays.

Dramatics, College Hill Station, Cincinnati, Ohio 45224; a publication of the National Thespian Society that prints some student written one-act plays.

Encore, Box 781, Elm Grove, Wisconsin; a bimonthly for community theatres interested in short plays suitable for such groups.

Evergreen Review, 80 University Place, New York, New York 10003; a provocative literary magazine which sometimes publishes short avant-garde plays.

Generation Magazine, 420 Maynard Street, Ann Arbor, Michigan; a publication interested in intellectual materials dealing with fine arts and in quality plays of any length.

Grade Teacher, 22 West Putnam Avenue, Greenwich, Connecticut 06830; uses a plays of about 1,000 words suitable for elementary school children.

The Instructor, F. A. Owen Publishing Co., Instructor Park, Dansville, New York 14437; prints plays of about 2,000 words for elementary children.

Players Magazine, University Theatre, Northern Illinois University, DeKalb, Illinois 60115; a quarterly devoted solely to American theatre that occasionally publishes outstanding new plays.

Plays, 8 Arlington Street, Boston, Massachusetts 02116; wants one-act plays for young people aged seven to seventeen, prefers short plays, and publishes about one hundred a year.

TDR, The Drama Review, New York University, 32 Washington Place, New York, New York 10003; frequently publishes plays of all lengths so long as they are genuinely original and of the highest quality.

Another specialized market, introduced in the list of periodicals, is children's theatre. Even to say that writers habitually overlook this market is an understatement. Children's theatre companies, publishers, and schools themselves need good new plays for young people. A rapidly growing market, it chronically lacks scripts of value and originality. Writing children's plays can serve a young dramatist with practice, an audience, and dollars. In addition to periodicals, the chief outlets for children's plays are such play-leasing companies as Baker's Plays, The Drama Shop, The Dramatic Publishing Company, Dramatists Play Service, Eldridge Publishing Company, Evans Plays, Samuel French, The Heuer Publishing Company, and The Anchorage Press (Cloverlot, Anchorage, Kentucky 40223).

The other three special markets worth noting are religious and social organizations, foreign theatres, and commissions. Literally thousands of religious groups are potential play markets, but they must be queried individually and usually given personal encouragement and technical help by the playwright himself. These provide a large and almost totally untapped reservoir of outlets. Also, foreign professional theatres are often more willing and more able to produce new plays than their American counterparts. Contacting these theatres, however, is best accomplished through personal acquaintances or agents. Every competent playwright's overseas travel itinerary might well include visits to such theatres in order to establish individual contacts. Lastly, writing on commission is a frequently possible and sometimes lucrative market possibility. But assignments go mostly to writers of some reputation, and they too come mainly from personal associations.

THE SUBMISSION DETAILS

Three items remain for a playwright to consider—letters, mailing, and biography. Like other business details, these may seem insignificant by comparison with the artistry of formulating and writing a drama, but such details sometimes mean the difference between securing a production and not. Thus, they affect the completion of a drama indirectly. Submission of scripts is an author's practical activity of offering his work to potential audiences.

A playwright will undoubtedly compose hundreds of letters during his career, but his letters of inquiry and his follow-up letters are likely to be among the most crucial. Every such letter should naturally be carefully written. After all, the writer *as a writer* is with his *letter* making a first and lasting impression on a potential producer. This is not to say that business letters should be decorated with verbal lace. They should be clear, interesting, fluent, and well proofread. Normally, letters of inquiry contain at least three paragraphs: (1) a statement of purpose, perhaps including the names of common acquaintances; (2) a description of what the writer wants, what his product is, and perhaps what rights he is offering for sale; and (3) an appeal for information or for acceptance. The best business letters get to the point quickly and yet are friendly. They should never carry a request that the writer knows cannot be fulfilled, nor should they be overly pressing. And everyone hates a letter full of gripes.

After receiving a letter of inquiry, most outlets will indicate whether or not the playwright should submit a script. If so, then a letter as well as the manuscript should get off immediately. They are best mailed simultaneously but separately. The letter goes first class, the manuscript at the special fourth class manuscript rate and so marked. If a postmaster is unfamiliar with this rate, one can refer him to Postal Regular 135.13. Every manuscript envelope, if mailed fourth class, should also carry the words *Return Requested.*

A 10″ × 13″ envelope is best for play manuscripts. It should contain not only the play but also a self-addressed envelope for return mailing. This envelope should have sufficient return postage clipped to it. Mailing envelopes look better with typed labels than with handwritten addresses. A writer reveals his working attitudes by his care in handling such details.

Sometimes it is wise for a writer to send a personal résumé with one of his follow-up letters, and a theatre company nearly always requests one when it plans a production. The writer would do well to have a single biographical sheet ready for this purpose. It should contain the following information:

- Personal data
  - Name
  - Addresses and phone numbers
  - Birth date and place
- Education
  - Schools
  - Degrees
- Publications (if any)
- Productions of the submitted play and others (if any)
- Other theatre experience
  - Position, company, and date

Neatness counts mightily in all the submission details. If a writer were to handle as many different manuscripts as do producers, directors, agents, editors, and teachers, he would soon discover that good looking, easy-to-read manuscripts are always impressive and surprisingly rare.

This treatment of audience as market can appropriately conclude with a reminder that a playwright should not expect the miracle of success, or even acceptance, with every submission. The marketing process brings more disappointment than pleasure. But unless a writer persists, his plays will decorate his own bookshelves rather than find life on a stage. He, like any other artist, must *make a place* for his work. Space for it does not otherwise exist.

Opportunities for unknown playwrights are increasing yearly, but the standards of craftsmanship and originality are also climbing. Difficult as it is for the unknown writer to believe, many people are desperately searching for well written plays. Marketing a play need not involve luck or magic. To get a production, a playwright should establish and maintain all the contacts he can manage with theatre people throughout the country. Even some of one's school friends turn out to occupy positions of responsibility in the theatre, somewhere! One friendship is better than a hundred letters to strangers, but friendships are often kept alive through letters.

The standard marketing practices followed by most professional writers are apt for playwrights too. Each market outlet needs careful study. No author should be discouraged by the first few rejections of an individual work. Submission must be a long, steady, and uninterrupted process. And each writer must face the fact that readers are slow, all of them. Every submission will require from one to six months. Therefore, multiple submissions are all right in theatre markets.

When all is said about salesmanship, however, the most significant point is that the artistic quality of a play is the chief stimulant, or deterrent, for its production. Finally, every playwright in our time should recognize that the only sure way to get produced is to join a theatre ensemble. Such associations are ever more desirable and possible. And the very future of American theatre depends on the growing willingness on the part of both companies and dramatists to join creative forces.

A final matter about marketing has to do with premature and improper submissions. A writer can be certain that his attempts to place bad scripts will earn him a bad name. It is not a good practice to send out a play just to see what happens. A writer had best submit only a finished play that he knows to be the best work he can do. Further, a writer ought to be certain that every submission is an appropriate one. It is senseless to send one type of play to a market outlet that is interested in another type. Inappropriate submissions waste a producer's time and unnecessarily disappoint the writer. Thus, the most important marketing idea of all is *always* to query an outlet with a letter and receive a reply before making a manuscript submission.

12: *CONCLUSION—A WAY OF LIFE*

Tell me where is fancy bred,
Or in the heart or in the head?
How begot, how nourished?
Reply, reply.

WILLIAM SHAKESPEARE
The Merchant of Venice

A playwright is an artist. He may be hard working or careless, serious or commercial, famous or lone. And being an artist makes him no better or worse a human being than a bricklayer or a woodcarver. But if he is to excel at his trade, he, like other craftsmen, should realize that even the basic techniques take years to master. He should understand that he will learn to write plays by writing plays, not by talking about those he intends to write. As every artist comes to know, practicing an art must be each artist's sole life purpose above all others. And real artists are too absorbed in their work to worry about their products' reception or their own reputation. As a working human, a playwright is committed to action, to a vision of life, to his craft, and to the creation of images.

ACTION

Drama requires action. Its action has three phases. A playwright and his fellow theatre artists involve themselves in the action of creation. Together they build an object of art that is at once poetic and theatrical. The object is made of words, vocal sounds, and physical actions. A playwright engages, thus, in an *act* of creativity. Second, a drama in itself, while it exists on a stage during a span of time, is an organized action. It is a pattern of situations leading to incidents leading to other situations, all occurring in the concrete present. A drama is concentrated and immediate activity of and by human beings. Like other forms of art, a play is a set of arranged details. A drama is a unity of presented acts. Finally, action arises, too, in the live participation of spectators. An audience uses the senses of seeing and hearing to perceive meaning in the dramatic object presented to them. Insofar as audience members feel, think, and express themselves, they are involved in human action—both their own and that within the play. Each spectator performs an act, on whatever level, of perception. To control action is, therefore, a playwright's main responsibility. He searches for discipline, structure, and meaning to satisfy the needs of the three phases of action that make drama possible. And he can best carry out his search by recognizing the nature and variety of human change, because human changes are action itself. By means of a playwright's treatment of action a play achieves value.

Action stands as both content and form in drama. To understand action in this regard is to recognize that, in its most simple yet most essential state, an action is a moral or ethical act. Writers and theorists from Sophocles and Aristotle to Hemingway and Sartre have understood and utilized this most basic of all principles. The structure of a play's action is its plot, its organization.

STRUCTURE

What does structure have to do with a moral act, with an ethical choice? Structure is the arrangement of all the human conditions antecedent to the act and therefore formative of it. Also, structure is the arrangement of all the human conditions that arise in consequence of the choice and therefore are resultant from it. Structure is as much a feature in non-story plays of vertical organization, such as *Prometheus Bound* and *Waiting for Godot*, as it is in story plays of horizontal organization, such as *Hamlet* and *The Good Woman of Setzuan*. The structure of a play's action depends upon probability, as Aristotle pointed out, but the probability is a logic of an individual play itself rather than of external concepts of dialectic. The logic in drama is the logic of the imagination not just the intellect, of the heart not merely the mind. There are no rules for structuring an action, only principles. And even the principles should not manipulate a play, but rather the play should *use* those principles appropriate to it. Structure in drama is crucial, but arbitrary form is deadly.

A playwright *constructs* a drama. He makes an object. In order to do so, he envisions a purposeful whole, selects appropriate materials, arranges them in some form, and expresses both form and content in words. For the playwright, a drama as object is the end of his activity. His materials are physical activities, sounds, words, thoughts, and characters. The form with which he works is plot—structured human action. And the style of the whole is both his own manner of rendering words and actors' manner of rendering physical actions and sounds. The structure of action, thus, is the systemization of morally differentiated human activities.

The principles of structure that each playwright employs—and every writer employs *some*—reveal his particular dreams, his philosophy, his vision. His vision, then, controls his selectivity. And his practices in selecting materials are as significant as the structural principles he accepts.

VISION

A playwright admits materials to a drama only because he chooses them. His choices—both in his art and in his life—comprise his vision. The writer's vision is a compound of his creative intuition and his reflective thought. A drama shows man suffering, acting, and reacting on the basis of convictions, of thoughts, and of ideas. If a character "comes to life" in a play, his ideas are inseparably connected to those of the author—whether positively or negatively, objectively or subjectively. Man cannot have a sense of right and wrong, of good and evil, of remorse and shame without some assumptions about morality or ethics. A man's ideas, convictions, and assumptions when combined with his sensory perceptions and his emotional sensitivities—these comprise his vision. And it controls the selection of materials for every play a writer constructs.

A playwright without ideas writes empty dramas, just as the playwright without imagination writes dull ones. A dramatist need not be a philosopher, but he must develop a philosophy. A dramatist need not be solely a thinker, as Eric Bentley pointed out, but he must think. Drama, at best, is not merely amusing, though every drama should entertain. A play is not merely a game; it is a spiritual compulsion. Whether a drama celebrates gods, whether it broods about man's fate, whether it illuminates the meaning or lack of meaning in life, a drama is a living demonstration of the fact that men live and die according to their ideas. The first business of a playwright is to deal with the attitudes, convictions, and actions—with the ideas—that give men peace in death or drive them to despair in life. All this can result only if a playwright cultivates intuitive and intellectual perceptions, only if he has a vision.

A vision is a center that implies circumference. For a poet—lyricist, novelist, or dramatist—vision proceeds from a central image to an ever-widening circumference, from idea to meaning, from knowledge to wisdom, from sight to insight. A playwright's vision of the world, of man, of life provides him with the conception of order he builds into his plays. His vision is the sense of sight he uses to select the materials of his dramas. A playwright's vision may arise primarily from his dreams, his reflective thoughts, or his intoxication with life. But a writer's vision needs some impetus from all three of these human facets of existence.

CRAFT

Too many would-be playwrights cannot write. Rather than wanting to *write* plays, they wish to *be* playwrights. Too many so-called professional dramatists never bothered to learn to formulate dramatic poems. They think writing a play means putting together a show. Too many avant-garde playwrights imagine that sensationalism substitutes for craftsmanship. To write plays, a playwright must learn to write.

Craft is a matter of style. But what is style to a writer? It is his *manner* of writing. It is much more than verbal dexterity. It is what the writer knows about life and about writing, how he goes about his work, and the final arrangement of words he makes—all three. Style is evinced in a drama, but it extends from the writer's knowledge and craftsmanship. Every author, in some manner or another, must handle three phases of poetic composition: invention, planning, and expression. First, he searches for and discovers material. He develops wisdom about choice of subjects, and he is exhaustive in accumulating materials. A playwright should, like any other competent writer, maintain the fertility of his imagination. Whatever the trouble or risk, discovery is crucial. Second, with what he finds or invents, the writer then formulates a structure for the whole. A play is more than a compilation of information. It is more than a group of characters talking. Thoughts, characters, and events contribute as parts to a whole. And in a play, the whole is a plot. Conversation is not drama. Drama is much more than mere dialogue. To think out a play before writing down its words is the difficult part of dramaturgy; arranging sequences of words is the easy part. But even that is harder than most beginners realize. The third phase of poetic composition, expression, involves putting words on paper one at a time. Style appears in a play's diction. And the best dialogue is clear, interesting, and appropriate. These are qualities that all writers think they are capturing all the time. But few do. Writing lucid English is hard work. Clarity and plainness of style are better in a drama than verbosity and ornamentation. And they are more difficult to achieve. Inventiveness, a sense of order, and a desire for clarity—these determine a writer's style.

Because writing is hard work, the writer needs to know himself in order to discipline himself. To write, a man needs to be irrepressibly confident, but without a feeling of self-importance. To write is to love the work of writing. But loving one's finished pieces is dangerous, and loving one's reputation is fatal. Only the working really matters. Working, as a human activity, makes style. And since a writer is a human at work, the man himself is the essence of the style. A playwright's self, his individuality, his personal voice create his unique style. The secret is to find the self, not the style. A writer finds himself in the working. Writing is not so much the command of the language as it is the discovery of self.

Excellence of craftsmanship means originality, emphasis, and economy. Originality, rather than perfection, is the chief mark of genius. Every artist is derivative, but the great artists are so only in the matter of learning certain skills. More than skillful rendering, excellence in literary work is a matter of exploring new country. A complete artist opens a new frontier. For a playwright, originality is deviation from conventionalized norms and established traditions. A true artist is an innovator, not an imitator. A truly original play sets forth new relevance.

Although originality of invention, form, and expression is the principal mark of excellence in art, it depends upon emphasis and economy. To arrest attention, to stress, to emphasize—first require selectivity. Emphasis contributes to originality not as accentuation of the obvious, but as the movement of vision to a neglected area of experience, as the illumination of a dark recess of existence. Progressively, emphasis is the increasingly explicit statement.

Economy, on the other hand, is implicitness. But rather than being the opposite of emphasis, economy is its complement. Economy means condensation, omission, and infolding. To condense is to make every particle of a play mean more than one thing and perform more than one function. Omission contributes to economy insofar as a writer cuts out the irrelevant and the obvious. Hemingway called it "leaving out." He explained that the more a writer knows the more he can leave out. A work of literature should be like an iceberg, only a small portion of it is visible. Infolding is another method of getting economy. It involves compression, hints, symbols, motifs, and obliquity. To condense several allusions into one word, one metaphor, or one speech is to be economical. Obscurity is the danger, not the goal of infolding. But the art of economy in writing is the art of implication.

About the craft of writing, a playwright can be certain of these things. The style reflects the man. It is better to be working than to want to work or to have worked. Craftsmanship spans the actions of discovery, formulation, and expression. But the discoveries of one artist, or some artistic school, are the commonplaces of later artists. To mimic another, to join a movement, to follow—all are fun, if the artist wants to make money or commit suicide. For a playwright-as-artist, living is a commitment to the new.

Conceptions about art need constant and complete overhauling. However universal some formative principles of play construction may be, they need redefinition by each generation, indeed by each writer. In this second half of the twentieth century, a redefinition of beauty in all art is underway. The best playwrights and theorists are rethinking the possibilities of drama. They are discovering a new pertinency for dramatic art. With vigor, they point to the importance of audacity in the theatre. Although a playwright inevitably must face traditional problems of material and structure, the new answers will shock many traditionalists. To practice the art of drama is surely to rethink traditions and perhaps to attack tradition itself. Regardless of the achievements of the past, those solutions can never work perfectly in the present. An artist should learn what he can about the principles identified by his predecessors, but he must make his own unique use of them. Artistry requires that he assimilate knowledge about his work to the degree that the principles are no longer a conscious checklist but rather a part of his selective imagination.

IMAGE

A playwright is a maker of images. Using his personal vision, he looks at his world and then constructs an object that has—if he is skillful and lucky—beauty, meaning, and value. A play is a creative image of life.

A playwright is a poet of dramas. And like all poets, he deals with being more than with thought. A poet makes images rather than philosophies. Although poets can learn from philosophers, the philosophers sit at the feet of the poets in order to see the primary images of their age. A philosopher conceives the universal; an historian records the particular; and a poet formulates the human. Not the infinite nor the infinitesimal but the finite is the province of poetry. Poetry is a finite image of life because man himself occupies that position in the universe. Logical ideas do not perfect the poet's vision, although they can improve it a little. Because life is concrete and complex, creative images cannot be made from axioms. Because man is contradictory and ambivalent, art presents his images better than mathematics. Because the human condition is feeble and frightening, reason cannot penetrate human experience—religious, sexual, aesthetic, or whatever is mystic—as well as artistic perception. A playwright's knowledge of craft, such as knowing most of the relatively logical principles in this book, must become a part of his subconscious. His craft simply contributes to his creative intuition.

For a playwright, problems abound in all the three areas with which he is chiefly concerned—with the process of writing, within each play as an object, and with the business of reaching an audience. Every writer struggles with himself in the matter of discipline, and most learn that writing must be a daily activity. Most playwrights could benefit by writing more than they do. Few write enough plays. Although the composition of a scenario is recognized by most playwrights as important to the construction of a play, too many fail to plan fully before writing. For many, the planning stage simply furnishes excuses for not really working, rather than being a period of careful labor and of putting materials on paper. Creativity in drama is not talking about playwriting; it requires discipline, imagination, craftsmanship, and work. Reaching an audience demands patience and persistence. Marketing a play is an activity of salesmanship both with one's products and one's self. But success for a playwright who wishes to be an artist is more a matter of making dramas of quality than a business of eliciting fat royalty checks.

Happiness for a playwright, despite the claims of the careerists, is writing plays. To take a set of human materials and to shape them into a drama gives the writer delight. It is a way for him to deal with the chaos and pain of life. Playwriting, like the practice of any art, can itself become a way of life. And for the real dramatic artist, it must become his manner of living.

Personal involvement with an art form has little to do with career, if career means wealth, fame, and glamour. Each of these has nothing to do with art, except that they may become obstacles to the artist. A person can write plays, can be a playwright in the fullest sense without being recognized on Broadway or in Hollywood. If professionalism means skill and dedication, the playwright should be as professional as possible. But if professionalism means writing by formula or if it means public acclaim, then it is unimportant. Not merely a career, playwriting is a commitment. It is a profession in that it is a way of thinking about life. By taking up playwriting as a profession, a person declares a calling. He assumes a special task, one requiring talent and know-how. If he operates in a society that values plays and if he is good enough at his work, he will be able to make a living from his profession.

When a person decides to write dramas, he evinces his belief in himself as alive and in drama as a thing of value. He demands unity and rejects the chaos of the world. He explores the moral order of life and seeks revelation about the nature of man. He rejects the world by reconstructing it. By being concerned enough to envision life clearly and wholly, by being rebellious enough to attack all forces of dehumanization, by being confident enough about his own talent and about life's value, a writer is able to work seriously. In this spirit, he can learn to make the intuitive images called dramas. More specifically, he himself takes action; he structures the action of each play; and he stimulates action in others. A playwright is committed to action, and action is life.

BIBLIOGRAPHY

Although this is not a comprehensive bibliography, these books can be useful and enjoyable to a playwright. Some deal directly with playwriting, others with dramatic theory, aesthetics, psychology, contemporary theatre, modern society, and writing mechanics. All contributed to the writing of this book. Any playwright will no doubt find other books that please or educate him about drama, but these are some of the most basic ones.

Alexander, Hubert G. *Language and Thinking: A Philosophical Introduction.* Princeton, New Jersey: D. Van Nostrand Company, Inc., 1967.

Aristotle. *Poetics.* Trans. Ingram Bywater. New York: Random House, 1954.

———. *Rhetoric.* Trans. Rhys Roberts. New York: Random House, 1954.

Artaud, Antonin. *The Theatre and Its Double.* Trans. Mary Caroline Richards. New York: Grove Press, Inc., 1958.

Baker, George Pierce. *Dramatic Technique.* Boston: Houghton Mifflin Company, 1919.

Barrett, William. *Irrational Man: A Study in Existential Philosophy.* New York: Doubleday & Company, Inc., 1958.

Bentley, Eric. *The Life of the Drama.* New York: Atheneum, 1965.

———. *The Theatre of Commitment: And Other Essays on Drama in Our Society.* New York: Atheneum, 1967.

Brecht, Bertolt. *Brecht on Theatre: The Development of an Aesthetic.* Trans. John Willet. New York: Hill and Wang, 1964.

Brockett, Oscar G. *The Theatre: An Introduction.* 2nd ed. New York: Holt, Rinehart and Winston, Inc., 1964.

Brook, Peter. *The Empty Space.* New York: Atheneum, 1968.

Brustein, Robert. *The Theatre of Revolt: An Approach to the Modern Drama.* Boston: Little, Brown and Company, 1964.

———. *The Third Theatre.* New York: Alfred A. Knopf, 1969.

Burke, Kenneth. *Grammar of Motives.* Englewood Cliffs, New Jersey: Prentice-Hall, Inc., 1945.

———. *The Philosophy of Literary Form.* Revised ed. New York: Vintage Books, 1957.

Busfield, Roger M., Jr. *The Playwright's Art: Stage, Radio, Television, Motion Pictures.* New York: Harper & Row, Publishers, 1958.

Cage, John. *Silence: Lectures and Writings.* Middletown, Connecticut: Wesleyan University Press, 1961.

Cain, Thomas H. *Common Sense about Writing.* Englewood Cliffs, New Jersey: Prentice-Hall, Inc., 1967.

Cameron, Kenneth M., and Theodore J. C. Hoffman. *The Theatrical Response.* New York: The Macmillan Company, 1969.

Camus, Albert. *The Rebel: An Essay on Man in Revolt.* Trans. Anthony Bower. New York: Alfred A. Knopf, Inc., 1956.

Cassirer, Ernst. *An Essay on Man: An Introduction to a Philosophy of Human Culture.* New Haven: Yale University Press, 1944.

———. *Language and Myth.* Trans. Susanne K. Langer. New York: Harper and Row, Publishers, 1964.

Clark, Barrett H., ed. *European Theories of the Drama.* Revised ed. New York: Crown Publishers, 1947.

Cole, Toby, ed. *Playwrights on Playwriting: The Meaning and Making of Modern Drama from Ibsen to Ionesco.* New York: Hill and Wang, 1961.

Crane, R. S., ed. *Critics and Criticism: Ancient and Modern.* Chicago: The University of Chicago Press, 1952.

———. *The Languages of Criticism and the Structure of Poetry.* Toronto: University of Toronto Press, 1953.

Croce, Benedetto. *Guide to Aesthetics.* Trans. Patrick Romanell. Indianapolis: The Bobbs-Merrill Company, Inc., 1965.

Dutch, Robert A., ed. *The Original Roget's Thesaurus of English Words and Phrases.* New ed. New York: St. Martin's Press, 1965.

Else, Gerald F. *Aristotle's Poetics: The Argument.* Cambridge, Massachusetts: Harvard University Press, 1957.

Esslin, Martin. *The Theatre of the Absurd.* New York: Doubleday & Company, Inc., 1961.

Fergusson, Francis. *The Idea of a Theater.* New York: Doubleday & Company, Inc., 1953.

Frenz, Horst, ed. *American Playwrights on Drama.* New York: Hill and Wang, 1965.

Gallaway, Marian. *Constructing a Play.* Englewood Cliffs, New Jersey: Prentice-Hall, Inc., 1950.

Gibson, William. *The Seesaw Log: A Chronicle of the Stage Production, with the Text of Two for the Seesaw.* New York: Bantam Books, 1962.

Goffman, Erving. *The Presentation of Self in Everyday Life.* New York: Doubleday & Company, Inc., 1959.

Gordon, Jesse E. *Personality and Behavior.* New York: The Macmillan Company, 1963.

Grebanier, Bernard. *Playwriting.* New York: Thomas Y. Crowell Company, 1961.

Grotowski, Jerzy. *Towards a Poor Theatre.* Holstebro, Denmark: Odin Teatrets Forlag, 1968.

Guthrie, Edwin R. *The Psychology of Human Conflict: The Clash of Motives Within the Individual.* Boston: Beacon Press, 1962.

Heffner, Hubert C., Samuel Selden, and Hunton D. Sellman. *Modern Theatre Practice: A Handbook of Play Production.* 4th. ed. New York: Appleton-Century-Crofts, 1959.

Ionesco, Eugène. *Notes and Counter Notes: Writings on the Theatre.* Trans. Donald Watson. New York: Grove Press, Inc., 1964.

Kitto, H. D. F. *Greek Tragedy: A Literary Study.* New York: Doubleday & Company, Inc., 1955.

———. *Poiesis: Structure and Thought.* Berkeley: University of California Press, 1966.

Koestler, Arthur. *The Act of Creation.* New York: Dell Publishing Company, Inc., 1964.

Kostelanetz, Richard. *Theatre of Mixed Means.* New York: Dial Press, 1968.

Langer, Susanne. *Feeling and Form: A Theory of Art.* New York: Charles Scribner's Sons, 1953.

Lawson, John Howard. *Theory and Technique of Playwriting.* New York: Hill and Wang, 1960.

Levin, Harry, *Contexts of Criticism.* Cambridge, Massachusetts: Harvard University Press, 1957.

Lewis, Emory. *Stages: The Fifty-Year Childhood of the American Theatre.* Englewood Cliffs, New Jersey: Prentice-Hall, Inc., 1969.

Macgowan, Kenneth. *A Primer of Playwriting.* New York: Doubleday & Company, Inc., 1962.

McCrimmon, James M. *Writing with a Purpose.* 4th ed. Boston: Houghton Mifflin Company, 1967.

McLuhan, Marshall. *Understanding Media: The Extensions of Man.* New York: McGraw-Hill Book Company, 1964.

Miller, Arthur. Introduction, *Collected Plays.* New York: The Viking Press, 1958.

Nietzsche, Friedrich. *The Birth of Tragedy.* Trans. Walter Kaufman. New York: Random House, Inc., 1967.

Olson, Elder, ed. *Aristotle's "Poetics" and English Literature: A Collection of Critical Essays.* Chicago: The University of Chicago Press, 1965.

———. "An Outline of Poetic Theory," in *Critics and Criticism: Ancient and Modern.* Ed. R. S. Crane. Chicago: The University of Chicago Press, 1952, pp. 546–66.

———. *The Theory of Comedy.* Bloomington, Indiana: Indiana University Press, 1968.

———. *Tragedy and the Theory of Drama.* Detroit: Wayne State University Press, 1961.

Pasolli, Robert. "The New Playwrights' Scene of the Sixties," *The Drama Review,* vol. 13, no. 1 (Fall, 1968), 150–62.

———. *The Open Theatre.* Indianapolis: The Bobbs-Merrill Company, Inc., 1969.

Plutchik, Robert. *The Emotions: Facts, Theories, and a New Model.* New York: Random House, Inc., 1962.

Polking, Kirk, and Natalie Hagen. *Writer's Market '71.* Cincinnati: Writer's Digest, 1971.

Pound, Ezra. *A B C of Reading.* New York: New Directions Publishing Corp., 1960.

Riesman, David, with Nathan Glazer and Reuel Denny. *The Lonely Crowd: A Study of the Changing American Character.* New York: Doubleday & Company, Inc., 1955.

Santayana, George. *Interpretations of Poetry and Religion.* New York: Harper & Brothers, 1957.

Sartre, Jean-Paul. *Literary and Philosophical Essays.* Trans. Annette Michelson. New York: Collier Books, 1962.

———. *What Is Literature?* Trans. Bernard Frechtman. New York: Harper & Row, Publishers, 1965.

Schechner, Richard. *Public Domain: Essays on the Theatre.* Indianapolis: The Bobbs-Merrill Company, Inc., 1969.

Simon, Bernard, ed. *Simon's Directory of Theatrical Services and Information.* 4th ed. New York: Package Publicity Service, 1969.

Singleton, Ralph H. *Style.* San Francisco: Chandler Publishing Company, 1966.

Sontag, Susan. *Against Interpretation and Other Essays.* New York: Dell Publishing Company, Inc., 1969.

Tolstoy, Leo. *What Is Art?* Trans. Aylmer Maude. London: Oxford University Press, 1930.

Weales, Gerald. *The Jumping-Off Place: American Drama in the 1960's.* London: The Macmillan Company, 1969.

Webster's Seventh New Collegiate Dictionary. Springfield, Massachusetts: G. & C. Merriam Company, Publishers, 1963.

Williams, George G. *Creative Writing.* Revised ed. New York: Harper & Row, Publishers, 1954.

INDEX

Q

R